Arthur Rizzi/Henri Viviand (Eds.)

Numerical Methods for the Computation of Inviscid Transonic Flows with Shock Waves

Notes on Numerical Fluid Mechanics

Volume 3

Manuscripts should be well over 100 pages. As they will be reproduced fotomechanically they should be typed with utmost care on special stationary which will be supplied on request. In print, the size will be reduced linearly to approximately 75 %. Figures and diagrams should be lettered accordingly so as to produce letters not smaller than 2 mm in print. The same is valid for handwritten formulae. Manuscripts (in English) or proposals should be sent to Prof. Dr. K. Förster, Institut für Aerodynamik und Gasdynamik, Pfaffenwaldring 21, D-7000 Stuttgart 80.

Arthur Rizzi/Henri Viviand (Eds.)

Numerical Methods for the Computation of Inviscid Transonic Flows with Shock Waves

A GAMM Workshop

With 121 Figures

Springer Fachmedien Wiesbaden GmbH

CIP-Kurztitelaufnahme der Deutschen Bibliothek

**Numerical methods for the computation of inviscid
transonic flows with shock waves:** a GAMM work-
shop/Arthur Rizzi; Henri Viviand (eds.). —

(Notes on numerical fluid mechanics; Vol. 3)
ISBN 978-3-528-08077-8   ISBN 978-3-663-14008-5 (eBook)
DOI 10.1007/978-3-663-14008-5

NE: Rizzi, Arthur [Hrsg.]; Gesellschaft für
Angewandte Mathematik und Mechanik

Produced by IVD, Industrie- und Verlagsdruck, Walluf b. Wiesbaden

ISBN 978-3-528-08077-8

ABOUT THIS WORKSHOP

1. FOREWORD

This is one in a series of workshops organized by the GAMM
Specialist Group for Numerical Methods in Fluid Mechanics
(GAMM-Fachausschuss für Numerische Methoden in der Strömungs-
mechanik) whose purpose is to bring together the small group
of researchers actively working on a sharply defined topic in
order to discuss in detail their problems and experiences, to
promote direct comparison and critical evaluation of algorithms,
and to stimulate new ideas for numerical methods in fluid
dynamics. The chairmen of this workshop were A. Rizzi of FFA,
Sweden, and H. Viviand of ONERA, France.

2. INTRODUCTION

Practically ten years have passed since it was first demonstrated that the nonlinear potential equation of mixed type which
governs inviscid transonic flow could be solved in a numerical
procedure. These years have seen an interest in the computation
of transonic flow that continues to grow because of the develop-
ing and ever-increasing ability of the numerical methods to
solve more and more complex flows and because of the great
practical use to which their solutions can be put. From the
question of whether we can solve the equations of transonic flow
we have now progressed to the question of how accurately can we
solve them. Any attempt to answer it must by necessity include
a collective comparison of the results obtained from the computational methods that are being applied today for the numerical
solution of inviscid steady transonic flow. Provided that the
results from the various computing procedures are in close agreement and that they ultimately converge to a unique solution as
the mesh is progressively refined, the outcome of such a comparison could be the determination of the accurate numerical

solution to each of several well-chosen test problems.

Theoretically, however, the question of which procedures lead
to the most accurate calculation of inviscid flow remains largely unanswered. Of the many factors involved in this question
perhaps three main aspects can be highlighted as sources of
error which are difficult to analyze and which have a large influence on the accuracy of the solution obtained by presently
existing methods:

   i)   choice and implementation of boundary conditions on
       the body and at "infinity" (which in practice means
       an artificial boundary at the finite outer limit of
       a given mesh)

  ii)   special treatment, if any, at the trailing edge of
       the airfoil and in the wake following thereafter

 iii)   procedures for the accurate computation of shock
       waves.

And these seem to be equally sensitive whether or not the
governing equations assume irrotationality. Unfortunately no
complete theory of boundary conditions exists to guide us on
items (i) and (ii). Even more unsettled is item (iii) since
some methods use conservative and others nonconservative difference schemes as well as a plethora of forms for artificial
viscosity.

It therefore seemed appropriate to hold a workshop on the computation of inviscid transonic flow with the aim of critically
comparing and evaluating the performance and success of current
methods on a number of controlled test problems. The purpose,
of course, was to gain some insight into the methods and
ultimately to yield a set of definitive solutions of inviscid
transonic flows which can stand as benchmark cases whose valid-
ity is established independently of experimental measurements.

3. STANDARD MESH SYSTEM

Of course an obvious factor which directly affects the accuracy
of the solution but which is difficult to evaluate is the type
of mesh and the number of grid points and their concentration.
Our hope was that this influence could be detected and perhaps
even measured by carefully controlling, and then systematically
varying the mesh system used for the test problems. Such a
study, however, required that everyone had access to a common
mesh generating program that is easily adapted and altered.
Furthermore, we could arrive at a fair comparison of the respective performance of the various methods only from solutions
which were computed on a common discretization of the flowfield.
Each participant, therefore was sent a simple computer program
together with the input cards necessary to generate the standard
mesh networks that we proposed for Problems A, C, D and G. This
was thought to be not only more convenient than sending the
nodal points on data cards, but also more flexible in that it
offered a possibility to adapt this grid to individual procedures.
The nature of this program and how to use it is documented in
Appendix A. For the internal flow of Problem B a sheared rectangular grid network is what we recommended as the standard mesh.
No mesh was proposed for Problem E.

We urged everyone who could to use the mesh we proposed. However, we realized that the problem of determining the number and
distribution of grid points which would constitute the most
economical discretization of space for a given accuracy is a
rather elusive one. And it is a problem we hoped that the Workshop would help to shed some light on. Consequently all the
participants were advised that our proposed mesh should not be
taken as the one having the optimum balance between resolution
and economy of computation, but instead just as one providing a
common reference. We therefore stressed to everyone the signi-
ficance of progressively refining any computational mesh
(Problem F) and recommended all to explore the influence of the
grid on the accuracy of the solution.

4. TEST PROBLEMS

The test problems, demoted A, B, C, D, E, F, and G, are described below. The participants were given the following
directives:

> Use the algorithm of your choice to solve either the
> small perturbation or full potential equation, or the
> Euler equations for the following transonic flow condi-
> tions (assume a perfect gas and take $\gamma = 1.4$). For
> problems A, B, C, D and G, we ask everyone to use the
> standard mesh (partially displayed below) so that a
> comparison can be made of results obtained upon a common
> discretization. In the event that you cannot use the
> mesh that we propose (for example your method is not
> based on a body-fitted system, or it uses a Poisson solver
> or multigrip technique, etc.) you are free to choose your
> own, but you should employ the same number of points
> (IL, JL) that we have in our mesh, or as close to those
> as possible. We urge everyone, however, to try his best
> to honour the goal of a common discretization. It is
> essential that no ambiguity arises from the definition
> of the test problems. For this reason those airfoils
> (Problems A and C), which have a thick trailing edge
> have been very slightly modified so as to make the trail-
> ing edge pointed (i.e. single valued).

A. (compulsory)   External two-dimensional flow past the
                      NACA 0012 airfoil at subsonic freestream speeds

I)   subcritical flow

     i) $M_\infty = .72$ $\alpha = 0^\circ$ ii) $M_\infty = .63$ $\alpha = 2^\circ$

II) supercritical flow

     i) $M_\infty = .8$ $\alpha = 0^\circ$ ii) $M_\infty = .85$ $\alpha = 0^\circ$

     iii) $M_\infty = .95$ $\alpha = 0^\circ$ iv) $M_\infty = .8$ $\alpha = 1.25^\circ$

     v) $M_\infty = .85$ $\alpha = 1^\circ$

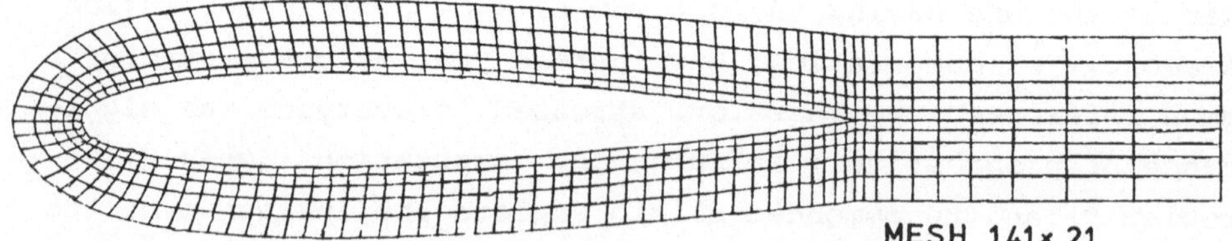

X

B. (compulsory)   Internal two-dimensional flow through a parallel
                    channel having a 4.2 % thick circular arc "bump"
                    on the lower wall. The ratio of static down-
                    stream pressure to total upstream pressure is
                    0.623512 (corresponding to M = 0.85 in isentropic
                    flow), and the distance between the walls is
                    2.073 times the chord length of the bump.

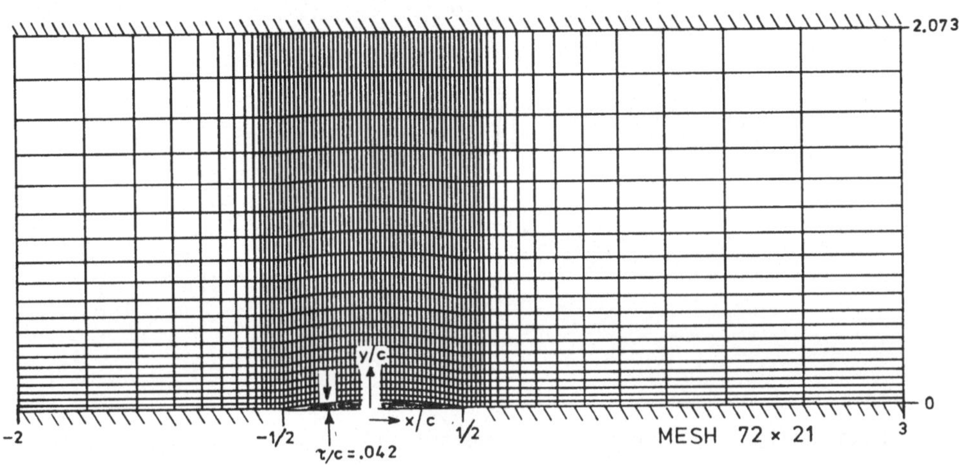

C. (optional)   External two-dimensional flow past the
                    RAE 2822 airfoil at subsonic freestream speeds

   I)   subcritical flow

        $M_\infty = .676 \quad \alpha = 1^\circ$

   II) supercritical flow

        $M_\infty = .750 \quad \alpha = 3^\circ$

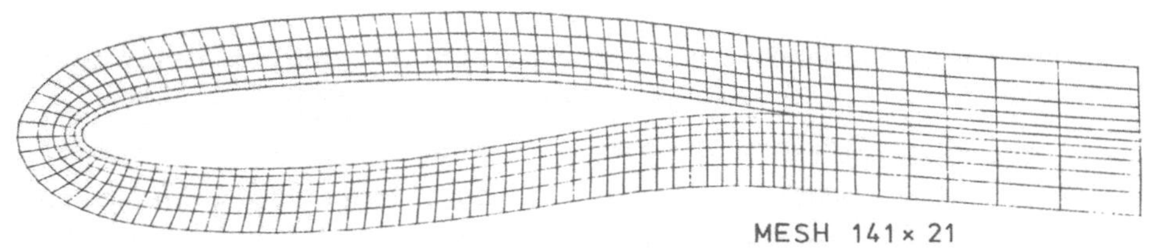

D. (optional)   External two-dimensional flow past the
                CAST 7 airfoil at subsonic freestream speeds

   I)   subcritical flow

        $M_\infty = .700 \quad \alpha = -1^\circ$

   II) supercritical flow

        $M_\infty = .760 \quad \alpha = \frac{1}{2}^\circ$

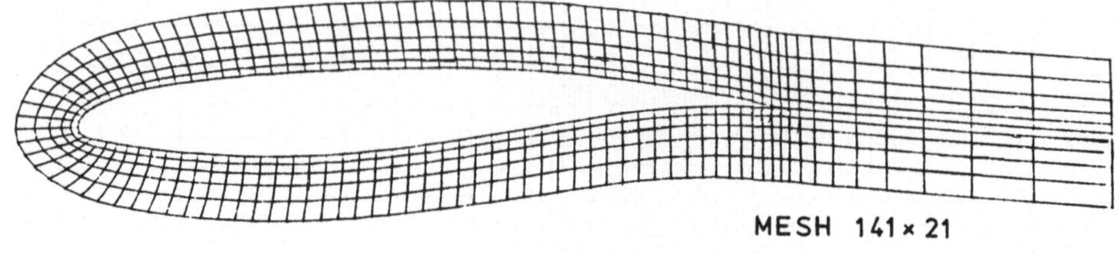

E. (optional)   External axisymmetric flow past a tangent ogive-
                cylinder at supersonic freestream speed $M_\infty = 1.2$.
                The shape of the body is defined by

$$0 \leq x/c < 1 \qquad y/c = -2.4383 + \sqrt{(2.6354)^2 - (x/c - 1)^2}$$

$$1 \leq x/c \qquad y/c = 0.1971$$

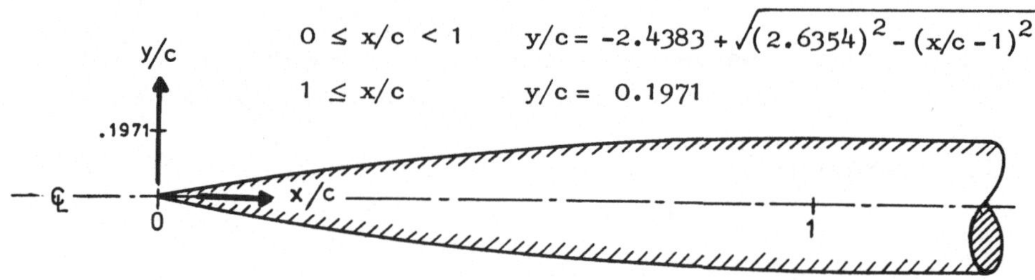

F. (optional)   Repeat any of the other test problems using the
                mesh system of your choice and progressively in-
                crease the number of grid points until you obtain
                what is, in your opinion, the accurate solutions
                to the problems you select.

G. (optional)   External two-dimensional flow past the Korn I
                  airfoil at subsonic freestream speeds

   I)   design conditions

$$M_\infty = .75 \qquad \alpha = 0.115^\circ$$

   II) off design

$$M_\infty = .75 \qquad \alpha = 1^\circ$$

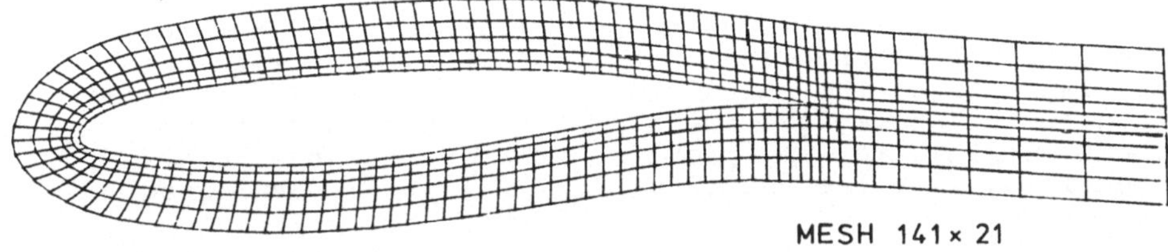

5. ORGANIZATION

This two-day Workshop, which took place at the Aeronautical Research Institute of Sweden (FFA) located in Bromma, a suburb of
Stockholm, was divided equally into an individual expository
phase and a collective evaluation phase. During the first day
each participant made an oral presentation covering the theoretical background of his method, and he discussed only selected
results which served to illustrate those features of his computational procedure which he found significant to the attain-
ment of an accurate solution. In addition Bertil Gustafsson
was invited to present a survey lecture on the analysis of
boundary conditions for aerodynamic flow problems. The written
manuscripts of these presentations constitute the main body of
this publication.

The next day was devoted to the group analysis and collective

comparison of all the computed results which we requested all
participants to submit one month prior to the Workshop (for the
most part dutifully observed, too). In order to facilitate the
comparisons we issued the following instructions for a unified
format of the results:

> Plot the coefficient of pressure C_p that is computed on
> all solid walls (with the negative values vertically up-
> ward) versus x/c (c = chord length) as well as C_p versus
> 10 y/c, all to the scale of 1 unit of abscissa and 1 unit
> of ordinate, both equal to 10 centimeters of the graph paper.
> (For the problem of the NACA 0012 airfoil with M_∞ = .95 ,
> α = 0° , extend the plot downstream along the x axis.)
> Furthermore, give your computed values of the lift coeffi-
> cient C_L and the drag coefficient C_D and state the
> amount of computing time, number of iterations, and the
> state of convergence (e.g. the residues) along with the
> initial data of the calculation. For Problem F, in addi-
> tion to the above, provide a plot of the mesh which you
> used. Lastly, for all problems treated with the Euler
> equations, plot the quantity $\left\{ \dfrac{p}{p_\infty} \middle/ \left(\dfrac{\rho}{\rho_\infty}\right)^\gamma - 1 \right\}$ evaluated
> at the wall.

> At your option you may also include contour plots of the
> various flowfield variables whose values and format should
> be:
>
> 1. scale: 1 chord length = 5 cm
>
> 2. isoMach: contour lines of constant Mach number M
>   starting at M = 0,5 and increasing with
>   increment M = 0.05
>
> 3. isobar: (for solutions of the Euler equations
>   only) contour lines of static pressure p
>   for the isentropic values of p which
>   correspond to the values of M in item 2
>   above, i.e. assuming isentropic expansion
>   from the reservoir conditions of the
>   problem. The two contour plots of M
>   and p should therefore coincide upstream
>   of all shocks, but will differ where the
>   field is nonisentropic.

The group discussion was structured around each of the Test
Problems A to G. We assigned a different chairman to each of
these problems whose task was to examine beforehand all the results submitted on his particular problem. During the Workshop
he opened the discussion of the solutions to his problem with a
survey of his findings and was responsible for leading a fruit-
ful discussion. We summarize the outcome of this group analysis
in our evaluation article.

6. PARTICIPANTS AND THEIR METHODS

PARTICIPANT/AFFILIATION	METHOD	PROBLEMS SOLVED
1. A. Jameson[x] Courant Inst., USA	TFP multi-grid-FVM	A,B,C,D,F,G
2. T. Holst[+] NASA-Ames	TFP-R-FDM	A,C,D,F,G
3. J.J. Chattot & C. Coulombeix[+] ONERA, France	TFP-R-FDM	A^*,B,C,D,F,G
4. J.P. Veuillot[+] & H. Viviand ONERA, France	TFP-PTD-FDM	A,B,C,D,F,G
5. W. Schmidt Dornier, Fed. Rep. Germany	TSP-R-FDM	A^*
6. L. Fuchs Royal Inst. Technology, Sweden	TSP-multi-grid-FDM	A^*
7. H. Deconinck & Ch. Hirsch Univ Brussels, Belgium	TFP-R-FEM	A^*,B
8. A. Eberle MBB, Fed. Rep. Germany	TFP-R-FEM	A,B
9. S. Jepps British Aerospace, England	TFP-R-FEM	A^*,B
10. T. Baker & M. Carr Aircraft Research Assoc., England	TFP-R-FEM	A,B,C,D,F,G
11. L. Carlson Texas A&M University, USA	TFP-R-FDM	A,B,C,D,F,G
12. R. Lock RAE, England	TFP-R-FDM	A^*,B,C,D,G
13. L. Zannetti Politecnico Torino, Italy	EE-TD-FDM	A^*,B
14. C. Sells RAE, England	EE-TD-FVM	A^*,C,D,G
15. A. Lerat[+] & J. Sides[+] ONERA, France	EE-PTD-FVM	A^*,B
16. J.P. Veuillot[+] & H. Viviand ONERA, France	EE-PTD-FDM	A,B,C,D,F,G

PARTICIPANT/AFFILIATION	METHOD	PROBLEMS SOLVED
17. A. Rizzi FFA, Sweden	EE-PTD-FVM	A,B,C,E,F

* incomplete

x in collaboration with D. Caughey, Cornell Univ., Wen Huei Jou,
Flow Research; Richard Pelz, Courant Institute; and John
Steinhoff, Grumman Aerospace.

+ did not attend in person

KEY

Model equation	Solution procedure	Discretization

TSP - transonic small per-
     turbation equation
	R - relaxation	FEM - finite
		     element

TFP - transonic full
     potential equation
	TD - time dependent	FDM - finite
		     difference

EE - Euler equations	PTD- pseudo-time-
	     dependent
		FVM - finite
		     volume

7. ACKNOWLEDGEMENTS

This Workshop could not have taken place without the help of
innumerable people. But of these there are some we wish to
single out for particular thanks:

The Aerodynamics Department of FFA (director Dr. G. Drougge)
provided generous support in all organizational matters.
The Theoretical Aerodynamics Division of ONERA for their
help in the development of the standard mesh and the overall
planning of the Workshop. And finally Inez Engström at FFA
for her meticulous preparation of the texts and figures of
the various FFA papers in these Proceedings.

(A. Rizzi, H. Viviand)

$$\underline{\text{Boundary Conditions for}}$$

$$\underline{\text{problems in aerodynamics}}.$$

Bertil Gustafsson

Department of Computer Sciences,
Uppsala University, Uppsala Sweden.

In this paper we will discuss some recent results which are of importance when solving aerodynamic problems. We will consider three different situations where the correct treatment of the numerical boundary conditions is not obvious: down stream boundaries where part of the flow is subsonic, the Rankine-Huginot conditions at a shock, and the boundary conditions at a solid wall.

Down stream boundary conditions

We consider the inviscid case where the flow is governed by a set of conservation laws. As an alternative these equations can be written in quasilinear form and we obtain a system of hyperbolic equations in two space dimensions.

$$(1) \qquad \underline{w}_t = A\underline{w}_x + B\underline{w}_y$$

We will first dicsuss the one dimensional case

$$(2) \qquad \underline{w}_t = A\underline{w}$$

and assume that the flow is subsonic at the boundaries $x = 0$, and $x = a$ This means that after linearization the proper model system is defined by a matrix A with eigenvalues $-u$ and $-u \pm c$, where c is the speed of sound. We assume that the velocity u is positive, so that there is two characteristics entering the domain at $x = 0$ and one characteristics entering the domain at $x = a$. Since A can be diagonalized, we simplify one more step, and consider the characteristic variable corresponding to the positive eigenvalue of A, i.e. we consider the scalar normalized equation

$$(3) \qquad \begin{cases} u_t = u_x & 0 \leq x \leq a, \quad 0 \leq t \\ u(x,0) = f(x) \end{cases}$$

We assume that we want to solve the steady state problem by integrating (3) by a marching technique. Since most difference approximations used are dissipative, we apply as our model approximation the Lax-Wendroff scheme (which is equivalent to the Mac-Cormack scheme for the equation (3)):

$$(4) \qquad u_j^{n+1} = (I + kD_o + \frac{k^2}{2} D_+D_-)u_j^n \quad , \quad j=1,2,\ldots,N-1$$

If boundary values are specified at $x = a$ the boundary conditions are

$$(5) \qquad \begin{cases} (hD_+)^p u_0^n = 0 \quad , \quad p \geq 1 \\ u_N^n = g \end{cases}$$

We have the following result for steady state calculations:

<u>Theorem 1</u>. For fixed h,k the solution to (4), (5) converge to $u_j = g$, $j=0,1,\ldots,N$ as $n \to \infty$. The convergence is exponentially fast, i.e.

$$||u^n - g|| \leq Ce^{-\alpha t_n}$$

where $\alpha > 0$ is independent of h,k. This is the normal situation. However, sometimes no data are available at the downstream boundary. If the specification of data is substituted by extrapolation or the use of one-sided differences, one might still get convergence, but the choice of initial data affects the limit function. With the boundary conditions

$$(6) \qquad \begin{cases} hD_+ u_0^n = 0 \\ hD_- u_N^n = 0 \end{cases}$$

we get

<u>Theorem 2</u>. For fixed h,k, the solutions to (4), (6) converge weakly as $n \to \infty$, i.e. the limit function u^∞ depends on the initial data.

This is of course an unsatisfactory situation. One way to overcome this non-uniqueness is to specify data at x = 0 instead.

$$(7) \qquad \begin{cases} u_0^{\,n} = g \\[2ex] hD_- u_N^{\,n} = 0 \end{cases}$$

This is not an unrealistic approach, since sometimes data for all
variables can be provided at a subsonic inflow boundary. ·

<u>Theorem 3</u>. For fixed h,k the solutions to (4), (7) converge to the
unique steady state $u_j^\infty \equiv g$, j=0,1,...,N if and only if N is odd.

In the first place we must be aware that even if N is odd, the convergence rate is very slow. If the marching process is considered as an
iteration procedure $u^{n+1} = Mu^n + v$, where M is a matrix of order N by
N, then it can be shown that the spectral radius has the form

$$\rho(M) \approx 1 - \sigma^N$$

where σ is a number less than one in magnitude.

One should also be aware of another interesting effect. It is possible
to prove that if $\partial f/\partial x = 0$ at x = a, the solutions to the difference
scheme will converge as $h \to 0, k \to 0$ to the solution of

$$(8) \qquad \begin{cases} u_t = u_x \\[1ex] u(x,0) = f(x) \\[1ex] u(a,t) = f(a) \end{cases}$$

on any finite time interval $0 \le t \le T$, and in the x-interval $\delta \le x \le 1$,
$\delta > 0$. Since the problem (8) is independent of the value g at the left
boundary, there seems to be a contradiction. However, this is not the
case, since in the beginning the transportation of the f(a) values
along the characteristics to the left will dominate the calculation. Due
to the dissipation in the scheme, the g-values will "dissipate" into the
domain, and after a long time $u_j^{\,n}$ will approach g.

The following figures show the result from a calculation made for
f(x) = sinx.

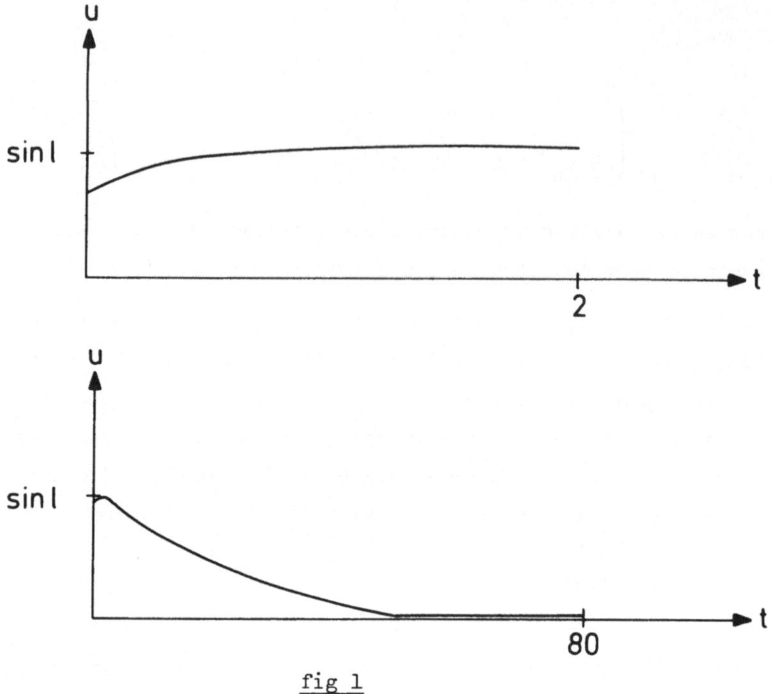

<u>fig 1</u>

The first curve represents the solution at $x = 2/3$ as a function of time. The second curve shows the same values, but the time interval is now extended until the real steady state is reached.

For a system of differential equations with constant coefficients of the form $u_t = Au_x$, the scalar results can be applied immediately if A is diagonalized.

From the results demonstrated above, which were derived in [3], it is clear that the true steady state solution can be obtained with reasonable computational effort only if correct data can be provided for as many variables as there are ingoing characteristics into the domain.

As an example we consider the simplest form of the one dimensional gas dynamic equations

$$(9) \qquad \begin{pmatrix} \rho \\ u \end{pmatrix}_t + \begin{bmatrix} u & \rho \\ c^2/\rho & u \end{bmatrix} \begin{pmatrix} \rho \\ u \end{pmatrix}_x = 0 \quad , \quad 0 \le x \le a$$

Here ρ is the density, u is the velocity and c is the local speed of sound. The linearized form of (9) is

$$
(10) \qquad \begin{pmatrix} \rho \\ u \end{pmatrix}_t + \begin{pmatrix} \hat{u} & \hat{\rho} \\ \hat{c}^2/\hat{\rho} & \hat{u} \end{pmatrix} \begin{pmatrix} \rho \\ u \end{pmatrix}_x = 0
$$

where $\hat{u}, \hat{\rho}, \hat{c}$ are assumed to be constants. To obtain as high convergence rate as possible, we want to specify the ingoing characteristic variables at each boundary without any coupling to the outgoing characteristic variable.

The advantage of this is seen by a simple application of the energy method for the problem

$$
\begin{pmatrix} u_1 \\ u_2 \end{pmatrix}_t + \begin{pmatrix} \lambda_1 & 0 \\ 0 & \lambda_2 \end{pmatrix} \begin{pmatrix} u_1 \\ u_2 \end{pmatrix}_\alpha = 0 \ , \quad \lambda_1 > 0, \quad \lambda_2 < 0
$$

with boundary conditions

$$
u_1(0,t) = S_1 u_2(0,t)
$$

$$
u_2(a,t) = S_2 u_1(a,t)
$$

We get with $\displaystyle \|u\|^2 = \int_0^a |u|^2 dx$

$$
\frac{d}{dt} \|u\|^2 = -2 \int_0^a (\lambda_1 u_1 u_{1_x} + \lambda_2 u_2 u_{2_x}) dx
$$

$$
= \lambda_1 u_1^2(0,t) + \lambda_2 u_2^2(0,t) - \lambda_1 u_1^2(a,t) - \lambda_2 u_2^2(a,t)
$$

$$
= (|\lambda_1|S_1^2 - |\lambda_2|)u_2^2(0,t) + (|\lambda_2|S_2^2 - |\lambda_1|)u_1^2(a,t) \ ,
$$

and from this follows that $\|u\|$ decreases as fast as possible when $S_1 = S_2 = 0$. The diagonal form of (10) is

$$
(11) \qquad \begin{pmatrix} \rho + \hat{\rho}/\hat{c}u \\ \rho - \hat{\rho}/\hat{c}u \end{pmatrix}_t + \begin{pmatrix} \hat{u}+\hat{c} & 0 \\ 0 & \hat{u}-\hat{c} \end{pmatrix} \begin{pmatrix} \rho + \hat{\rho}/\hat{c}u \\ \rho - \hat{\rho}/\hat{c}u \end{pmatrix}_x = 0
$$

Assuming that the values $\hat{\rho}, \hat{u}, \hat{c}$ are the steady state values, we
get the boundary conditions

$$\begin{cases} \rho + \frac{\hat{\rho}}{c} u = \hat{\rho} + \hat{\rho}\hat{u}/\hat{c} & \text{at} \quad x = 0 \\[3mm] \rho - \frac{\hat{\rho}}{c} u = \hat{\rho} - \hat{\rho}\hat{u}/\hat{c} & \text{at} \quad x = a \end{cases}$$

Unfortunately these conditions require more data than those available in
general.

The specification of the characteristic variables is a special case of
the so called absorbing boundary conditions derived by Engquist and
Majda. In [1] general systems of first order *are* treated, in [2] the small
disturbance equation for transonic flow. The aim is to construct conditions such that the amplitudes of waves reflected from artificial
boundaries are minimized. The result is an hierarchy of conditions which
all but the first contain both time- and space-derivatives. They all
have in common that certain combinations of function values and their
derivatives must be known at the boundaries. For example, the second
approximation for a system of type (1) has the form

$$w_t + bw_y = 0$$

where w denotes an ingoing characteristic variable. Higher order
approximations of the totally absorbing conditions contain higher order
derivatives both in time and space. In general we must expect that the
steady state solutions depend on the initial data.

This is also the case for the type of conditions derived by Hedström
[5]. He analyzed the one dimensional fully non-linear gas dynamic
equations, and arrived at conditions of the form

$$(12) \qquad \begin{cases} \rho_t + \frac{\rho}{c} u_t = 0 & \text{for} \quad x = 0 \\[3mm] \rho_t - \frac{\rho}{c} u_t = 0 & \text{for} \quad x = a \end{cases}$$

for the system (9). He also showed that weak shocks are reflected as
much weaker shocks when (12) is used.

Rudy and Strikwerda [6] considered the two dimensional gas dynamic
equations with the pressure p as one of the dependent variables. Under
the assumption that $p = \hat{p}$ is known at the outflow boundary, they

propose the condition

$$p_t - \rho c u_t + \alpha(p - \hat{p}) = 0$$

The value of α is determined by an analysis of the linear one-dimensional equations, such that the solution decays as fast as possible. The result is

$$\alpha = 0.28 \frac{\hat{c}^2 - \hat{u}^2}{\hat{c}a}$$

and their experiments exhibit a high convergence rate.

The blunt body problem

In this section we will first briefly describe the solution method presented in [4] for the blunt body problem. We consider the symmetric case for zero angle of attack according to the figure

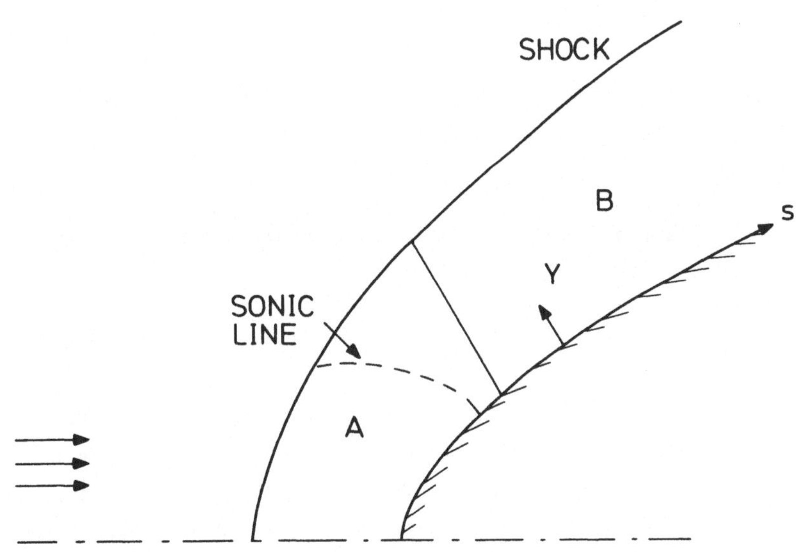

fig 2

7

The physical cordinates are s and $\underline{Y}$, but we introduce a new coordinate
in the Y-direction such that the shock stand-off distance δ in the old
coordinate system is normalized to one:

$$y = Y/\delta$$

If $\underline{w}$ denotes the vector containing the dependent variables velocity in
s-direction u, velocity in Y-direction v, enthalpy h and pressure p,
we obtain a system of the form

(13) $$\underline{Lw}_s + \underline{Mw}_y + \underline{f} = 0$$

where the matrices L and M depend on $\underline{w}$ and δ . (12) is approximated
using central differences over the region A in fig 2, and the non-linear
system of equations obtained is solved by Newton's method. δ is one of the
unknowns in this system, and since it is determined by the Newton procedure, the variation of δ is independent of the real time dependent
equations. We will now discuss the boundary conditions at the shock and
at the body in more detail.

The computational domain A is defined such that the component u is
supersonic at the outflow boundary. This is important as we have seen
from the discussion in the first part of this paper. Since the flow is
transonic, the system (13) has two different characters: there are either
two or four real characteristics. The boundary conditions for (13) do not
depend on the type of the system. At the body there is one condition

$$v = 0 \quad at \quad y = 0$$

and at the shock there are the four Rankine-Huginot conditions which can
be expressed as

(14) $$\underline{w} = \underline{R}(\underline{w}_\infty,\delta) \quad at \quad y = 1,$$

where $\underline{w}_\infty$ denotes the known free stream values. Since δ is un-
determined, these five conditions are required to determine the solution
uniquely. For the difference scheme four extra boundary conditions are
required. These are defined by letting the central differences be substituted by one-sided differences at the boundaries. The important thing is
to pick the right four equations. In the non-hyperbolic part there is no
characteristic pointing out of the domain at either boundary. However, in
the hyperbolic part the directions of the characterstics are according to
fig 3.

fig 3

Hence, the boundary treatment is defined as follows:

$\underline{y = 0}$ v = 0 and one-sided differences for the equations no 1, 3, 4 in (13).

$\underline{y = 1}$ Rankine-Huginiot equations and one-sided differences for equation number 4 in (13).

If for example all four extra conditions are applied at y = 0, the result is an oscillating solution, and the oscillations are more severe the longer the region A is extended into the hyperbolic part of the domain. It should be noted that region A should be cut off as soon as possible anyway, since a marching procedure is more effective in the hyperbolic part. The trouble is, that the position of the sonic line is not known a priori.

Finally we will make some comments on the outflow boundary treatment. As was mentioned above, it is essential that the flow perpendicular to the boundary is supersonic, so that we can use extrapolation of all variables (or upstream differencing). However, since the system (13) is hyperbolic and time-like in the s-direction, it is a rather strange procedure to end a calculation by extrapolation.   We consider the model problem

$$u_s + \tau u_y = 0$$

After a Fouriertransformation of the difference approximation we get

(15a) $u^{n+1} - u^{n-1} + 2i\lambda \sin(\omega\Delta y)u^n = 0$, $n = 1,2,\ldots,N-1$

$\lambda = \tau\Delta s/\Delta y$

9

with the boundary conditions

$$u^0 = a$$

(15b) $$(hD_-)^p u^N = b$$

It is possible to prove

<u>Theorem 4</u>. The solutions to (15) satisfy for all ω, λ an estimate

(16) $$||u^n|| \leq K(|a|+|b|)$$

where K is independent of ω and λ.

In particular, we note that λ can take any value, which is not the case if (15a) is used as a marching procedure.

References:

1. B. Engquist and A. Majda: Absorbing boundary conditions for the numerical simulation of waves. Math. Comp., Vol 31 (1977), 629-651.

2. B. Engquist and A. Majda: Numerical radiation boundary conditions for unsteady transsonic flow. To appear in J. Comp. Phys.

3. B. Gustafsson and H.-O. Kreiss: Boundary conditions for time dependent problems with an artificial boundary. J. Comp. Phys., Vol 30 (1979), 333-351.

4. B. Gustafsson and P. Wahlund: Finite difference methods for computing the steady flow about blunt bodies. To appear in J. Comp. Phys.

5. G. W. Hedström: Nonreflecting boundary conditions for nonlinear hyperbolic systems. To appear in J. Comp. Phys.

6. D. Rudy and J. Strikwerda: A non-reflecting outflow boundary condition for subsonic Navier-Stokes calculations. ICASE report No 79-2 (1979).

ACCELERATED FINITE-VOLUME CALCULATION OF TRANSONIC POTENTIAL FLOWS

Antony Jameson
New York Univ.
New York, N. Y.

D. A. Caughey
Cornell Univ.
Ithaca, N. Y.

W. H. Jou
Flow Research, Inc.
Kent, Washington

John Steinhoff
Grumman Aerospace Corp.
Bethpage, N. Y.

Richard Pelz
New York University
New York, N. Y.

Abstract

A fully-conservative finite-volume algorithm is applied to the calculation of transonic potential flows past isolated airfoils and through two-dimensional channels. The difference equations are solved by a multi-grid technique which uses an alternating-direction-implicit method as a smoothing algorithm. The finite-volume formulation provides a very powerful framework within which to treat flows past complicated geometries, while the multi-grid/alternating-direction scheme provides rapid convergence of the solution to very small residuals. Results of calculations using the algorithm are presented for the isolated airfoil and transonic channel test cases provided by the Symposium organizers.

I. Introduction

The finite-volume methods of Jameson and Caughey[1,2,3] provide a general framework within which it is relatively easy to calculate the transonic potential flow past essentially arbitrary geometrical configurations. Their use of only local properties of the difference-grid generating transformations essentially decouples the solution of the transonic flow equations from the grid generating step, so that minor modifications of a universal algorithm can be applied in any boundary-conforming coordinate system. Although the initial variants of these methods used line relaxation to solve the difference equations, the multi-grid/alternating-direction-implicit (MAD) scheme of Jameson[4] is used in the present work to provide high rates of convergence to very small residuals. The scheme is applied in three coordinate systems for the present calculations (1) a C-type mesh generated by weakly shearing a parabolic mapping for airfoil calculations, (2) an O-type mesh generated by weakly shearing a conformal mapping to the exterior of a near-circle for airfoil calculations, and (3) a weakly sheared Cartesian system for channel flows. In the following sections, the fully-conservative finite-volume method will be briefly reviewed, and the application of the MAD algorithm will be described. The construction of the three mesh systems generated for the application of the technique to airfoil and channel calculations will then be described. Finally, results will be presented for the airfoil and channel-flow test cases provided by the Symposium organizers.

II. Analysis

A. Finite-Volume Scheme

The equations of steady, inviscid, isentropic flow can be represented as follows. Let x,y be Cartesian coordinates and u,v be the corresponding components of the velocity vector q. Then the continuity equation can be written as

$$(\rho u)_x + (\rho v)_y = 0, \tag{1}$$

where ρ is the local density. This is given by the isentropic law

$$\rho = (1 + \frac{\gamma-1}{2} M_\infty^2 (1 - q^2))^{\frac{1}{\gamma-1}}, \tag{2}$$

where γ is the ratio of specific heats, and M_∞ is the freestream Mach number. The pressure p and the speed of sound a follow from the relations

$$p = \rho^\gamma /(\gamma M_\infty^2), \tag{3}$$

and

$$a^2 = \rho^{\gamma-1}/M_\infty^2. \tag{4}$$

Consider now a transformation to a new set of coordinates X,Y. Let the Jacobian matrix of the transformation be defined by

$$H = \begin{Bmatrix} x_X & x_Y \\ y_X & y_Y \end{Bmatrix}, \tag{5}$$

and let h denote the determinant of H. The metric tensor of the new coordinate system is given by the matrix $G = H^T H$, and the contravariant components of the velocity vector U,V are given by

$$\begin{Bmatrix} U \\ V \end{Bmatrix} = H^{-1} \begin{Bmatrix} u \\ v \end{Bmatrix} = G^{-1} \begin{Bmatrix} \phi_X \\ \phi_Y \end{Bmatrix}, \tag{6}$$

where ϕ is the velocity potential. Eq.(1), upon multiplication by h, can then be written

$$(\rho h U)_X + (\rho h V)_Y = 0. \tag{7}$$

The fully-conservative finite-volume approximation corresponding to Eq.(7) is constructed by assuming separate bilinear variations of the independent and dependent variables within each mesh cell. Numbering the cell vertices as illustrated in Figure 1, and assuming that the local coordinates $X_i = \pm 1/2$, $Y_i = \pm 1/2$ at the vertices, the local mapping can be written

$$x = 4 \sum_{i=1}^{4} x_i(1/4 + X_i X)(1/4 + Y_i Y), \tag{8}$$

with similar formulas for y and ϕ. At a cell center, this
transformation yields formulas for derivatives such as

$$x_X = 1/2(x_2 - x_1 + x_3 - x_4). \tag{9}$$

If we introduce the averaging and differencing operators

$$\mu_x f_{i,j} = 1/2(f_{i+1/2,j} + f_{i-1/2,j})$$
$$\hat{\delta}_x f_{i,j} = (f_{i+1/2,j} - f_{i-1/2,j}) \tag{10}$$

then the transformation derivatives, evaluated at the cell centers, can
be expressed by formulas such as

$$x_X = \mu_Y \hat{\delta}_X x$$
$$x_Y = \mu_X \hat{\delta}_Y x \tag{11}$$

with similar expressions for the derivatives of y and the potential.
Such formulas can be used to determine ρ, h, U, and V at the center of
each cell using Eqs.(2),(5), and (6). Eq.(7) is represented by conser-
ving fluxes across the boundaries of auxiliary cells whose faces are
chosen to be midway between the faces of the primary mesh cells. This
can be represented as

$$\mu_Y \hat{\delta}_X(\rho hU) + \mu_X \hat{\delta}_Y(\rho hV) = 0. \tag{12}$$

This formula can also be obtained by applying the Bateman variational
principle that the integral of the pressure

$$I = \int p \, dx \, dy$$

is stationary, and approximating I by a simple one-point integration
scheme in which the pressure at the center of each grid cell is
multiplied by the cell area. For subsonic flow the finite-volume method
can equally well be regarded as a finite element method with isoparametric bi-linear elements. The extension to treat transonic flows is
accomplished by adding an artificial viscosity to introduce an upwind
bias. The use of the one-point integration scheme leading to Eq.(12)
has the advantage of requiring only one density evaluation per mesh
point, but also has the undesirable effect of tending to decouple the
solution at odd- and even-numbered points of the grid, and suitable
recoupling terms can be added to improve the stability of the solution.
If we represent the influence coefficients of the terms containing
ϕ_{XX} and ϕ_{YY} in the expanded form of Eq.(12) as

$$A_X = \rho h(g^{11} - U^2/a^2)$$
$$A_Y = \rho h(g^{22} - V^2/a^2) \tag{13}$$

where g^{ij} are the elements of G^{-1}, then the compensated
equation can be written

$$\mu_Y \, \delta_X(\rho hU) + \mu_X \, \delta_Y(\rho hV) - \epsilon \delta_{XY}(A_X + A_Y) \, \delta_{XY}\phi = 0, \qquad (14)$$

where $0 \leq \epsilon \leq 1/2$. In practice $\epsilon = 1/2$ is generally used. An alternative method of obtaining the recoupling terms is to use a higher-order integration scheme which takes account not only of the pressure at the center of each cell, but also its x- and y- derivatives.

The scheme is stabilized in supersonic regions by the explicit addition of an artificial viscosity, chosen to emulate the directional bias introduced by the rotated difference scheme of Jameson.[5] We define

$$\hat{P} = \rho h \sigma/a^2 (U^2 \, \delta_{XX} + UV \, \delta_{XY})\phi$$

$$\hat{Q} = \rho h \sigma/a^2 (UV \, \delta_{XY} + V^2 \, \delta_{YY})\phi \qquad (15)$$

where the switching function

$$\sigma = \max(0., 1 - (M_c/M)^2). \qquad (16)$$

is non-zero only for values of the local Mach number M greater than some critical Mach number M_c. Then, after defining

$$P_{i+1/2,j} = \begin{cases} \hat{P}_{i,j} & \text{if } U \geq 0, \\ \hat{P}_{i-1,j} & \text{if } U < 0, \end{cases} \qquad (17)$$

with a similar shift for Q, we represent Eq.(14) as

$$\delta_X(\mu_Y(\rho hU) + P) + \delta_Y(\mu_X(\rho hV) + Q)$$
$$- \epsilon \delta_{XY}(A_X + A_Y) \, \delta_{XY}\phi = 0. \qquad (18)$$

The resulting difference equations are solved using a generalization of the multi-grid/alternating-direction-implicit (MAD) scheme of Jameson.[3] Let

$$L^h \phi = 0 \qquad (19)$$

represent Eq.(18) on a grid with a spacing proportional to h. Given an initial estimate $\phi^{(n)}$ of the solution to Eq.(19), the basis of the multiple grid method is to calculate an improved estimate $\phi^{(n+1)}$ on a coarser grid according to

$$L^{2h}\phi^{(n+1)} = f \qquad (20)$$

where

$$f = L^{2h}\phi^{(n)} - I_h^{2h} L^h \phi^{(n)}, \qquad (21)$$

and I_h^{2h} is a collection operator which averages the residuals over the fine mesh points in the neighborhood of each coarse mesh point. The fine grid solution is then improved according to

$$\phi^{(n+1)} = \phi^{(n)} + I_{2h}^{h}(\phi^{(n+1)} - \phi^{(n)}), \qquad (22)$$

where I_{2h}^{h} is an interpolation operator.

The success of the multiple-grid method generally depends upon the use of a relaxation method to rapidly eliminate high frequency errors on any given grid. Since point and line relaxation schemes do not necessarily provide the necessary smoothing of all high-frequency errors on non-uniform grids, a generalized alternating-direction scheme is used for our calculations. We introduce the difference operator

$$S = \alpha_0 + \alpha_1 \delta_x^- + \alpha_2 \delta_y^- \qquad (23)$$

where δ_x^-, δ_y^- represent one-sided difference operators, and we solve

$$(S - A_x \delta_x^2)(S - A_y \delta_y^2)C^{(n+1)} = SL^h\phi^{(n)} \qquad (24)$$

This scheme can be considered a discrete approximation to the time-dependent equation

$$\beta_0\phi_t + \beta_1\phi_{xt} + \beta_2\phi_{yt} = A_x\phi_{xx} + A_y\phi_{yy} \qquad (25)$$

where the coefficients β_0, β_1, and β_2 are related to the parameters α_0, α_1, and α_2. The advantage of using the operator defined in Eq.(23) in place of the simple constant of conventional ADI schemes is that Eq.(25) remains hyperbolic when the signs of A_x or A_y change. To ensure consistency, the δ_x^2 in Eq.(24) is replaced by an upwind operator when $A_x < 0$, and the δ_x^- in S is chosen to be upwind.

The simple fixed multiple-grid strategy proposed by Jameson[4] is used. An initial estimate for the solution on the finest grid is used to start the iteration. The high frequency error in this is removed by a single sweep of the ADI scheme, and the smoothed residual is passed to the next coarsest grid. This process is repeated until the coarsest grid is reached. An additional ADI sweep is made on the coarsest grid, and the predicted corrections to the solution are interpolated onto the next finest grid. This process is repeated until the corrections have been interpolated onto the finest grid, completing one multiple-grid cycle. Only the residuals on the finest grid are monitored, and the iteration ceases when the average residual over the field is less than some specified tolerance.

Further details regarding the difference approximation, artificial viscosity, and iterative scheme are contained in References 1 and 4.

B. Grid Generation

An important advantage of the finite-volume method is its decoupling of the solution procedure from the grid-generation step. This permits the grid to be generated in any convenient manner, and allows application of an essentially universal algorithm to any problem for which a boundary-conforming coordinate system can be generated.

A convenient procedure is to generate the boundary-conforming grid using a sequence of simple, analytically-defined transformations. Conformal mappings can frequently be used as building-blocks in this procedure, and are attractive because of their many desirable properties (such as orthogonality). The finite-volume calculations to be described were performed on three different coordinate grids, the construction of which will be described in this section.

Parabolic Airfoil Grid

The C-grid generation sequence for airfoil-shaped bodies is illustrated in Figure 2. Let x,y be Cartesian coordinates in the streamwise and vertical directions, respectively. The origin of the x-y system is chosen to be approximately halfway from the profile leading edge to its center of curvature. This results in the profile's surface being mapped to a slowly varying bump

$$Y_1 = S(X_1) \tag{26}$$

by the square-root transformation

$$x + iy = 1/2(X_1 + iY_1)^2. \tag{27}$$

Finally, the shearing transformation

$$\widetilde{X} = X_1$$
$$\widetilde{Y} = Y_1 - S(X_1) \tag{28}$$

results in a nearly-orthogonal transformation, since $S(X_1)$ is slowly-varying. Stretching transformations of the form

$$\widetilde{X} = X, \qquad\qquad\qquad |x| \leq \hat{x}$$
$$\widetilde{X} = \pm(\hat{x} + (|x| - \hat{x})/(1 - ((|x| - \hat{x})/(1-\hat{x}))^2)^a), \qquad \hat{x} \leq |x| \leq 1 \tag{29}$$

and

$$\widetilde{Y} = bY/(1 - Y^2)^a \tag{30}$$

are introduced, and a uniform grid on

$$-X_{lim} \leq x \leq X_{lim}$$
$$0 \leq Y \leq Y_{lim} \tag{31}$$

is transformed back to the physical plane to give the Cartesian coordinates of the mesh points. Values of $X_{lim} = Y_{lim} = .90$, $a = 1.0$, $b = .50$, and $\hat{x} = .45$ were used for the calculations to be presented. The choice of $\hat{x} = X_{lim}/2$ ensures that the airfoil trailing edge is at a grid point on all grids containing a multiple of four cells in the X-direction.

<u>Circle Airfoil Grid</u>

A second grid system for airfoil calculations is provided by weakly
shearing a conformal map of the profile to a near-circle (see Figure 3).
A Joukowsky mapping of the profile according to

$$x + iy = .25((X_1 + iY_1) + 1/(X_1 + iY_1)) \tag{32}$$

is followed by a shearing of the radial coordinate according to Eq.(28)
to make the system boundary conforming. A stretched radial coordinate
is introduced according to Eq.(30), and a uniform polar grid on

$$0 \leq \tilde{X} \leq 2\pi$$

$$\tag{33}$$

$$0 \leq Y \leq Y_{lim}$$

is then mapped back to the physical plane to give the Cartesian
coordinates of the mesh points.

<u>Channel Grid</u>

The grid generation sequence for the channel geometry is similar to
the latter stages of those for the airfoil. Let y_{max} be the
channel height. Then the shearing transformation

$$\tilde{X} = x$$

$$\tag{34}$$

$$\tilde{Y} = (y - S(x))/(y_{max} - S(x))$$

provides a boundary-conforming coordinate system. The stretching of
Eqs.(29)-(30) is again used, and values of X_{lim} and Y_{lim}
are chosen so that the upper channel wall and selected values of x
form the boundaries of the computational domain.

C. Boundary Conditions

Two types of boundary conditions must be specified to determine
solutions for the potential flow problems considered herein. The
no-flux condition must be enforced across any solid boundaries (such as
the airfoil and channel surfaces); and appropriate far-field boundary
conditions must be specified at the necessarily finite limits of the
computational domain. In addition, for the airfoil calculations, a
discontinuity in potential across some branch cut must be incorporated
if the airfoil has lift.

The solid-surface boundary conditons are quite easy to enforce in
boundary-conforming coordinate systems because the difference scheme is
formulated in terms of the contravariant components of the velocity.
The appropriate condition is that the out-of-plane component be zero.

This is incorporated by reflection of the normal-flux contributions for
a given auxiliary cell on the boundary.

A disadvantage of the finite-volume schemes is the need to truncate
the usually infinite domains of aerodynamic interest to finite
computational regions. This is in contrast to methods in which the
equation can be analytically transformed with suitable stretching
functions so that the difference mesh extends to infinity. (See, e.g.,
References 5, 6, and 7.) In both of the analyses treated here, a
reduced potential is introduced to describe the perturbations upon an
otherwise uniform stream. This potential is set to values appropriate
for a compressible vortex of circulation Γ

$$\phi = \Gamma/2\pi \arctan(y\sqrt{1 - M_\infty^2}/x) \tag{35}$$

on the farfield boundaries of the computational domain for the airfoil
calculations.

For the channel flow calculations, the reduced potential is set to
zero at the upstream boundary of the domain, while the streamwise
perturbation velocity is set to zero at the downstream boundary.

III. Results

A. Computational Considerations

Most of the results to be presented here have been calculated on
meshes containing 128x32 mesh cells in the X and Y directions,
respectively, and also 256x64 mesh cells. The rapid convergence rate
of the MAD scheme allows reduction of the residuals to very small
values. The solutions to be presented here have been converged to
residuals (representing the unbalanced flux in each cell, or the
differential operator multiplied by the cell area) of 10^{-9} unless
otherwise specified.

Calculations on a 128x32 cell grid require about 2 minutes of CPU
time on the CDC 6600 and use 230K(octal) of high-speed core storage.
Solutions of accuracy adequate for engineering purposes can be obtained
in a fraction of this time, since the surface pressure coefficients are
converged to four significant digits after about ten multi-grid cycles
in most cases.

B. Selected Results

The results of calculations for the test cases provided by the
Symposium organizers will now be presented. Results of calculations on
the two different mesh systems will be presented for the airfoil test
cases, and results of predictions for the flow through the two-
dimensional channel will be presented at a Mach number near that
proposed by the Symposium organizers.

In the test cases for subsonic flows containing shock waves of moderate strength, results of the finite-volume calculations on the two grid systems agree quite well. At larger angles of attack or higher freestream Mach numbers, the shock waves become too strong for the potential flow approximation to be reasonably accurate. In this situation, the shock wave on the upper surface tends to move back to the trailing edge, and satisfactory results are not obtained on the circle grid because of the interaction between the shock wave and the singularity in the grid at the trailing edge. The solution obtained on the parabolic grid is also open to question, as for example, in the case of the NACA 0012 airfoil at 1.25 degrees angle of attack and 0.80 freestream Mach number. For this case the shock wave appearing on the airfoil upper surface has moved to the trailing edge, and the inability of the numerical scheme to resolve the details of the flow in the vicinity of the stagnation point immediately downstream of the shock probably results in an incorrect value for the circulation. The trailing edge must be a stagnation point in this flow since the only alternative configuration which satisfies the Kutta-Joukowski Condition would be for the shock to turn the flow parallel to the lower surface; this solution is not realizable, however, because the maximum turning angle through an oblique shock at the Mach number which exists at the upper surface trailing edge is less than the included angle of the profile trailing edge. The calculated solution is apparently possible because the corner at the trailing edge is not resolved by the numerical scheme, and is effectively replaced by a cusp, with the result that a stagnation point is not obtained, and it becomes possible to match the lower surface pressure to the pressure behind the shock wave on the upper surface. This suggests the possibility that there may be another solution with less circulation and a shock wave forward of the trailing edge, which would be consistent with the need for a stagnation point to be present at a trailing edge with a finite corner angle. A nonuniqueness associated with the presence of a shock wave at the trailing edge has, in fact, been found in the solution for a Joukowsky airfoil by Steinhoff and Jameson, and is presented in a separate note in this volume.

The third set of results is for the transonic flow through the two-dimensional channel, calculated on grids containing 64x16 grid cells in the X and Y directions. It is easily verified that no solution exists at the freestream Mach number provided by the Symposium organizers when the flow is constrained to be isentropic, since the throat of the nozzle is too small to pass the required mass flux. The required mass flux could be passed only if the flow were uniformly sonic across the channel width, but this flow is inconsistent with the normal momentum equation at the lower channel wall which has a finite curvature at the throat. Results are therefore presented at the highest freestream Mach number at which it has been possible to achieve a converged solution with the present scheme to date. The figure shows the lower and upper wall surface pressure distributions in the vicinity of the bump on the lower wall at a freestream Mach number of 0.8435. It is clear from the value of the minimum pressure coefficient on the upper wall that the channel is not quite choked at this Mach number, but it is very nearly. Inspection of the complete flowfield indicates that the supersonic pocket reaches nearly 60 percent of the way across the channel.

IV. Conclusions

Results of transonic potential flow calculations using the geometrically-general finite-volume method have been presented for airfoil and channel geometries. The potential equation is differenced in strong conservation form, and the difference equations are solved by a multi-grid/alternating-direction-implicit algorithm which allows convergence of the solutions to very small residuals in reasonable computing times. Results are presented for the airfoil and channel flow test cases provided by the sponsors of the Symposium.

V. Acknowledgements

This work was supported in part by the Office of Naval Research under Contracts N00014-77-C-0032, N00014-77-C-0033, and N00014-78-C-0079, and by NASA under Grants NSG-1579 and NGR-33-016-201. Most of the calculations were performed at the DOE Mathematics and Computing Laboratory under Contract EY-76-C-02-3077.

VI. References

1. Jameson, Antony and Caughey, D.A., "A Finite-Volume Method for Transonic Potential Flow Calculations," Proc. of AIAA 3rd Computational Fluid Dynamics Conference, pp. 35-54, Albuquerque, N.M., June 27-29, 1977.

2. Caughey, D.A. and Jameson, Antony, "Numerical Calculation of Transonic Potential Flow about Wing-Body Combinations," AIAA Journal Vol. 17, pp. 175-181, February 1979.

3. Caughey, D.A. and Jameson, Antony, "Progress in Finite-Volume Calculations for wing-Fuselage Combinations," AIAA Paper 77-677 (Also, AIAA Journal, to appear).

4. Jameson, Antony, "A Multi-Grid Scheme for Transonic Potential Calculations on Arbitrary Grids," Proc. of AIAA 4th Computational Fluid Dynamics Conference, pp 122-146, Williamsburg, Va., July 23-25, 1979.

5. Jameson, Antony, "Iterative Solution of Transonic Flows over Airfoils and Wings including Flows at Mach 1," Comm. Pure and Appl. Math. Vol 27, pp. 283-309, 1974.

6. Jameson, Antony and Caughey, D.A., "Numerical Calculation of the Transonic Flow past a Swept Wing," New York University ERDA Report COO-3077-140, June 1977.

7. Caughey, D.A. and Jameson, Antony, "Accelerated Iterative Calculation of Transonic Nacelle Flowfields," AIAA Journal Vol. 15, pp.1474-1480, October 1977.

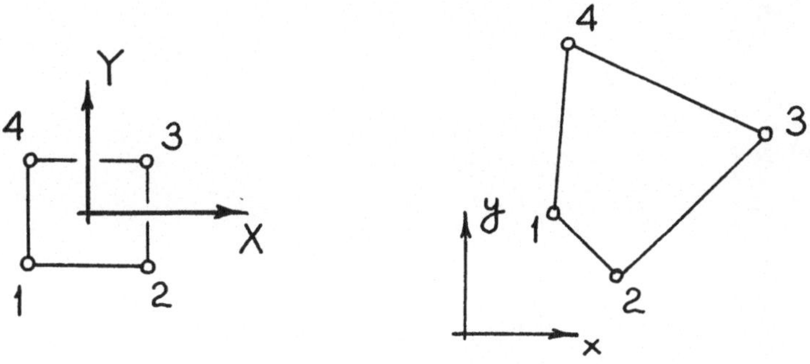

Figure 1.   Mesh cell in Computational and Physical Domains.

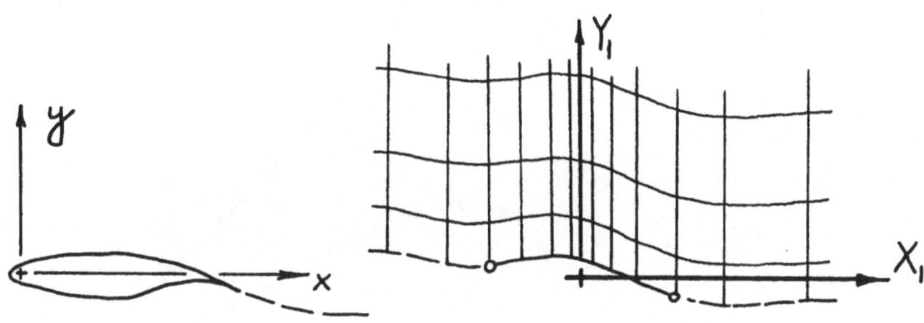

Figure 2.   Parabolic Grid generating sequence.

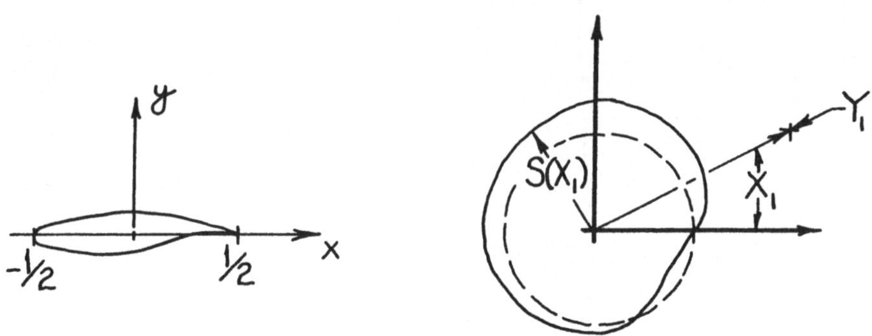

Figure 3.   Circle Grid generating sequence.

NACA 0012
MACH 0.630
GRID 256X64
CL 0.3343
FLO42
ALPHA 2.000
RESO.278E-09
CD 0.0001
CM -0.0026
O-type grid
CP
1.200 1.000 0.800 0.600 0.400 0.200 0.000 -0.200 -0.400 -0.600 -0.800 -1.000 -1.200 -1.400
X/C
0.000 0.200 0.400 0.600 0.800 1.000

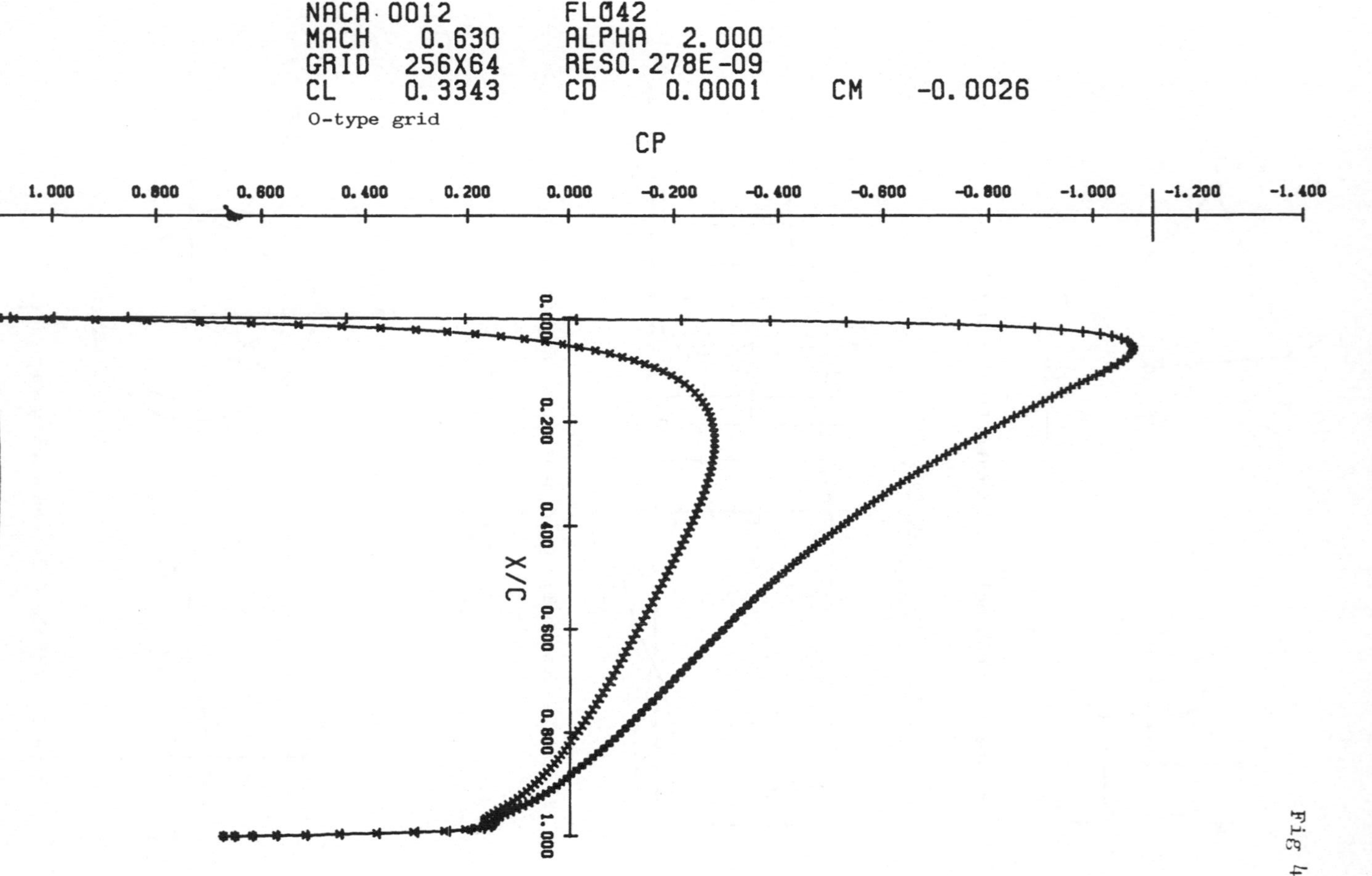

Fig 4.

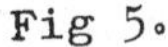

Fig 5.

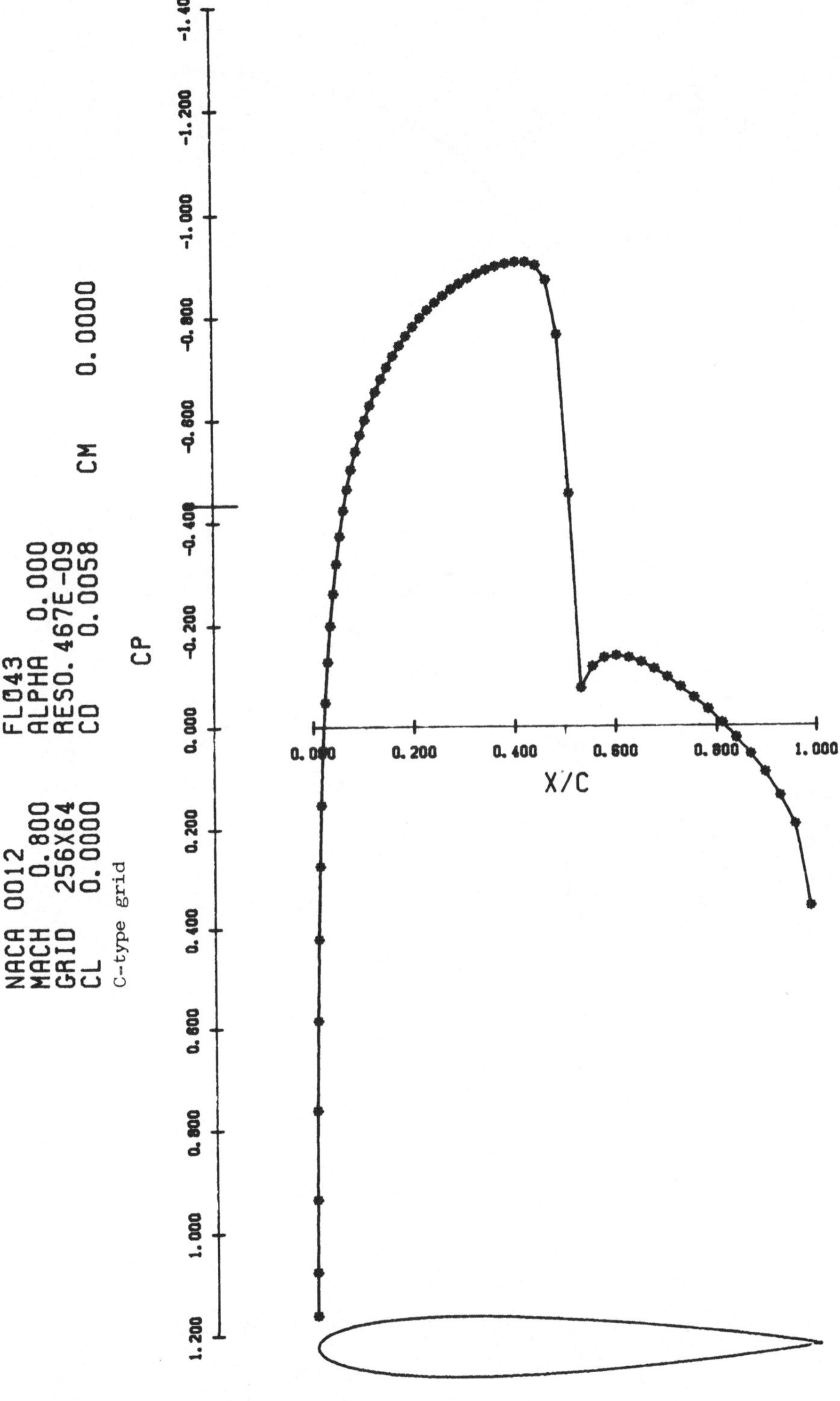

NACA 0012 FL043
MACH 0.800 ALPHA 0.000
GRID 256X64 RESO. 467E-09
CL 0.0000 CD 0.0058 CM 0.0000
C-type grid
CP
X/C
0.000 0.200 0.400 0.600 0.800 1.000
-1.400 -1.200 -1.000 -0.800 -0.600 -0.400 -0.200 0.000 0.200 0.400 0.600 0.800 1.000 1.200

NACA 0012   FL043
MACH   0.800 ALPHA 1.250
GRID 256X64 RESO.310E-09
CL   1.0912 CD   0.0874   CM   -0.3330
C-type grid
CP
X/C
1.200 1.000 0.800 0.600 0.400 0.200 0.000 -0.200 -0.400 -0.600 -0.800 -1.000 -1.200 -1.400
0.000 0.200 0.400 0.600 0.800 1.000

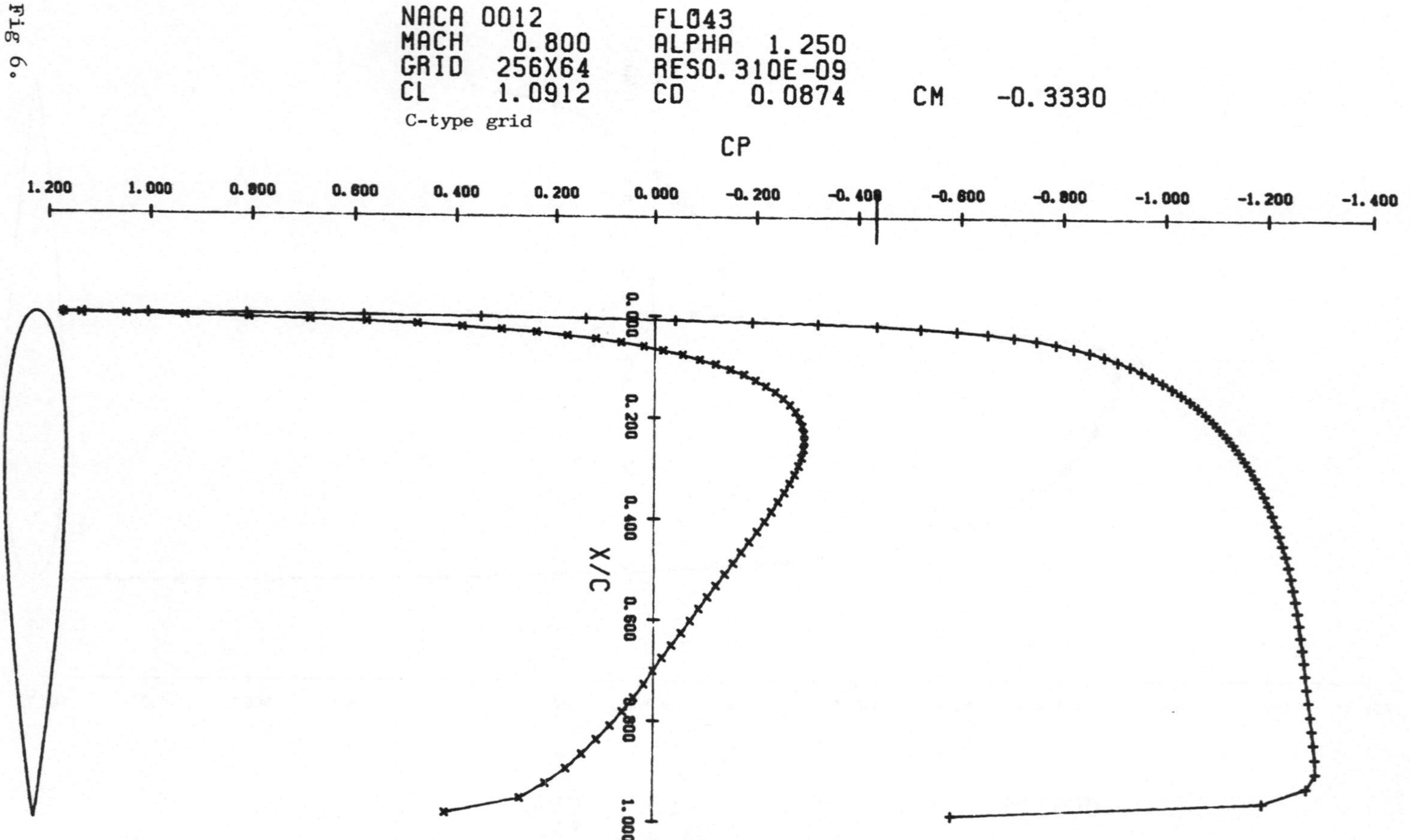

Fig 6.

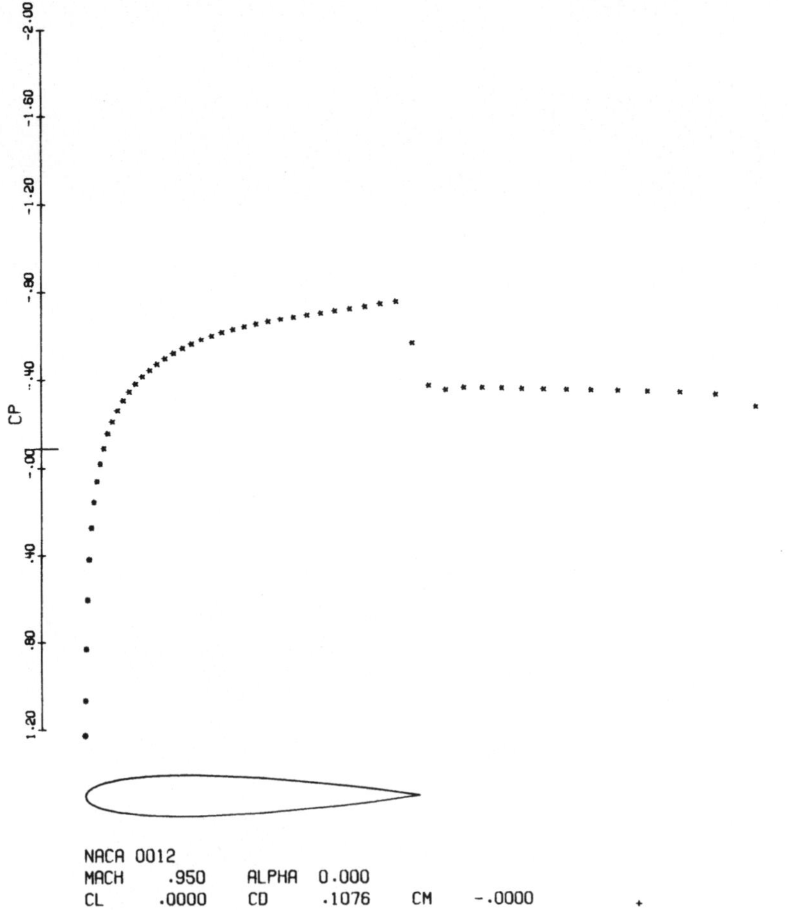

C-type grid   128 x 32   Fig 7.

4.00
2.00
0.00
-2.00
-4.00
-6.00
-8.00
-10.00
-12.00
LOG (ERROR)
0.00
10.00
20.00
30.00
40.00
50.00
60.00
WORK (X10^1)
CHANNEL TEST CASE
MACH   0.8435
RES1   0.00062633   RES2   0.00000000
WORK   498.55   RATE   0.9742
NX   64   NY   16

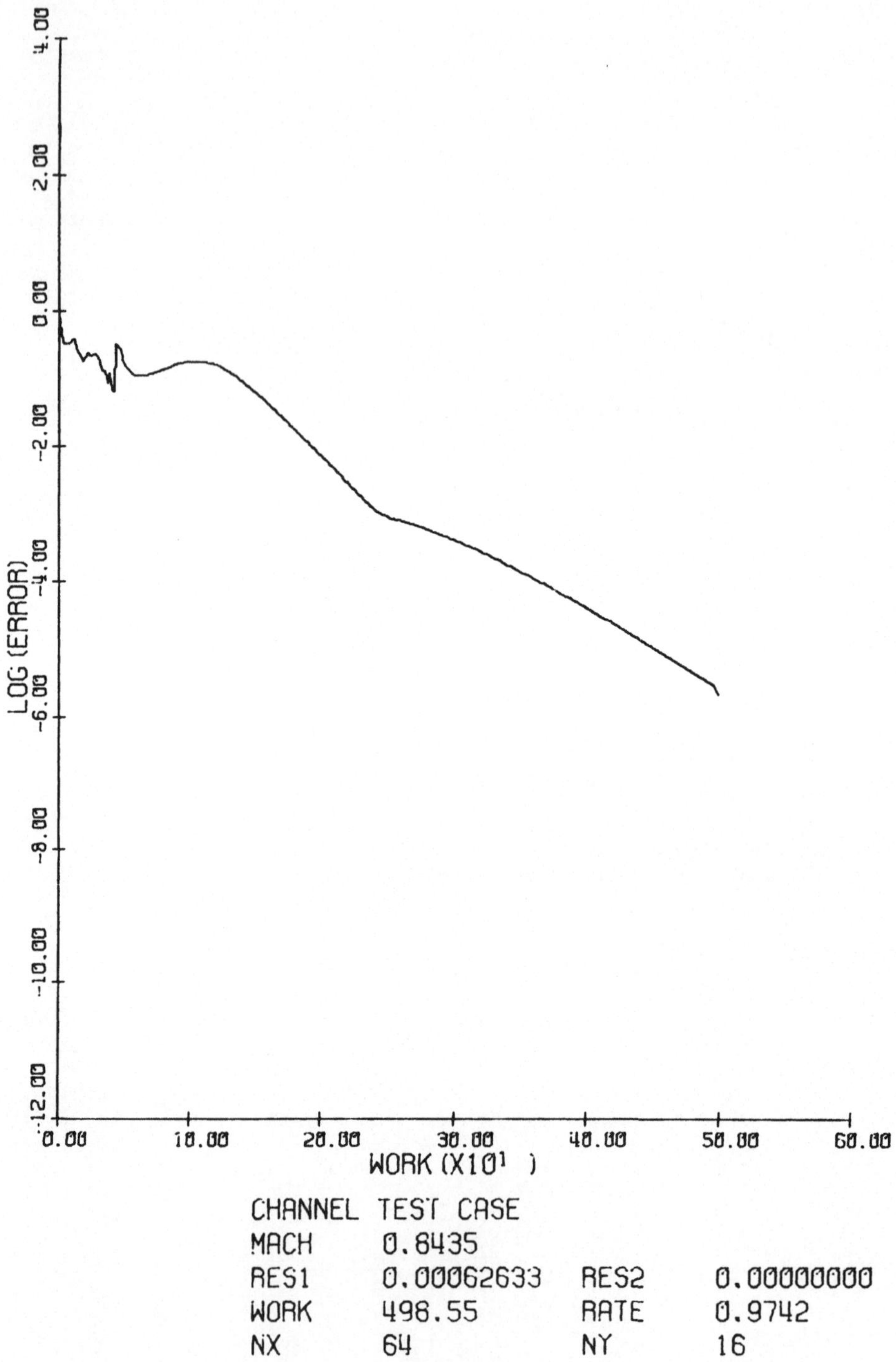

Fig 8.

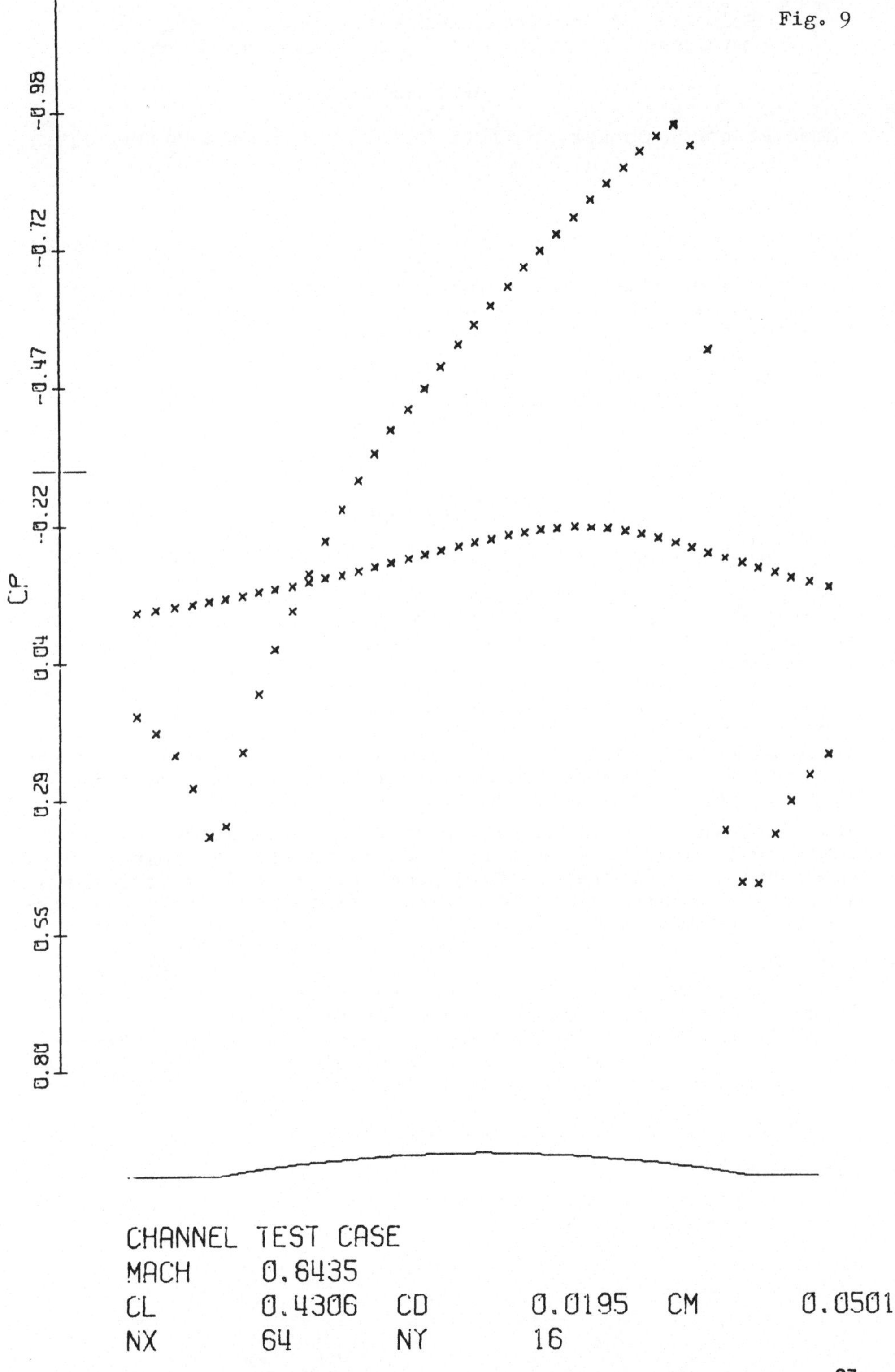
CP
-0.98
-0.72
-0.47
-0.22
-0.0
0.04
0.29
0.55
0.80
CHANNEL TEST CASE
MACH   0.8435
CL   0.4306   CD   0.0195   CM   0.0501
NX   64   NY   16

Solution of the Transonic Full Potential Equation
in Conservative Form Using An Implicit Algorithm

Terry Holst

Ames Research Center, Moffett Field, California 94035, U.S.A.

ABSTRACT

Numerical solutions of the full potential equation in conservative form are presented. The iteration scheme used is a fully implicit approximate factorization technique called AF2 and provides a substantial improvement in convergence speed relative to standard successive line overrelaxation algorithms. The spatial differencing algorithm is centrally differenced in both subsonic and supersonic regions with an upwind evaluation of the density coefficient in supersonic regions to maintain stability. This effectively approximates "rotated differencing" and thereby greatly improves the reliability of the present algorithm. The solutions presented in this paper have been selected from the GAMM Workshop on Numerical Methods for the Computation of Inviscid Transonic Flow with Shock Waves.

1. INTRODUCTION

Several implicit approximate factorization (AF) algorithms for solving the transonic small-disturbance equation (refs. 1 and 2) and the full potential equation (refs. 3-6) have been presented in recent years. One of the most recently introduced AF schemes is the so-called AF2 scheme of reference 1. For the AF2 algorithm, significant improvement in convergence speed has been experienced relative to the standard transonic solution procedure, successive-line overrelaxation (SLOR). A brief summary of the full potential form of the AF2 algorithm will be presented herein. For more details see references 4-6.

Stability in the present full potential formulation for supersonic regions of flow is maintained by an upwind evaluation of the density coefficient. This procedure is effectively the same as the addition of the upwind-differenced artificial viscosity term introduced in reference 7. Use of the upwind-biased density coefficient greatly simplifies the solution procedure and effectively allows the simple two- and three-banded matrix forms of the AF scheme to be retained over the entire mesh, even in supersonic regions. Other studies (refs. 8-9) have used similar spatial differencing schemes for many different applications to further substantiate this differencing procedure as being both reliable and flexible.

2. FULL POTENTIAL EQUATION ALGORITHM

Governing Equations

The full potential equation written in strong conservation-law form is given by

$$(\rho\phi_x)_x + (\rho\phi_y)_y = 0 \tag{1a}$$

$$\rho = \left[1 - \frac{\gamma - 1}{\gamma + 1}\left(\phi_x^2 + \phi_y^2\right)\right]^{1/(\gamma-1)} \tag{1b}$$

where the density (ρ) and velocity components (ϕ_x and ϕ_y) are nondimensionalized by the stagnation density (ρ_s) and the critical sound speed (a_*), respectively; x and y are Cartesian coordinates; and γ is the ratio of specific heats.

Equation (1) is transformed from the physical domain (Cartesian coordinates) into a computational domain by using a general independent variable transformation. This general transformation, indicated by $[(\xi,\eta) \to (x,y)]$, maintains the strong conservation-law form of equation (1). The full potential equation written in the computation domain (ξ-η coordinate system) is given by

$$\left(\frac{\rho U}{J}\right)_\xi + \left(\frac{\rho V}{J}\right)_\eta = 0 \tag{2}$$

$$\rho = \left[1 - \frac{\gamma - 1}{\gamma + 1}\left(U\phi_\xi + V\phi_\eta\right)\right]^{1/(\gamma-1)} \tag{3}$$

where

$$U = A_1\phi_\xi + A_2\phi_\eta, \quad V = A_2\phi_\xi + A_3\phi_\eta, \quad J = \xi_x\eta_y - \xi_y\eta_x \left.\phantom{\begin{matrix}1\\1\end{matrix}}\right\}$$
$$A_1 = \xi_x^2 + \xi_y^2, \quad A_2 = \xi_x\eta_x + \xi_y\eta_y, \quad A_3 = \eta_x^2 + \eta_y^2 \tag{4}$$

U and V are the contravariant velocity components along the ξ and η directions, respectively; A_1, A_2, and A_3 are metric quantities; and J is the Jacobian of the transformation.

Grid Generation

Because only wraparound meshes have been used to date, the workshop standard mesh could not be adapted to the present computer code without undue difficulties. As a result, the wraparound grid has been retained for all calculations. The automatic grid generation scheme introduced by Thompson et al. (ref. 10) has been used for the generation of these wraparound meshes. Basically, this grid generation scheme uses numerically generated solutions of Laplace's equation (or in some cases Poisson's equation) to establish regular and smooth finite-difference meshes around arbitrary bodies. Details of the present scheme can be found in reference 4.

Spatial Differencing

A finite-difference approximation to equation (2), suitable for both subsonic and supersonic flow regions, is given by

$$\overleftarrow{\delta}_\xi\left[\tilde{\rho}_i\left(\frac{U}{J}\right)_{i+1/2,j}\right] + \overleftarrow{\delta}_\eta\left[\overline{\rho}_j\left(\frac{V}{J}\right)_{i,j+1/2}\right] = 0 \tag{5a}$$

$$\tilde{\rho}_i = \left[(1 - \nu)\rho\right]_{i+1/2,j} + \nu_{i+1/2,j}\,\rho_{i+k+1/2,j} \tag{5b}$$

$$\overline{\rho}_j = \left[(1 - \nu)\rho\right]_{i,j+1/2} + \nu_{i,j+1/2}\,\rho_{i,j+\ell+1/2} \tag{5c}$$

where

$$k = \begin{cases} -1 & \text{when } U_{i+1/2,j} > 0 \\ 1 & \text{when } U_{i+1/2,j} < 0 \end{cases} \qquad \ell = \begin{cases} -1 & \text{when } V_{i,j+1/2} > 0 \\ 1 & \text{when } V_{i,j+1/2} < 0 \end{cases} \qquad (6)$$

and

$$\nu_{i+1/2,j} = \begin{cases} \max[(M_{i,j}^2 - 1)C, 0] & \text{for } U_{i+1/2,j} > 0 \\ \max[(M_{i+1,j}^2 - 1)C, 0] & \text{for } U_{i+1/2,j} < 0 \end{cases} \qquad (7)$$

the operators $\overleftarrow{\delta}_\xi(\ )$ and $\overleftarrow{\delta}_\eta(\ )$ are first-order-accurate backward-difference operators in the ξ and η directions, respectively; $M_{i,j}$ is the local Mach number; C is a user specified constant (usually between 1.0 and 2.0); and the quantities U and V are computed by standard second-order-accurate finite-difference formulas. The density (ρ) is computed in a straightforward manner from the second-order-accurate, discretized version of equation (3) and is stored at half points in the finite-difference mesh (i.e., i + 1/2, j + 1/2). Values needed at i + 1/2,j or i,j + 1/2 are obtained by using simple averages.

Use of the density coefficients given by equations (5b) and (5c) is equivalent to the addition of an appropriately differenced artificial viscosity term (refs. 4 and 5)

$$-\Delta\xi\left(\nu\rho_\xi\frac{|U|}{J}\right)_\xi - \Delta\eta\left(\nu\rho_\eta\frac{|V|}{J}\right)_\eta \qquad (8)$$

This effectively maintains an upwind influence in the differencing scheme for supersonic regions anywhere in the finite difference mesh for any orientation of the velocity vector, thus approximating a rotated differencing scheme.

The scheme given by equation (5) is centrally differenced and second-order-accurate in subsonic regions. In supersonic regions, the differencing is a combination of the second-order-accurate central differencing used in subsonic regions and the first-order-accurate upwind differencing resulting from the upwind evaluation of the density. As the flow becomes increasingly supersonic, the scheme is increasingly retarded in the upwind direction.

AF2 Iteration Scheme

The AF2 fully implicit approximate factorization scheme is given by

$$\text{Step 1} \qquad \left[\alpha - \overrightarrow{\delta}_\eta \bar{\rho}_j^n\left(\frac{A_3}{J}\right)_{i,j-1/2}\right] f_{i,j}^n = \alpha\omega L\phi_{i,j}^n \qquad (9)$$

$$\text{Step 2} \qquad \left[\alpha\overleftarrow{\delta}_\eta \mp \alpha\beta\overrightarrow{\delta}_\xi - \overleftarrow{\delta}_\xi \tilde{\rho}_i^n\left(\frac{A_1}{J}\right)_{i+1/2,j} \overrightarrow{\delta}_\xi\right] C_{i,j}^n = f_{i,j}^n \qquad (10)$$

where the n superscript is an iteration index, α is an acceleration parameter (see ref. 3 for a discussion of α), ω is a relaxation parameter (set equal to 1.8 for all cases), $L\phi_{i,j}^n$ is the nth iteration residual operator (defined by equation (5a)), and $f_{i,j}^n$ is an intermediate result stored at each point

in the finite-difference mesh. In step 1, the f array is
obtained by solving a simple bidiagonal matrix equation for each
ξ = constant line. The correction array ($C_{i,j}^n = \phi_{i,j}^{n+1} - \phi_{i,j}^n$)
is then obtained in the second step from the f array by solving
a tridiagonal matrix equation for each η = constant line. Note
that with the AF2 scheme the η-direction difference approximation is split between the two steps. This generates a $\phi_{\eta t}$-type
term, which is useful to the iteration scheme as timelike dissipation. (The iterative process is considered as an iteration
in pseudotime. Thus, the time derivative is introduced by
( )$^{n+1}$ - ( )$^n \sim$ ( )$_t$.) The split η term also places a sweep
direction restriction on both steps, namely, outward (away from
the airfoil) for the first step and inward (toward the airfoil)
for the second step (ref. 6). No sweep restrictions are placed
on either of the two sweeps due to flow direction.

A $\phi_{\xi t}$-type term has been added inside the brackets of step 2
(see eq. (10)), to provide time-dependent dissipation in the
ξ direction. The parameter β is fixed at a value of 0.3 in
subsonic regions and specified as needed in supersonic regions
(usually between 1.0 and 5.0). The double arrow notation in
equation (10) on the δ_ξ-difference operator indicates that the
difference is always upwind, which on the upper surface is a
backward difference and on the lower surface is a forward difference. The sign is chosen in such a way that the addition
of $\phi_{\xi t}$ increases the magnitude of the second sweep diagonal.
Further details about the AF2 algorithm are discussed in
references 4-6.

3.   COMPUTED RESULTS

The full potential equation algorithm just presented has been
coded into a transonic airfoil analysis computer code (TAIR).
All of the workshop airfoil test problems have been computed
with TAIR, and several selected cases are discussed in this
section. Because TAIR has been written for use with a wraparound grid, the standard workshop grid was not used. Instead
a wraparound grid with an equivalent number of mesh points was
selected (105 × 28). Unless otherwise noted, the outer boundary
for each grid was located at a radius of six chords from the
airfoil leading edge. In addition to the standard C_p vs x/c
and C_p vs 10 y/c plots, the lift, drag, and quarter-chord
pitching-moment coefficients obtained from surface pressure
integration are included for each case. The number of iterations and CPU time (Ames CDC 7600 computer, FTN compiler, OPT=2)
required for convergence is also indicated in each case as
standard output. For all cases in which the predicted shock
strength occurred in the full potential regime (i.e., for cases
in which the maximum local Mach number was near or less than
1.3), two additional solutions on refined meshes have been
included. The refined mesh calculations contain two and four
times as many grid points as the standard mesh (i.e., 149 × 39
and 209 × 56). To accommodate the fine grid calculations,
several arrays of storage were moved to the less efficient
large core memory on the 7600 computer. As a result the CPU
time per iteration per grid point for these calculations is
roughly 25% larger than for the in core, standard grid calculations. For all cases the initial solution was that of freestream flow. Iteration was continued until the maximum residual

dropped by four orders of magnitude, which generally represents
a very "tight" level of convergence. For instance, in many
strong shock calculations this level of convergence represents
three times the number of iterations required for plottable
accuracy.

Of the 13 workshop cases proposed, six are considered to lie
within the full potential regime. These cases are (1) NACA
0012, $M_\infty = 0.72$, $\alpha = 0°$; (2) NACA 0012, $M_\infty = 0.63$, $\alpha = 2°$;
(3) NACA 0012, $M_\infty = 0.8$, $\alpha = 0°$; (4) RAE 2822, $M_\infty = 0.676$,
$\alpha = 1°$; (5) CAST 7, $M_\infty = 0.7$, $\alpha = -1°$; and (6) KORN I, $M_\infty = 0.75$,
$\alpha = 0.115°$. These solutions are either subcritical or super-
critical with shock waves weak enough to be reasonably approximated by the isentropic, irrotational full potential formulation.
The full potential solutions for these six cases should agree
well with Euler solutions. The full potential solutions for
the remaining seven strong shock cases will be incorrectly predicted relative to the Euler solutions. In each strong shock
case the full potential shock strength and position will be
stronger and further downstream than the shock strength and
position for a corresponding Euler solution. This is a well
known result, at least when conservative form is used. Nevertheless, with this information, even the strong shock full potential solutions can be quite useful. For instance, in a
transonic cruise design application, a solution with even a
moderate strength shock would probably be intolerable because
of the resulting wave drag. These cases could easily and efficiently be eliminated from consideration by using the "pessi-
mistic" full potential formulation. As the proper weak shock
solution is approached in the design process, the full potential
formulation would provide an accurate solution. This type of
design procedure implemented with the Euler equations would be
quite expensive.

The first case discussed concerns the Korn airfoil at $M_\infty = 0.75$
and $\alpha = 0.115°$, which is the design condition. The pressure
coefficient distributions for this case on all three meshes
(105×28, 149×39 and 209×56) are shown in figure 1. Instead of
the expected shock-free condition, a weak shock is predicted at
about 70% of chord. An indication of shock strength is given
by the wave drag (C_D), which for this case is too small to
register above truncation error. It is interesting to note
that the shock strength decreases and the lift coefficient
approaches the design value as the mesh is refined. Usually
solutions computed on the intermediate mesh provide adequate
accuracy (relative to the fine mesh) while solutions on the
coarse mesh generally contain small differences.

The second case discussed herein involves an NACA 0012 airfoil
at a freestream Mach number of 0.95 and zero degrees angle of
attack. Besides the standard grid solution (105×28), two additional solutions with outer computational boundary locations at
12 and 24 chords have been computed, both on 149×56 meshes.
The pressure coefficient distributions for this series of calculations is shown in figure 2. The axis along the chord of
the airfoil has been extended several chords downstream of the
airfoil trailing edge. Immediately obvious is the existence of
a two-shock system, i.e., a so-called "fishtail shock." The
first shock is an oblique shock (supersonic-to-supersonic) at

the airfoil trailing edge.
The second shock is a normal
shock (supersonic-to-
subsonic) and exists downstream of the airfoil trailing edge. As the outer
boundary is extended, the
second shock location moves
further downstream in such
a way that the final position is difficult to predict.
An additional mesh refine-
ment study for a fixed outer
computational boundary produced no effect on the
normal shock location.
Several features of this
flow field do seem to be
invariant with outer boundary position, namely, the
airfoil surface pressure
distribution, the position
of the oblique shock, and
the pressure level down-
stream of the oblique shock.
The solutions of figure 2
seem to be qualitatively
correct; however, obtaining
anything more quantitative
for this sensitive strong

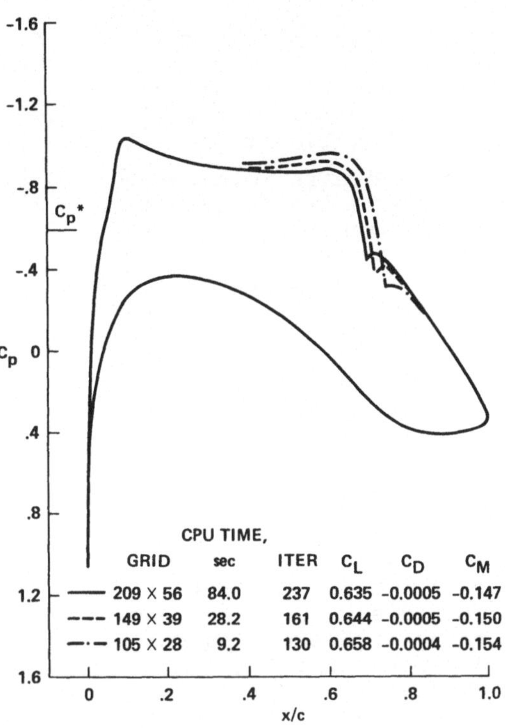

Figure 1. – Pressure coefficient
distributions (KORN I airfoil,
$M_\infty = 0.75$, $\alpha = 0.115°$).

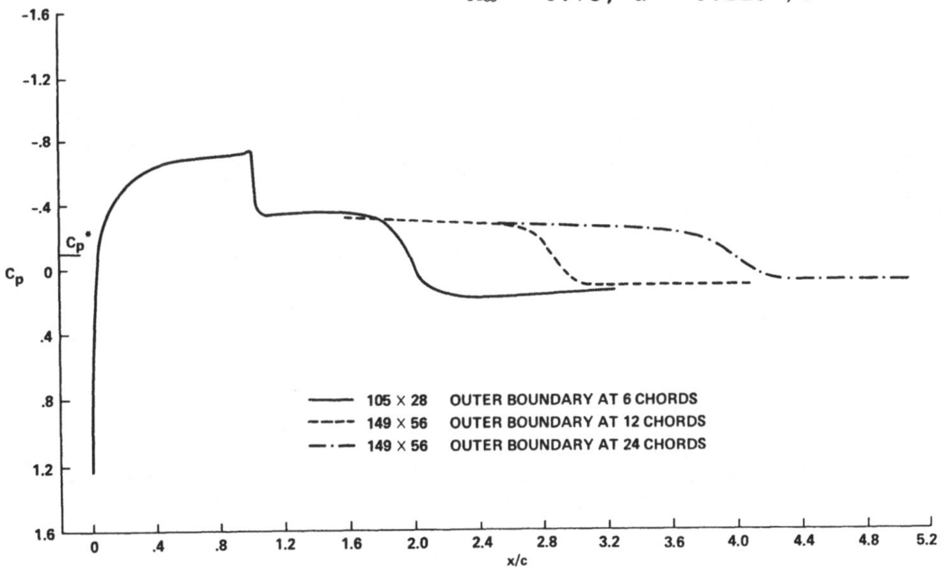

Figure 2. – Pressure coefficient distributions with x/c axis
extended downstream of airfoil trailing edge (NACA 0012
airfoil, $M_\infty = 0.95$, $\alpha = 0°$).

shock solution (maximum local Mach number is 1.44) from the present full potential formulation will be very difficult.

The third case involves the NACA 0012 airfoil at a freestream Mach number of 0.85 and one degree angle of attack. The pressure coefficient distribution for this calculation is shown in figure 3. A very strong shock at the airfoil trailing edge upper surface and a moderately strong shock at 65% of chord on the lower surface are clearly evident. Convergence histories for this calculation are shown in figure 4, where the lift coefficient (C_L), the number of supersonic points (NSP), and the maximum residual ($|R|_{max}$) are all plotted versus iteration number. This calculation was one of the more difficult because just over 200 iterations were required to reduce the maximum residual by four orders of magnitude. However, as indicated by the C_L and NSP curves, crude convergence is obtained in about 60 iterations. At this point the maximum residual has just begun to drop. Therefore, for this case, a four-order-of-magnitude reduction in $|R|_{max}$ represents an extremely conservative convergence criterion. This behavior is quite characteristic of the AF2 iteration scheme for cases involving strong shock waves (see refs. 5 and 6 for more discussion on this point). In contrast, the convergence history for a subcritical calculation (NACA 0012 airfoil, $M_\infty = 0.63$, $\alpha = 2°$) is shown in figure 5. For this particular case the $|R|_{max}$ drops much faster than in the previous case even though the lift evolves at about

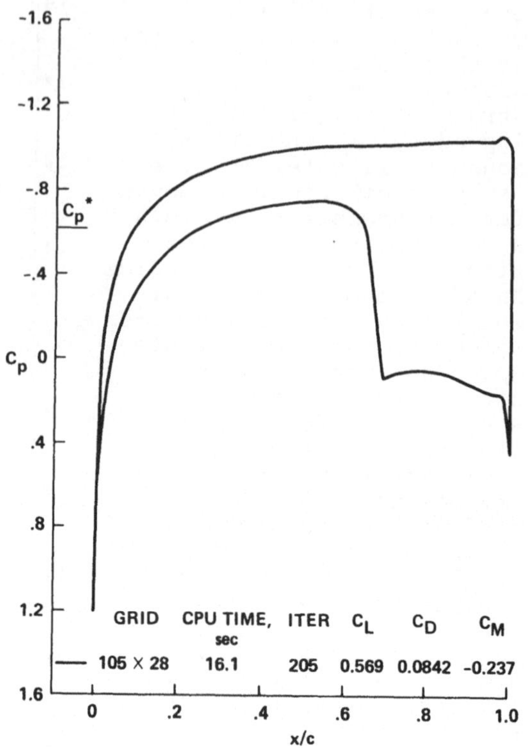

Figure 3. – Pressure coefficient distribution (NACA 0012, $M_\infty = 0.85$, $\alpha = 1°$).

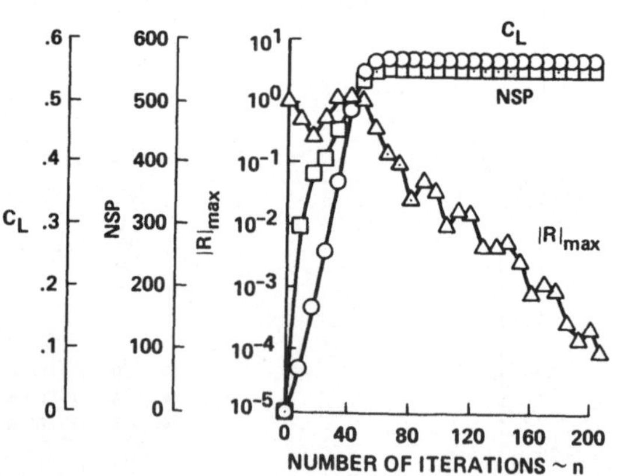

Figure 4. – Convergence history (NACA 0012 airfoil, $M_\infty = 0.85$, $\alpha = 1°$).

the same rate. For this subcritical case a four-order-of-magnitude reduction in $|R|_{max}$ represents only a slightly conservative convergence criterion.

4. CONCLUDING REMARKS

All of the 13 workshop airfoil cases have been computed with the present conservative full potential equation formulation. Six of these cases are considered to lie within the full potential regime, i.e., either subcritical or supercritical weak shock solutions. For each of these six cases, solutions on successively finer meshes have been computed. Solutions computed on the intermediate mesh (149×39)

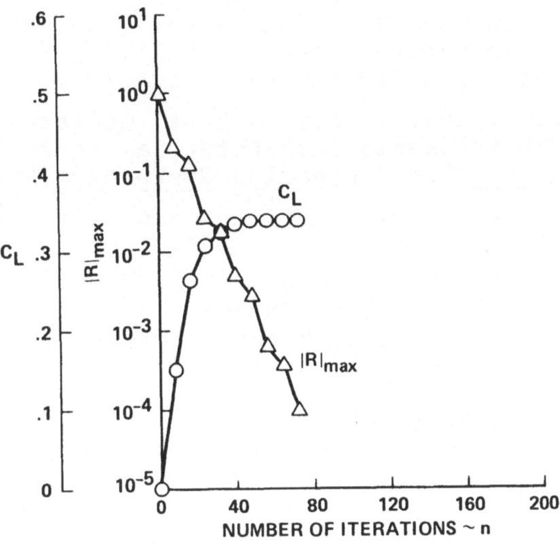

Figure 5. – Convergence history (NACA 0012 airfoil, M_∞ = 0.63, α = 2°).

usually provided adequate accuracy (relative to the fine mesh) while the coarse mesh generally produced solutions with small discrepancies. The remaining seven strong shock calculations were easily computed with the present conservative full potential formulation. Although discrepancies between these solutions and corresponding Euler solutions will exist, valuable qualitative information about these strong shock solutions can be efficiently and routinely obtained from the present full potential formulation.

ACKNOWLEDGMENT

The author expresses gratitude to Reese Sorenson for writing the graphics routine used to plot the results presented for this workshop.

REFERENCES

1. Ballhaus, W. F. and Steger, J. L., "Implicit Approximate Factorization Schemes for the Low-Frequency Transonic Equation," NASA TM X-73,082, 1975.

2. Ballhaus, W. F., Jameson, A., and Albert, J., "Implicit Approximate Factorization Schemes for the Efficient Solution of Steady Transonic Flow Problems," AIAA Journal, Vol. 16, No. 6, 1978, pp. 573-579.

3. Holst, T. L. and Ballhaus, W. F., "Fast Conservative Schemes for the Full Potential Equation Applied to Transonic Flows," NASA TM-78469, 1978. (Also AIAA Journal, Vol. 17, No. 2, Feb. 1979, pp. 145-152.)

4. Holst, T. L., "An Implicit Algorithm for the Conservative, Transonic Full Potential Equation Using an Arbitrary Mesh," AIAA Paper 78-1113, July 1978.

5.   Holst, T. L. and Albert, J., "An Implicit Algorithm for the Conservative Transonic Full Potential Equation with Effective Rotated Differencing," NASA TM-78570, April 1979.

6.   Holst, T. L., "A Fast, Conservative Algorithm for Solving the Transonic Full-Potential Equation," _AIAA Fourth Computational Fluid Dynamics Conference Proceedings_, July 1979.

7.   Jameson, A., "Transonic Potential Flow Calculations Using Conservation Form," _AIAA Second Computational Fluid Dynamics Conference Proceedings_, June 1975, pp. 148-155.

8.   Hafez, M. M., Murman, E. M., and South, J. C., "Artificial Compressibility Methods for Numerical Solution of Transonic Full Potential Equation," AIAA Paper 78-1148, July 1978.

9.   Eberle, A., "Transonic Potential Flow Computations by Finite Elements: Airfoil and Wing Analysis, Airfoil Optimization," Lecture held at the DGLR/GARTEUR 6 Symposium, Transonic Configurations, Bad Harzburg, Germany, June 1978.

10.   Thompson, J. F., Thames, F. C., and Mastin, C. M., "Automatic Numerical Generation of Body-Fitted Curvilinear Coordinate System for Field Containing Any Number of Arbitrary Two-Dimensional Bodies," _Journal of Computational Physics_, Vol. 15, 1974, pp. 299-319.

RELAXATION METHOD FOR THE FULL-POTENTIAL EQUATION

by Jean-Jacques CHATTOT and Colette COULOMBEIX

Office National d'Etudes et de Recherches Aérospatiales (ONERA)
92320 Châtillon (France)

♦

INTRODUCTION

In this paper a brief and fragmentary account is made of our contribution to the GAMM workshop on Numerical Methods for the Computation of Inviscid Transonic Flow with Shock Waves, since the main objective is the comparison, during the actual workshop, of the results obtained by various methods. Emphasis however is put in the first paragraph on the basic assumptions underlying the mathematical modelling of transonic flow, using the full-potential equation. In particular the semi-conservative form of the equation, used in the method, is derived. In the second part, the discretization schemes and the solution algorithm are sketched and reference is given to a more detailed paper. A sample of results is presented in the last paragraph, and the sensitivity of the numerical solution to the space discretization is shown due to an insufficient mesh concentration of the proposed mesh in the nose region.

1. MATHEMATICAL MODEL

The full-potential equation is used as mathematical model for steady transonic flows of perfect fluid. In cartesian coordinates in two space dimensions the following formulation holds :

$$(1) \qquad \begin{aligned} &\frac{\partial \rho u}{\partial x} + \frac{\partial \rho v}{\partial y} = 0 \\[1mm] &\rho = \left[(\gamma - 1)\, M_\infty^2 \left(h_i - \frac{q^2}{2} \right) \right]^{\frac{1}{\gamma - 1}} \\[1mm] &(u, v) = \overrightarrow{\text{grad}}\ \Phi \end{aligned}$$

where :

ρ is the density

$(u, v) = \vec{q}$ is the velocity vector

γ is the ratio of specific heats ($\gamma = 1,4$ in this application)

$h_i = \dfrac{1}{(\gamma - 1) M_\infty^2} + \dfrac{1}{2}$ is the stagnation enthalpy

M_∞ is the Mach number of the unperturbed flow

Φ is the full-potential variable.

All the dimensional quantities are normalized by the free-stream conditions and a reference length (chord of the profile).

System (1) expresses the conservation of mass and energy. The pressure p and the pressure coefficient C_p are related to the density by the isentropic law :

$$(2) \qquad p = \frac{1}{\frac{\gamma}{2} M_\infty^2} \rho^\gamma$$

$$C_p = \frac{\rho^\gamma - 1}{\frac{\gamma}{2} M_\infty^2}$$

the assumption of irrotational-isentropic flow implies a deviation from the Euler solution whenever shock waves are present, since the chosen model does not conserve momentum accross discontinuities. However, for a wide class of airfoils of practical interest, the shock waves are weak and it is commonly accepted that for Mach numbers upstream of the shocks up to 1.3–1.4 the isentropic shock polar and the Rankine-Hugoniot shock polar are in good agreement, and hence the solution to equation (1) is a good approximation to the Euler system. It remains to be seen, and this is one of the concerns of this workshop, that this is indeed the case.

Equation (1) is in conservation law form and admits several weak solutions to a given boundary value problem. Some of these solutions contain expansion shocks which are approximations to discontinuities of Euler equations through which the entropy decreases. Hence they do not model physical discontinuity surfaces, for which the entropy always increases, and must be rejected. It has been shown [1] that a uniqueness condition can be formulated for equation (1) to replace the entropy condition of the Euler system lost with the approximation of irrotational-isentropic flow as :

$$(3) \qquad \rho u \left[\frac{\partial}{\partial x} \left(\rho u^2 + p \right) + \frac{\partial}{\partial y} \left(\rho u v \right) \right] + \rho v \left[\frac{\partial}{\partial x} \left(\rho u v \right) + \frac{\partial}{\partial y} \left(\rho v^2 + p \right) \right] \geqslant 0$$

the elimination of expansion shocks and the capture of compression shocks is obtained by *Jameson* [2] by the introduction of an upwind bias in supersonic, a generalization of the upwind scheme of *Murman* and *Cole* [3] to the case of an arbitrary flow direction. This can be interpreted as an artificial viscosity term added to the equation as, ref. [2] :

$$\frac{\partial}{\partial x} \left(\rho u \right) + \frac{\partial}{\partial y} \left(\rho v \right) + T = 0$$

$$(4) \qquad T = -\frac{\partial}{\partial x} \left(\Delta x \, \mu \, |u| \, \frac{\partial \rho}{\partial x} \right) - \frac{\partial}{\partial y} \left(\Delta y \, \mu \, |v| \, \frac{\partial \rho}{\partial y} \right)$$

$$\mu = \max \left(0, \ 1 - \frac{a^2}{q^2} \right)$$

where a is the speed of sound defined by $\qquad a^2 = \dfrac{\rho^{\gamma - 1}}{M_\infty^2}$

It must be noted that T is in gradient or conservation form which preserves the jump conditions in the limit of decreasing mesh sizes, $\Delta x, \Delta y \longrightarrow 0$ the artificial viscosity term is proportional to Δx, Δy and vanishes with μ in subsonic where the scheme is second order accurate. In supersonic, for the upper Mach number range of 1.3–1.4, μ is order one (0(1)) and the scheme is only first order accurate.

Recently a slightly different formulation was proposed in [4 and 5] to introduce the penalization. This approach is called the artificial compressibility or artificial density method. It consists of including the term T

in the formula to compute the density, in such a way that the equation retains the very simple form (1). Let $\tilde{\rho}$ be a modified density. The equation to be solved is :

$$(5) \quad \frac{\partial \tilde{\rho} u}{\partial x} + \frac{\partial \tilde{\rho} v}{\partial y} = 0$$

$$\text{with} \quad \tilde{\rho} = \rho + \Delta x \, \mu \, \frac{\rho}{a^2} |u| \frac{\partial^2 \phi}{\partial x^2} + \Delta y \, \mu \, \frac{\rho}{a^2} |v| \frac{\partial^2 \phi}{\partial y^2}$$

It can be easily shown that (4) and (5) are equivalent in the leading terms. The full advantage of this can be appreciated in the actual coding of the method, if, as is done in the present work, system (5) is discretized using centered schemes, not only for the mass conservation equation, but also when computing the artificial density. This is in contrast to the schemes of *Holst* and *Hafez* [4, 5], in which upwind schemes are used for $\tilde{\rho}$. It must be noted also that the discretization and the associated matrix are similar at subsonic and supersonic points, and along coordinate lines tridiagonal matrices are obtained at all points, as is the case with (4) in subsonic flow only.

The equation is solved in a curvilinear coordinate system. The mass conservation equation, written in strong conservation law form as indicated by *Viviand* [6], and the artificial density, now assume the form :

$$(6) \quad \frac{\partial}{\partial \xi}\left(\frac{\tilde{\rho} U}{J}\right) + \frac{\partial}{\partial \eta}\left(\frac{\tilde{\rho} V}{J}\right) = 0$$

$$\tilde{\rho} = \rho + \Delta \xi \, \mu \, \frac{\rho}{a^2} |u| \frac{\partial^2 \phi}{\partial \xi^2} + \Delta \eta \, \mu \, \frac{\rho}{a^2} |v| \frac{\partial^2 \phi}{\partial \eta^2}$$

U and V are the contravariant components of the velocity, $U = \vec{q} \cdot \overrightarrow{\text{grad}} \, \xi$, etc. J is the Jacobian of the transformation. This easily extends to three dimensions. In fact we are using the semi-conservative form of system (6) which consists in writing the metric coefficients in front of the partial derivatives as :

$$(7) \quad A \frac{\partial}{\partial \xi}\left(\tilde{\rho}\, \frac{\partial \phi}{\partial \xi}\right) + B\left[\frac{\partial}{\partial \xi}\left(\tilde{\rho}\, \frac{\partial \phi}{\partial \eta}\right) + \frac{\partial}{\partial \eta}\left(\tilde{\rho}\, \frac{\partial \phi}{\partial \xi}\right)\right] + C \frac{\partial}{\partial \eta}\left(\tilde{\rho}\, \frac{\partial \phi}{\partial \eta}\right) + G = 0$$

$$A = \nabla \xi \cdot \nabla \xi \quad , \quad B = \nabla \xi \cdot \nabla \eta \quad , \quad C = \nabla \eta \cdot \nabla \eta \quad , \quad G = \nabla^2 \xi \, \tilde{\rho} \frac{\partial \phi}{\partial \xi} + \nabla^2 \eta \, \tilde{\rho} \frac{\partial \phi}{\partial \eta}$$

$\tilde{\rho}$ is the same as in (6).

2. DISCRETIZATION SCHEMES AND RELAXATION METHOD

The discretization schemes, as mentioned previously, are all centered schemes, similar to those used in the three-dimensional version of the method, presented in [7]. The boundary conditions and their discretization are also treated as in [7]. For reasons of convenience, fictitious points are introduced inside the boundary, as mirror images of the points nearest to the boundary. Thus, two versions of the workshop mesh are available, one of which satisfies this requirement. This is shown figure 1. Some results are also computed on a sheared parabolic mesh used in [2] and shown figure 2.

The relaxation method is of the approximate factorization type, first developped by *Ballbaus* and *Steger* [8] to solve the small perturbation transonic potential equation.

The iteration process can be described by :

$$(8) \qquad \left(\alpha - A\frac{\overleftarrow{\partial}}{\partial\xi}\,\tilde{\rho}\,\frac{\overrightarrow{\partial}}{\partial\xi}\right)\left(\alpha - C\frac{\overleftarrow{\partial}}{\partial\eta}\,\tilde{\rho}\,\frac{\overrightarrow{\partial}}{\partial\eta}\right)\left(\Phi^{n+1} - \Phi^n\right) = \alpha\,\omega\,\mathbb{R}(\Phi^n)$$

α is the acceleration parameter $\qquad \alpha > 0$

ω is the relaxation parameter $\qquad 0 \leq \omega \leq 2$

$\mathbb{R}$ is the residual of the equation, corresponding to the left hand side of (7).

It can be noted that this decomposition does not allow for an implicit treatment of the cross-derivative terms and hence improvements can be sought in this direction, see [7].

3. SAMPLE OF RESULTS

Some results are presented hereafter. To give some indication of the sensitivity of the solution of the space discretization, two mesh systems of 141×21 points have been used. The computation of the flows past a NACA 0012 airfoil at $M_\infty = 0.63$ and $\alpha = 2°$ yields a pressure coefficient distribution shown on figure 3. This result is obtained with the mesh proposed for the workshop.

The same case has been computed with a mesh system which gives a higher concentration of points near the leading edge where the large gradients are located, and both results are compared on figure 4. This last result compares favourably with the theoretical data of *Sells* reported in [9] who uses a mesh generated by conformal mapping of the profile into a circle. The second example concerns the computation of the transonic flow past the Korn airfoil at $M_\infty = 0.75$ and $\alpha = 0.115°$. This is the design condition for a shock-free flow [10]. Figure 5 shows the pressure distribution obtained with the workshop mesh, and this is compared with the sheared-parabolic mesh on figure 6. The "peaky" pressure distribution is better represented with the second mesh, however a weak compression wave still exists downstream of the supersonic region. A last result is presented with a finer grid composed of 201×25 points. The numerical solution seems to converge to the analytical solution as the mesh is refined but due to storage limitation it was not possible to use an even finer grid. This last result is shown along with the exact solution on figure 7.

The examples presented indicate the importance of the space discretization. Even in the case of a purely subsonic flow, first example, the accuracy of the solution near the leading edge strongly depends on the mesh distribution in the nose region. This seems to be even more so in the case of transonic flow, since the expansion waves emanating from the upstream part of the supersonic zone are reflected by the sonic line and the body and largely determine, by a complex interaction process, the structure of the recompression region leading the flow from supersonic back to subsonic through either a shock wave or a smooth recompression.

REFERENCES

[1] CHATTOT, J.J., "Condition d'unicité pour les écoulements irrotationnels stationnaires de fluide parfait compressible", C.R.Ac. Sc., Paris, t. 268-A (1978), p. 111-113.

[2] JAMESON, A., "Transonic flow calculations", VKI Lecture Series 87, March 15-19, 1976.

[3] MURMAN, E.M., COLE, J.D., "Calculation of plane steady transonic flows", AIAA Journal, vol. 9, 1971, p. 114-121.

[4] HOLST, T.L., BALLHAUS, W.F., "Conservative implicit schemes for the full potential equation applied to transonic flows", NASA TM-78469, March 1978.

[5] HAFEZ, M.M., SOUTH, J.C., MURMAN, E.M., "Artificial compressibility methods for numerical solution of transonic full potential equation", AIAA paper 78-1148, July 1978.

[6] VIVIAND, H., "Conservative forms of gas dynamics equations", La Rech. Aérosp., No 1, (1974), p. 65-68.

[7] CHATTOT, J.J., COULOMBEIX, C., da SILVA TOMÉ, C., "Calculs d'écoulements transsoniques autour d'ailes", La Rech. Aérosp., 1978-4, p. 143-159.

[8] BALLHAUS, W.F., STEGER, J.L., "Implicit approximate factorization schemes for the low frequency transonic equation", NASA, TM X-73, 082, 1975.

[9] LOCK, R.C., "Test cases for numerical methods in two-dimensional transonic flows", AGARD Report No 575, Nov. 1970.

[10] GARABEDIAN, P.R., KORN, D.G., "Numerical design of transonic airfoils", Numerical solution of partial differential equations. II — Academic Press, New York, 1971, pp. 253-271.

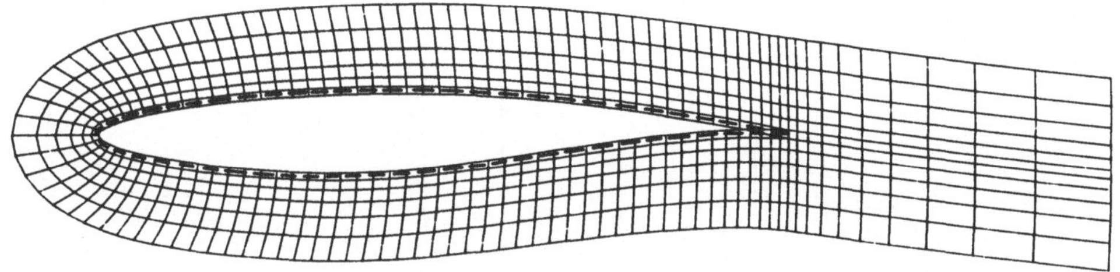

*Fig. 1 — Workshop mesh (partial view) RAE 2822 profile —
141 x 21 points.*

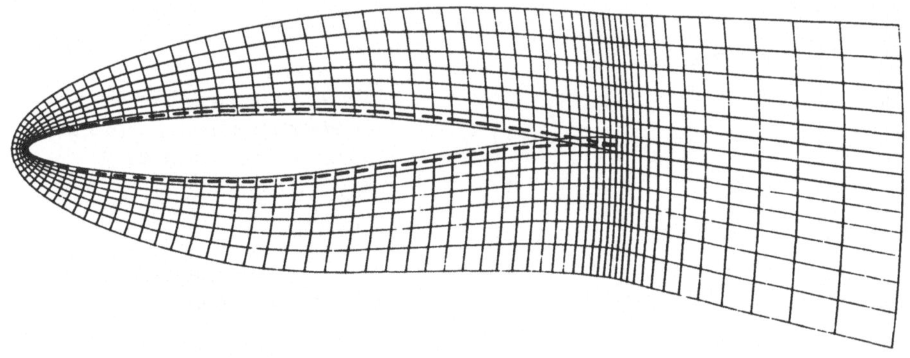

*Fig. 2 — Sheared parabolic mesh of ref. [7] (partial view) RAE 2822 profile
141 x 21 points.*

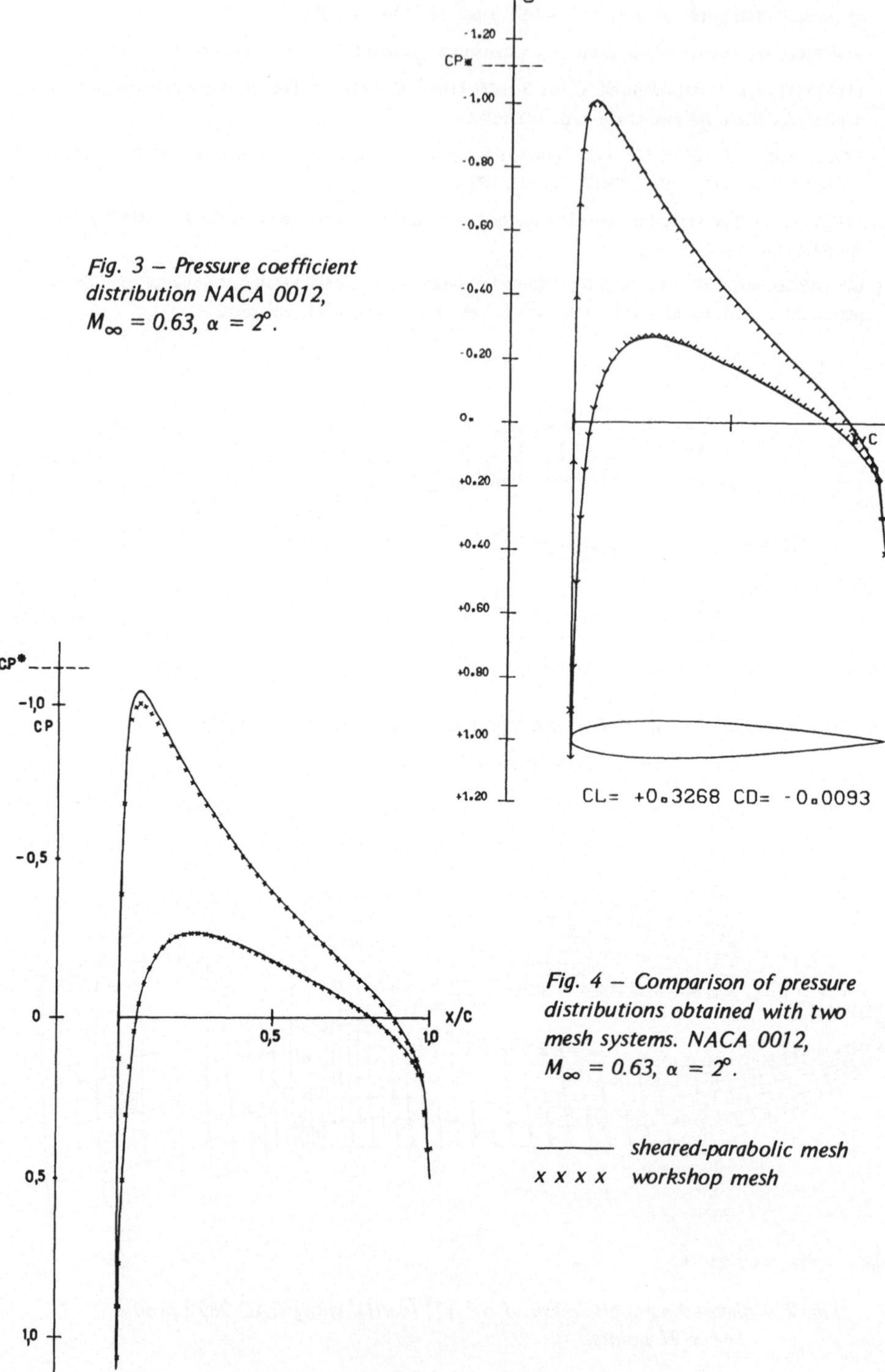

Fig. 3 – Pressure coefficient distribution NACA 0012, $M_\infty = 0.63$, $\alpha = 2°$.

CL= +0.3268 CD= -0.0093

Fig. 4 – Comparison of pressure distributions obtained with two mesh systems. NACA 0012, $M_\infty = 0.63$, $\alpha = 2°$.

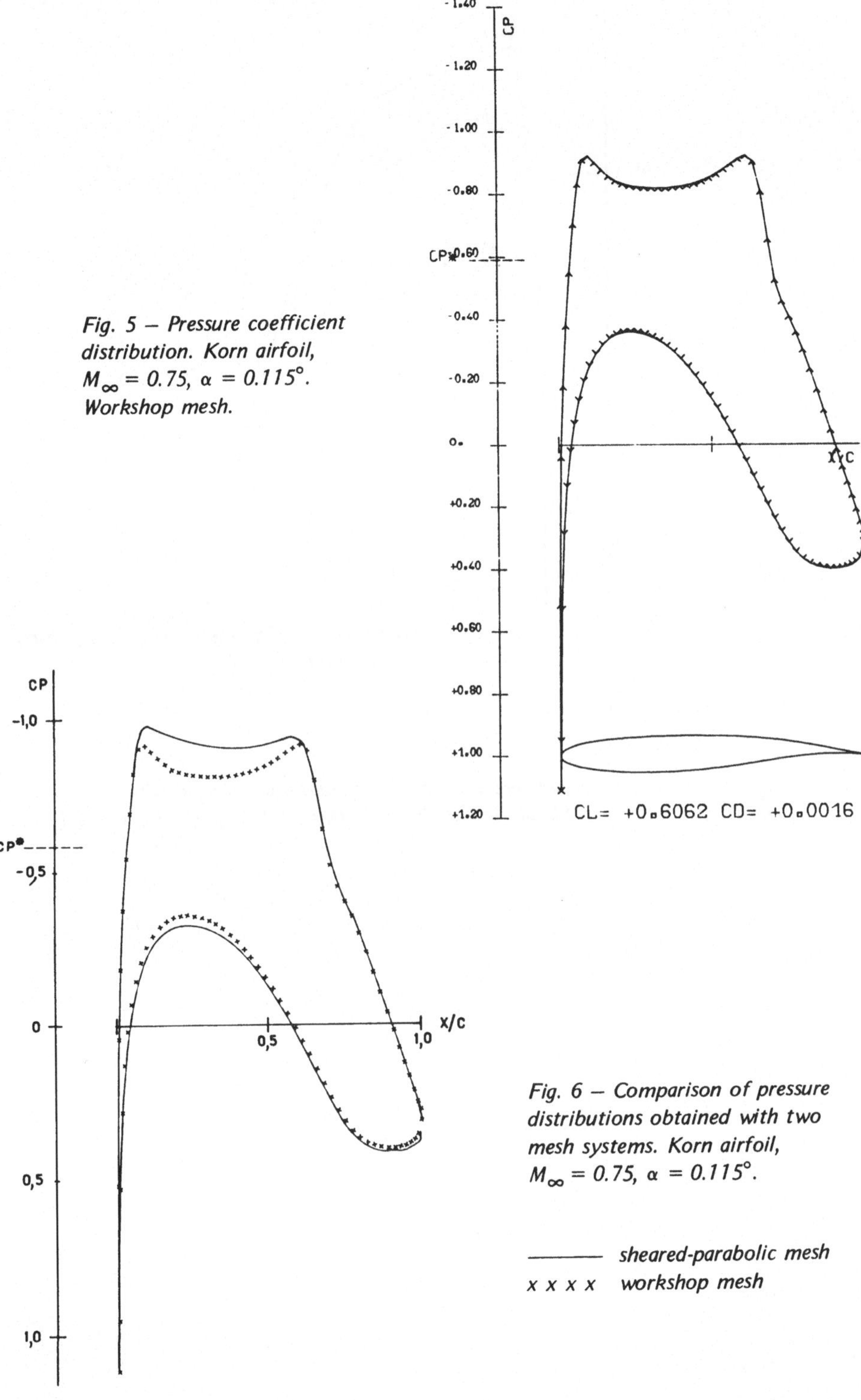

Fig. 5 — Pressure coefficient distribution. Korn airfoil, $M_\infty = 0.75$, $\alpha = 0.115°$. Workshop mesh.

Fig. 6 — Comparison of pressure distributions obtained with two mesh systems. Korn airfoil, $M_\infty = 0.75$, $\alpha = 0.115°$.

——— sheared-parabolic mesh
x x x x workshop mesh

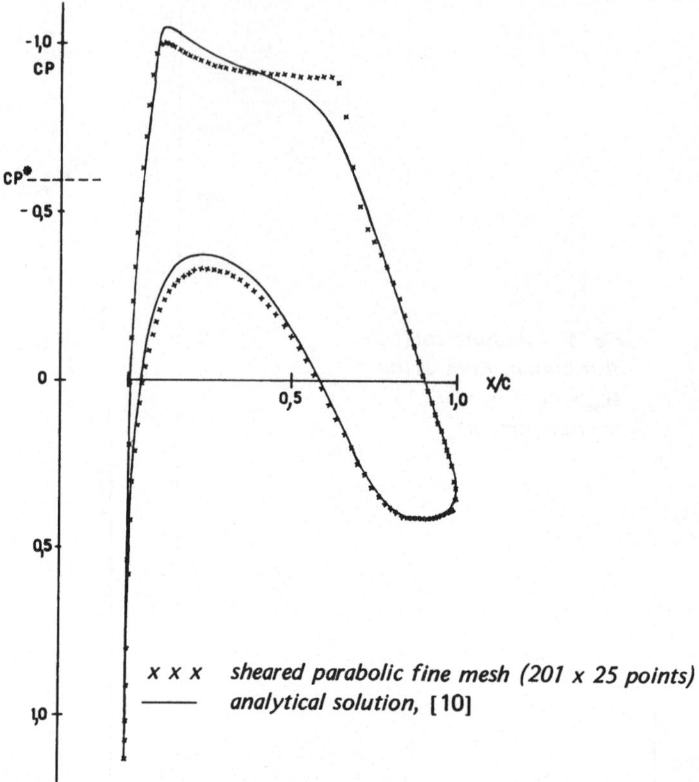

Fig. 7 — Comparison of pressure coefficient distributions obtained with a fine mesh with the exact analytical solution. Korn airfoil, $M_\infty = 0.75$, $\alpha = 0.115°$.

COMPUTATION OF STEADY INVISCID TRANSONIC FLOWS
USING PSEUDO-UNSTEADY METHODS

by Jean-Pierre Veuillot and Henri Viviand

Office National d'Etudes et de Recherches Aérospatiales (ONERA)
92320 Châtillon (France)

1 - THE PSEUDO-UNSTEADY SYSTEMS

The calculations presented at this workshop have been carried out by means of two pseudo-unsteady methods which have already been presented in details in Refs. [1] and [2]. Therefore only a general description of these methods is given here.

The pseudo-unsteady approach follows from the obvious remark that, if only steady flows are of interest, then the transient solution and the unsteady equations need not have a physical meaning. In particular it is possible, in view of reducing the computing cost, to make use of first integrals of the steady equations. Thus, for iso-energetic steady flows, the unsteady energy equation can be replaced by the steady Bernoulli relation :

$$h + \frac{1}{2}\vec{V}^2 = H_o \qquad \text{(const.), with} \qquad h = h(p,\rho) \qquad (1)$$

where h is the specific enthalpy, p the pressure, ρ the density and $\vec{V}$ the velocity vector. This is valid at steady state even when shock waves are present. The use of Bernoulli's relation (1) together with the unsteady continuity and momentum equations is not new (e.g. Refs.[3],[4],[5]) ; it leads to the first method used, which will be called method H.

In the case of steady iso-energetic and homentropic flows, we can made use not only of (1), but also of the uniformity of the specific entropy S :

$$S(p,\rho) = S_o \qquad \text{(const.), or} \qquad p = p(\rho, S_o) \qquad (2)$$

so that p and $V(=|\vec{V}|)$ are known functions of ρ . This allows to eliminate one more scalar unsteady equation, but the problem arises in the choice of the unsteady equations to be retained. We have shown,[1],[2], that the natural choice, for two-dimensional flows, is to discard the tangential (intrinsic) momentum equation, and to retain the normal (intrinsic) momentum equation. The latter can be written in the following form, after taking relations (1) and (2) into account :

$$\frac{\partial\theta}{\partial t} + \omega = 0 \qquad , \qquad \omega = \frac{\partial v}{\partial x} - \frac{\partial u}{\partial y} \qquad (3)$$

Work perfomed with the financial support of DRET.

45

where θ is the angle of the velocity vector with the x-axis, and ω
is the vorticity (x, y are cartesian or cylindrical coordinates
resp. for plane or axisymmetric flows ; $u = V\cos\theta$ and $v = V\sin\theta$ are the
corresponding velocity components). Equation (3), together with the
continuity equation, form the basis for our second method, to be called
method H-S. However, for stability reasons discussed in Refs.[1],[2], the
time-derivative term of the continuity equation is modified, and the
following pseudo-unsteady form of this equation is used :

$$ -k \, \frac{\rho_*}{a_*} \, \frac{\partial V}{\partial t} + \text{div} (\rho \vec{V}) = 0 \qquad (k = \text{const.} , > 0) \tag{4}$$

where ρ_* and a_* are the density and the sound speed at sonic conditions.
Steady solutions of the system (3), (4), represent irrotational flows.
Moreover, this system is in divergence form, if ω is written as shown
in Eq. (3), and it is easily verified that the corresponding weak solutions
satisfy the conservation of mass and of tangential velocity across a discontinuity surface. Therefore this system can be used to compute transonic
flows with shock waves under the isentropic shock approximation. This pseudo-
unsteady method for potential flows seems to be new, but similar methods
have also been studied by Essers [6],[7].

For numerical solution, the pseudo-unsteady systems of methods H and
H-S are written in the following conservative form :

$$ \frac{\partial f}{\partial t} + \frac{\partial F}{\partial x} + \frac{\partial G}{\partial y} + \varepsilon \, \frac{v}{y} \, g = 0 \tag{5}$$

where $\varepsilon = 0$ or 1 resp. for plane or axisymmetric flows, and f, F, G, g ,
are column-matrices ;

<u>in method H :</u>

$$ f = g = \begin{pmatrix} \rho \\ \rho u \\ \rho v \end{pmatrix} \quad , \quad F = \begin{pmatrix} \rho u \\ \rho u^2 + p \\ \rho u v \end{pmatrix} \quad , \quad G = \begin{pmatrix} \rho v \\ \rho u v \\ \rho v^2 + p \end{pmatrix} \tag{6}$$

(p is a known function of ρ and V from Eq. (1))
<u>in method H-S :</u>

$$ f = \begin{pmatrix} -k\rho_* V/a_* \\ \theta \end{pmatrix} , \quad F = \begin{pmatrix} \rho u \\ v \end{pmatrix} , \quad G = \begin{pmatrix} \rho v \\ -u \end{pmatrix} , \quad g = \begin{pmatrix} \rho \\ 0 \end{pmatrix} \tag{7}$$

(ρ is a known function of V through Eqs. (1) and (2)).

These two systems have been shown to be hyperbolic, and the characteristic cones have been determined. An approximate C.F.L. type stability
criterion for explicit schemes has been established.

2 - NUMERICAL SCHEME

The system (5) is discretized directly in the physical plane (i.e. without performing a coordinate transformation) in arbitrary curvilinear meshes, by means of a predictor-corrector scheme based on the MacCormack scheme. Let f_M^n be the approximate numerical value of f at the mesh point M and at the time $n\Delta t$; the two-step scheme used can be written :

$$f_M^{\widetilde{n+1}} = f_M^n - \Delta t \left(\delta_x F + \delta_y G + \varepsilon \, v \, g/y \right)_M^n \tag{8a}$$

$$f_M^{n+1} = \frac{1}{2} \left\{ f_M^n + f_M^{\widetilde{n+1}} - \Delta t \left(\widetilde{\delta}_x F + \widetilde{\delta}_y G + \varepsilon \, v \, g/y \right)_M^{\widetilde{n+1}} \right\} \tag{8b}$$

where $\delta_x \Phi$, $\widetilde{\delta}_x \Phi$ (resp. $\delta_y \Phi$, $\widetilde{\delta}_y \Phi$) are first order finite difference approximations to $\partial\Phi/\partial x$ (resp. $\partial\Phi/\partial y$) obtained by assuming a linear dependency of Φ on x and y over some triangle MPQ (P and Q being two neighbouring mesh points) ; such an approximation is given by the general formula :

$$\frac{\partial\Phi}{\partial x} = \frac{y_{MP}\,\Phi_{MQ} - y_{MQ}\,\Phi_{MP}}{y_{MP}\,x_{MQ} - y_{MQ}\,x_{MP}} + O\left(|\overrightarrow{MP}| + |\overrightarrow{MQ}| \right) \tag{9}$$

with $y_{MP} = y(P) - y(M)$, $\Phi_{MQ} = \Phi(Q) - \Phi(M)$, etc..; the formula for $\partial\Phi/\partial y$ is obtained by interchanging x and y .

The mesh points being now identified by two indices in the usual way, M being the point (k , ℓ), then, as an example, P and Q can be chosen to be the points ($k+1$, ℓ) and (k , $\ell-1$) for the definition of δ_x , δ_y , and the points ($k-1$, ℓ) and (k , $\ell+1$) for the definition of $\widetilde{\delta}_x$, $\widetilde{\delta}_y$. Four other variants are possible, as in the classical MacCormack scheme.

Instead of using the same time step at all points, we can use the maximum local value as given by the stability criterion, since the consistency in time is not necessary. The convergence can thus be much accelerated, especially in problems in which the mesh size varies greatly over the computation domain.

In regions where the flow properties exhibit strong gradients, numerical instabilities are likely to occur and a third smoothing step is added to the scheme (8), in the form :

$$f_M^{n+1(s)} = f_M^{n+1} + X_1 \frac{\Delta t}{4\Delta X} \left(|f_{ME}|\,f_{ME} + |f_{MW}|\,f_{MW} \right)^n \tag{10}$$

$$+ X_2 \frac{\Delta t}{4\Delta Y} \left(|f_{MN}|\,f_{MN} + |f_{MS}|\,f_{MS} \right)^n$$

where the superscript (δ) denotes the smoothed value, and $f_{ME} = f(E) - f(M)$, etc... The points E, W, N, S are the four neighbours of M, i.e. respectively ($k+1, \ell$), ($k-1, \ell$), ($k, \ell+1$) and ($k, \ell-1$). The constants ΔX, ΔY are measures of mesh size in the directions of the mesh lines ℓ = const. and k = const. respectively. The positive coefficients X_1, X_2, which can be variable, are of the order of unity, so that the scheme remains of second order accuracy if the mesh varies smoothly enough, i.e. if $|\overrightarrow{ME} + \overrightarrow{MW}| = O(\Delta X^2)$ and $|\overrightarrow{MN} + \overrightarrow{MS}| = O(\Delta Y^2)$.

3 - BOUNDARY CONDITIONS

At a boundary point P, the system (5) is replaced by an equivalent set of equations which are chosen to be the compatibility relations (of the pseudo-unsteady hyperbolic system used) associated with the characteristic planes through P parallel to the boundary. Such a compatibility relation can be used only if it corresponds to information coming from inside the computation domain, and this is known from the sign of the eigenvalue associated with the characteristic plane considered. The number of boundary conditions to be imposed is equal to the number of the missing compatibility relations.

A full discussion according to the different types of boundaries is given in Refs. [1] and [2]; here we shall only state the boundary conditions imposed in the test problems treated.

At a subsonic upstream boundary (fluid entering the computation domain with normal Mach number less than one), the orientation of the velocity vector is specified ; furthermore in method H the entropy distribution is specified (in method H-S , we have $S \equiv S_o$ everywhere ; it is not a boundary condition). One compatibility relation is used to compute the density.

At a subsonic downstream boundary (fluid leaving the domain with normal Mach number less than one), the static pressure is imposed. In method H, two compatibility relations and Eq. (1) allow to compute the density and velocity components ; in method H-S , one compatibility relation allows to compute the velocity angle θ.

At a solid impermeable wall, the slip conditon $\vec{V} \cdot \vec{n} = 0$ is imposed. In method H, two compatibility relations are used to compute the density and the tangential velocity ; in method H-S , one compatibility relation is used to compute the density.

In the computation of flows past profiles with a C-type mesh, the mesh line which leaves the trailing-edge and extends downstream is treated as a cut, i.e. as a double boundary line. A mesh point P on this line splits into two boundary points, say P_1 and P_2. The boundary conditions are simply $f(P_1) = f(P_2)$; together with the compatibility relations which can be used at P_1 and at P_2, they constitute a set of relations which allow to determine $f(P)$.

The compatibility relations used at a boundary point are discretized by the same scheme (8) as for inner points, except for some necessary modification of the finite-difference spatial operators at one of the two steps, since mesh points are available only on one side of the boundary.

4 - NUMERICAL APPLICATIONS

We now present some of the results obtained for the workshop test problems.

Figures 1 and 2 concern the channel flow problem and show iso-Mach contour maps of the potential solution given by method H-S (fig. 1) and of the Euler solution given by method H (fig. 2). The latter has been calculated in the workshop mesh of (72 × 21) points, and the former in a quasi-orthogonal mesh also of (72 × 21) points and very close to the workshop mesh. The two solutions are quite different ; in particular it can be noted that the sonic line reaches the upper wall in the potential solution whereas it goes up only to 41% of the channel height in the Euler solution. The shock position and strenght are also quite different. This difference can be explained mostly in terms of the mass flow rate, the values of which for the potential and the Euler solutions are found to be respectively 2.031 and 2.021 (referred to sonic conditions and to chord of the circular arc profile) ; the latter value corresponds to an upstream Mach number of 0.835, instead of 0.85 for the potential solution. Indeed, because of the total pressure loss across the shock in the Euler solution, and since the imposed exit pressure is the same in both solutions, the calculated mass flow rate is smaller for the Euler solution than for the potential solution, corresponding to unchoked flow in the first case, and to practically choked flow in the second case.

The figures 3 to 10 are relative to the NACA 0012 airfoil for the following three sets of free-stream conditions (M = Mach number, α = incidence) :

a) M = 0.85, α = 0°   (figs. 3 to 6)
b) M = 0.8, α = 1.25°   (figs. 7 and 8)
c) M = 0.95, α = 0°   (figs. 9 and 10)

Calculations have been made using different meshes, namely the workshop standard mesh with (141 × 21) points and a sheared-parabolic mesh given to us by J.J. Chattot and C. Coulombeix (see their paper in this volume), either with (141 × 21) points or with (189 × 25) points. In the following as well as on the figures, these three meshes will be referred to as WS, SP1 and SP2 respectively. The relative positions of the external boundaries of the WS and SP2 meshes can be compared on figure 3 where only half the computation domains are shown. The external boundary of the SP1 mesh is practically identical to that of the SP2 mesh.

Figure 4 compares the pressure distributions of the potential solutions obtained in the WS and SP2 meshes for the case a). We think that the noticeable difference between the two solutions is essentially due to the difference between the external boundaries ; since the supersonic region is rather large, the WS mesh may not extend far enough from the profile. However, the two solutions agree very well in the nose region, despite the fact that the WS mesh is much coarser there than the SP2 mesh. In the shock region, these two meshes give approximately the same resolution in the direction parallel to the wall. The theoretical isentropic jumps for normal shocks are indicated by horizontal dashes on all the C_p curves. It can be seen that the numerical solution is in good agreement with these exact jumps.

The Euler solutions for case a) in the WS mesh and in the SP1 mesh
are compared on figure 5. The calculation using the SP2 mesh yields a solution -not shown here- which is very close to the WS solution with exactly
the same shock position. The comparison of these three Euler solutions tends
to show that the difference between the external boundaries of the WS mesh
and of the SP1 (or SP2) mesh has no influence here, this hypothesis being
in agreement with the fact that the supersonic zone is much less extended
that in the potential solutions. The slight difference in shock position
between the WS (or SP2) solution and the SP1 solution could be a result of
the mesh at the shock being coarser for the latter ; indeed, with the same
values of the artificial viscosity coefficients X_1 , X_2 (see Eq. (10)),
the effective amount of artificial dissipation increases as the square of
the mesh size.

The value at the wall of the quantity Z defined by :

$$Z = \frac{P}{P_\infty} \left(\frac{\rho_\infty}{\rho}\right)^\gamma - 1 = e^{(S - S_\infty)/C_v} - 1$$

where the subscript ∞ refers to free-stream values, is shown on figure
6. It can be seen that the value of Z on the leading-edge (which should be
strictly zero) is very sensitive to the mesh size in this region. The WS
mesh being much coarser than the SP1 mesh in the nose region, a much larger
artificial dissipation is generated in this region ; further downstream, up
to the shock, the flow gradients being much smaller, the artificial dissipation is negligible and Z remains constant, equal to about 10^{-2} with the WS

mesh, and to about 10^{-3} with the SP1 mesh. The calculated shock jumps are
in good agreement with the Rankine-Hugoniot jumps.

The solutions obtained for case b) in the WS mesh and in the SP2
mesh are shown on figure 7 (potential solution), and on figure 8 (Euler solution). The influence of the position of the external boundaries on the
potential solutions is very important, much more than in the previous case
a). This influence exists also for the Euler solutions, but to a much smaller
degree. Also the general difference between the potential solutions and the
Euler solutions is very striking. The two Euler solutions exhibit a weak
shock on the lower surface, contrary to the potential solutions which give
a minimum for the pressure close to the critical value C_p^* in the WS mesh,
and slightly above this value in the SP2 mesh.

Case c) has been calculated only with method H. The corresponding
Euler solution is shown on figure 9 (pressure distribution on the profile
and on the downstream axis) and on figure 10 (isobar contour map). This
solution has been obtained with a special mesh, of the type of the WS mesh,
but with (71 × 23) points for a half-space. Because of the high free-stream
Mach number, the outer boundary must be placed at very large distances ;
here the lateral part of this boundary is at a distance of 12 chords from
the axis, the downstream part is at a distance of 9 chords from the trailing-
edge, and the upstream part is at a distance of 6 chords from the leading-
edge. In fact it would be necessary to carry out calculations in several
meshes extending more and more outwards to make sure that the outer boundary
is far enough. A lambda-shock configuration is seen to exist, with the oblique shock starting at the trailing-edge and the normal shock located at 1.01
chords from the trailing-edge.

Finally the results obtained for the RAE 2822 profile at a Mach number $M = 0.75$ and an incidence $\alpha = 3°$ are shown on figure 11 (potential solution in the WS mesh) and on figure 12 (Euler solutions in the WS and SP2 meshes). These meshes are similar to the corresponding meshes for the NACA 0012 with the same number of points and approximately the same outer boundaries. The large differences which can be observed between the potential and the Euler solution are related to the existence of large supersonic regions and of strong shocks (the maximum Mach number is about 1.7 in the potential solution, and 1.57 and 1.42 in the Euler solutions. Of course with such high values, the potential approximation is no more justified.

REFERENCES

1 - Viviand, H., et Veuillot, J.P., Méthodes pseudo-instationnaires pour le calcul d'écoulements transsoniques, Publication ONERA No 1978-4, Oct. 1978.

2 - Veuillot, J.P., and Viviand, H., A pseudo-unsteady method for the computation of transonic potential flows, AIAA Paper No 78-1150 and AIAA J., Vol. 17, N° 7, July 1979. also ONERA T.P. No. 1978-47, 1978.

3 - Ivanov, M.Ya., Solution of two and three-dimensional problems of transonic flow over bodies, U.S.S.R. Comput. Math. and Math. Phys., Vol. 15, No 5, 1975.

4 - Magnus, R., and Yoshihara, H., Unsteady transonic flows over an airfoil, AIAA J., Vol. 13, No. 12, Dec. 1975.

5 - Gottlieb, D., and Gustafsson, B., On the Navier-Stokes equations with constant total temperature, Studies in Appl. Math., 55, 1976, pp. 167-185.

6 - Essers, J.A., Time-dependent methods for mixed and hybrid steady flows, von Karman Institute Lecture Series on Computational Fluid Dynamics, 13-17 March 1978.

7 - Essers, J.A., Quasi-natural numerical methods for the computation of inviscid potential or rotational transonic flows, Appl. Math. Modelling, Vol. 3, Feb. 1979, pp. 55-66.

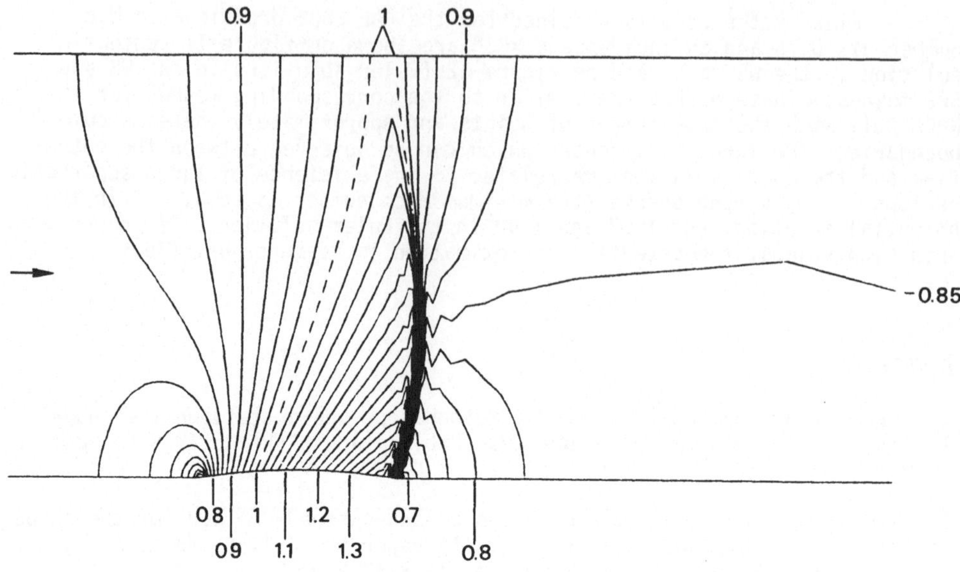

Fig. 1 — Iso-Mach lines for channel flow. Potential solution.

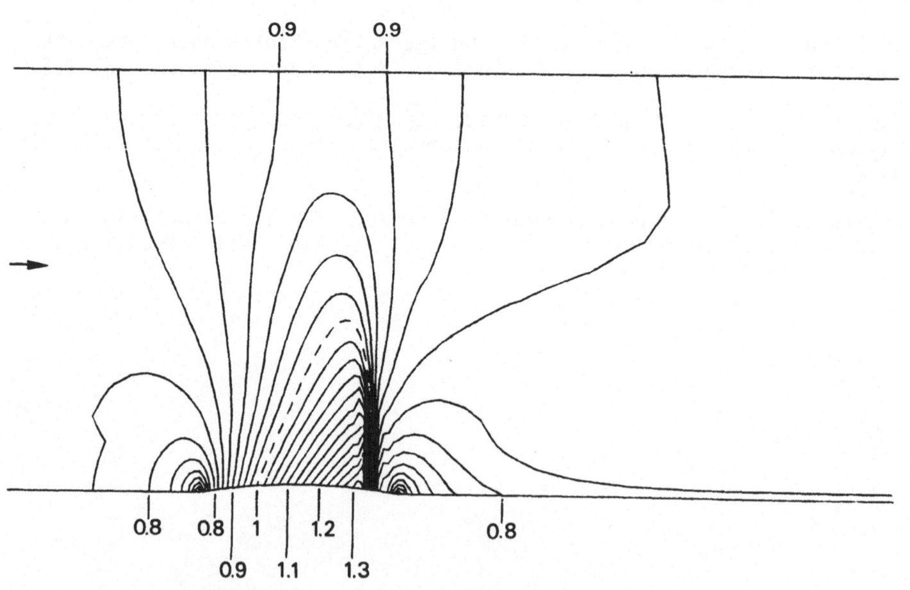

Fig. 2 — Iso-Mach lines for channel flow. Euler solution.

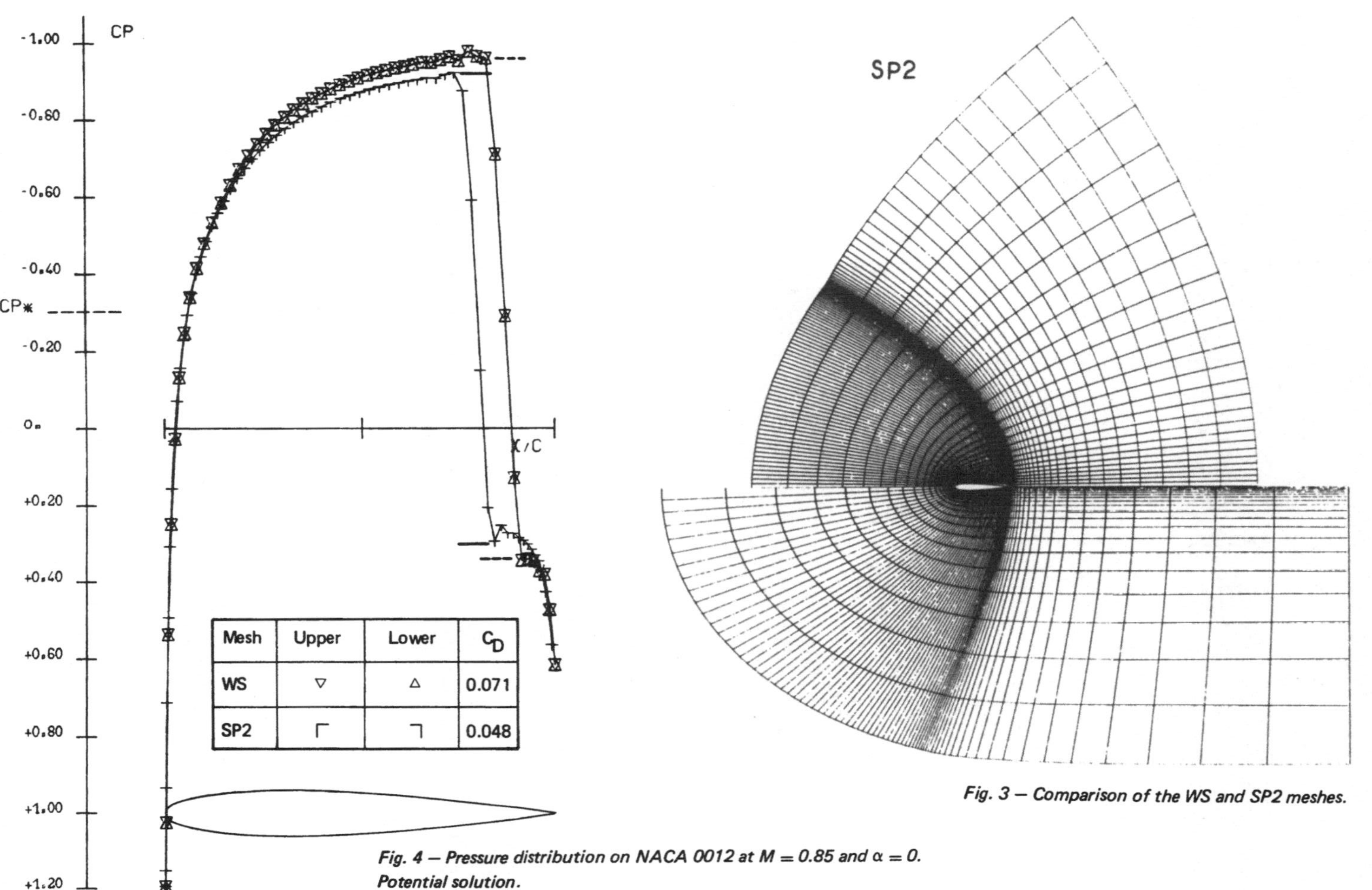

Mesh	Upper	Lower	C_D
WS	▽	△	0.071
SP2	Γ	⌐	0.048

Fig. 3 — Comparison of the WS and SP2 meshes.

Fig. 4 — Pressure distribution on NACA 0012 at M = 0.85 and α = 0.
Potential solution.

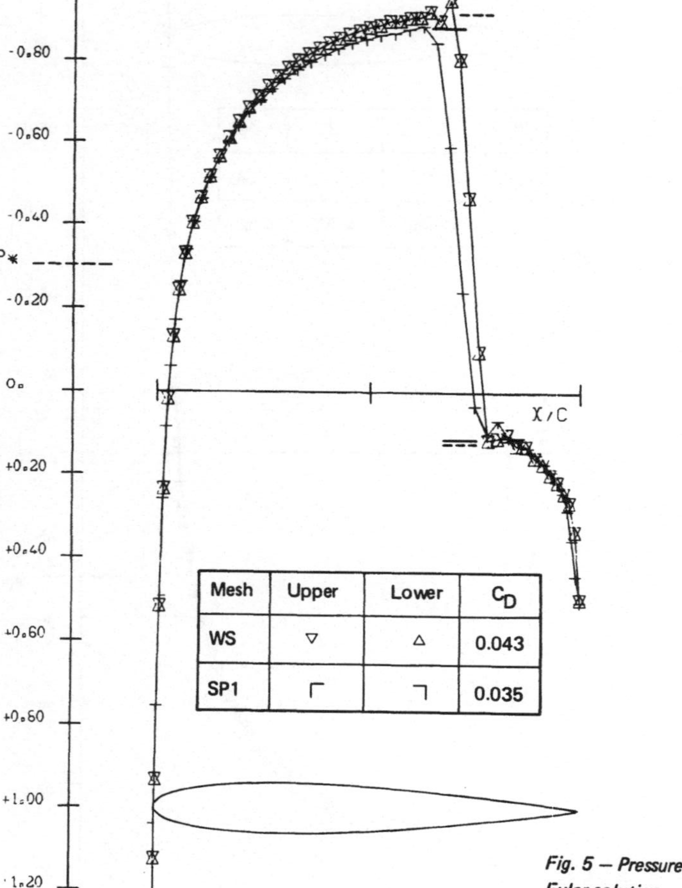

*Fig. 5 — Pressure distribution on NACA 0012 at M = 0.85 and α = 0.
Euler solution.*

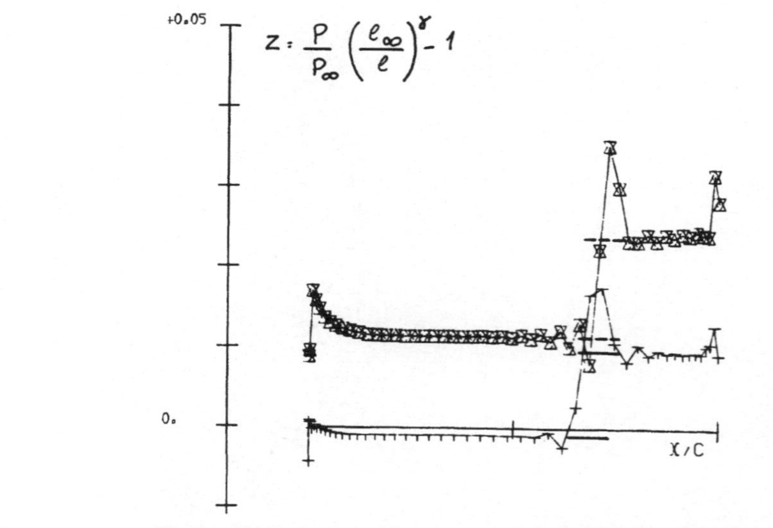

*Fig. 6 — Distribution of p/ρ^γ on NACA 0012 at M = 0.85 and α = 0.
Euler solution.*

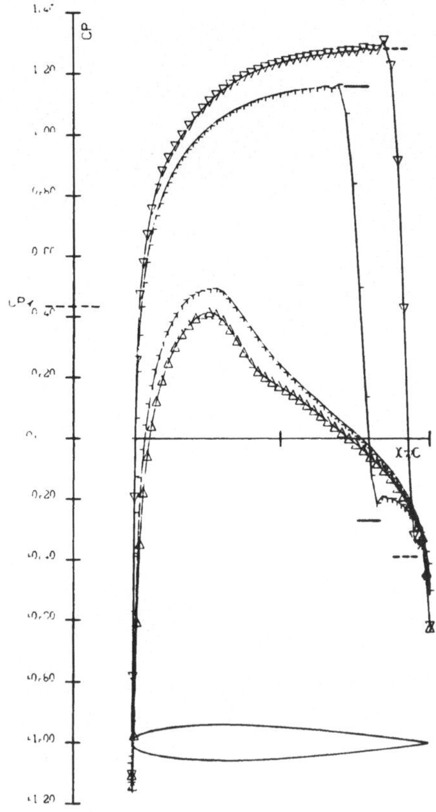

Mesh	Upper	Lower	C_D	C_L
WS	▽	△	0.072	0.898
SP2	⌐	¬	0.037	0.552

Fig. 7 — Pressure distribution on NACA 0012 et M = 0.80 and α = 1.25°. Potential solution.

Mesh	Upper	Lower	C_D	C_L
WS	▽	△	0.029	0.371
SP2	⌐	¬	0.019	0.292

Fig. 8 — Pressure distribution on NACA 0012 at M = 0.80 and α = 1.25°. Euler solution.

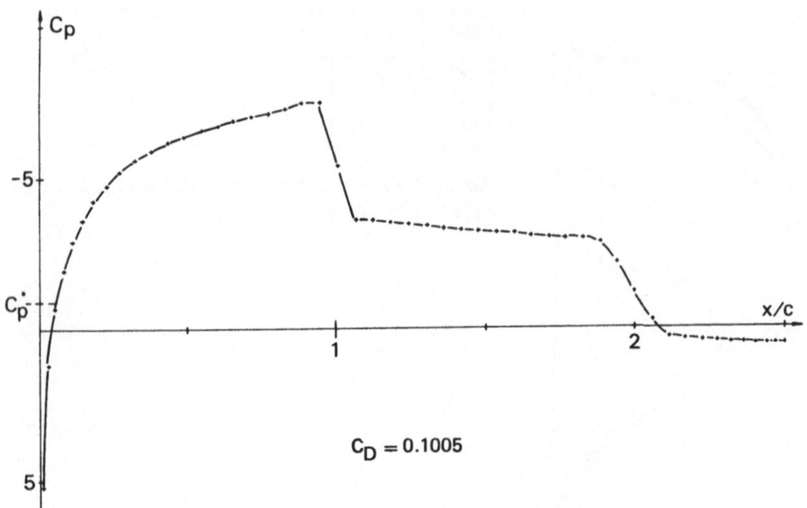

Fig. 9 — Pressure distribution on NACA 0012 and on the axis at $M = 0.95$ and $\alpha = 0$. Euler solution.

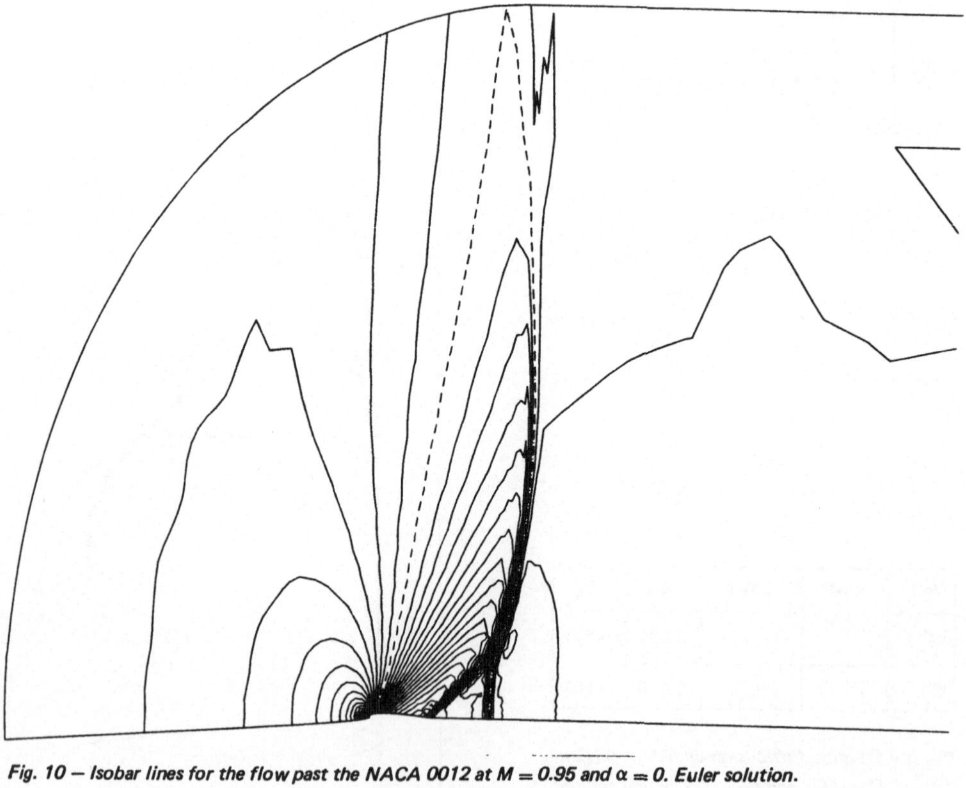

Fig. 10 — Isobar lines for the flow past the NACA 0012 at $M = 0.95$ and $\alpha = 0$. Euler solution.

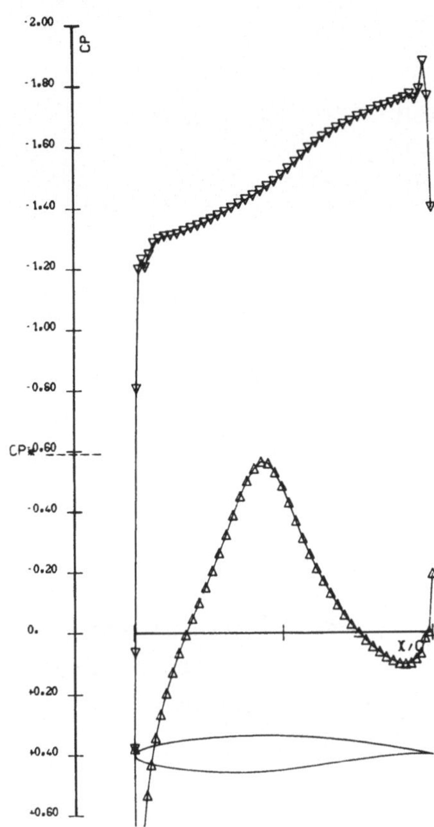

Mesh	Upper	Lower	C_D	C_L
WS	▽	△	0.150	1.406

Fig. 11 — Pressure distribution on RAE 2822 at $M = 0.75$ and $\alpha = 3°$. Potential solution.

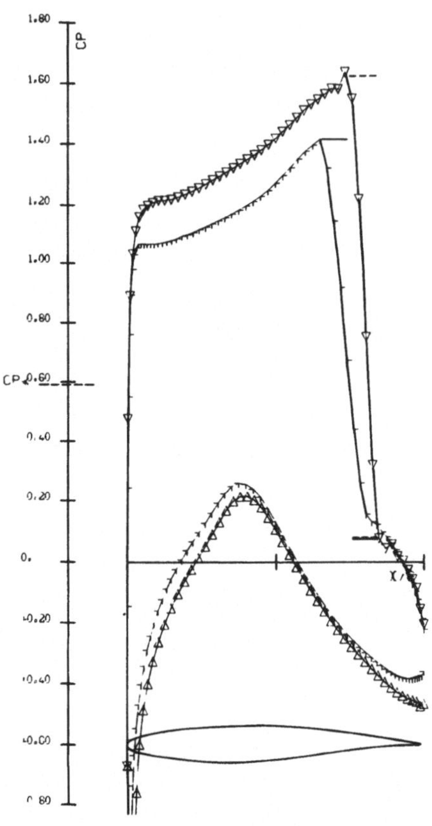

Mesh	Upper	Lower	C_D	C_L
WS	▽	△	0.072	1.248
SP2	⌐	¬	0.043	0.991

Fig. 12 — Pressure distribution on RAE 2822 at $M = 0.75$ and $\alpha = 3°$. Euler solution.

TRANSONIC FLOW COMPUTATION BY A MULTI-GRID METHOD

Laszlo J Fuchs
Department of Gasdynamics
The Royal Institute of Technology, Stockholm

INTRODUCTION

Steady-state flow problems are solved as a rule by some iterative technique. Relaxation methods are the most common ones. The multi-grid (MG) technique can be considered as a variant of a usual relaxation strategy even though it differs conceptually from the latter. The multi-grid method (MGM) has been first applied to transonic problems by Brandt and the present author. A refined and efficient version of this method is described in details by Fuchs [1]. A short review of the MGM is given here. Other applications and developments of the MGM are described by Brandt [2] and the references in that paper.

The discretization of the inviscid transonic equation in any form is type-dependent (to simulate well the mixed elliptic-hyperbolic PDE). Here, the small perturbation transonic equation is used with a new far-field boundary condition. The commonly used first order accurate boundary conditions on the surface of the airfoil (e.g. [3]) are replaced by second order approximations. The results computed on a coarse grid, using these conditions, are compared with those obtained on a finer net using a first order accurate method. We conclude that the second order method may help to reduce the total number of computational points. The far-field conditions eliminates in many cases the need to make coordinate transformations. This in turn allows a more efficient solution of the finite-difference equations.

The transonic small perturbation equation is given by

$$L(\varphi) = [K - (\gamma+1)\varphi_x]\varphi_{xx} + \varphi_{yy} = 0 \tag{1}$$

where x and y are the scaled Cartesian coordinates such that $y = \delta^{1/3}y^*$. x and y^* are coordinates scaled by the chord length. δ and M_∞ are the airfoil thickness ratio and the free-stream Mach number, respectively. The transonic similarity parameter is given by

$$K = (1 - M_\infty^2)/M_\infty^{2p}\delta^{2/3} \tag{2}$$

Here p is a free parameter choosen in such a way that equation (1) is a good approximation to the full potential transonic equation over a wide range of Mach numbers. The pressure coefficient is given by

$$C_p = -2M_\infty^{-q}\delta^{2/3}\varphi_x \tag{3}$$

where q is another free parameter. Murman and Cole [3] used p=q=0, while Langley [4] (using the same values as Krupp) takes p=1/2 and q= 3/4. The later values are adopted for the present computations as well.

The perturbation potential φ vanishes at infinity (uniform flow). At solid boundaries φ satisfies the condition that the flow is tangential to the boundaries. For the profile this is approximated by

$$\varphi_y\Big|_{\text{profile}} = F'(x) \tag{4}$$

applied at y=0. $\delta F(x)$ is the shape of the profile.

Equation (2) is approximated by the well known type-dependent (conservative) finite-differences as proposed by Murman and Cole [3,5] and Murman [6]. The usual central differences approximating φ_{yy} and which are used at the inner field points cannot be used on the profile, where condition (4) must be satisfied. There φ_{yy} is approximated by

$$\delta_y^2 \varphi \;=\; \frac{2}{h_y^2}\,[\varphi_{i,2}-\varphi_{i,1}-h_y\,F'(x_i)] \;=\; \tag{5}$$

$$=\; \varphi_{yy} - \frac{h_y}{3}\,\varphi_{yyy} + O(h_y^2)$$

This is a first order approximation. Second order accuracy can be obtained by using the differential equation (1) and (4)

$$\varphi_{yyy} \;=\; -[K-(\gamma+1)\varphi_x]F'''(x)+(\gamma+1)F''(x)\varphi_{xx} \tag{6}$$

or

$$\varphi_{yy} \;=\; \delta_y^2\,\varphi \;+\; \frac{h_y}{3}\,\varphi_{yyy} + O(h_y^2) \tag{7}$$

where φ_x and φ_{xx} are approximated by finite differences.
It should be noted that the often used method, inserting a so called "false" net line at $y=-h_y$ and computing φ at these points by central differences, leads to a first order scheme. Similarly the approximation suggested by Krupp (and reproduced by Langley [4]) is first order accurate.

The far-field perturbation potential can be written (see e.g. [7]) as the sum of contributions from the lifting effects, the airfoil thickness and the non-linearity of the governing equation. The contribution to the far-field potentials due to airfoil thickness and non-linearities can be written as a sum of potentials $\varphi^{(n)}$, i.e.

$$\varphi(x,y) \;=\; \sum_{i=0}^{\infty} \varphi^{(n)}(x,y) \tag{8}$$

such that

$$K\varphi_{xx}^{(n+1)} + \varphi_{yy}^{(n+1)} \;=\; (\gamma+1)\varphi_x^{(n)}\varphi_{xx}^{(n)} \tag{9}$$

It can be shown [8] that $\varphi_x^{(n+1)} \sim [\varphi_x^{(n-1)}]^2$ up- and downstream far away from the airfoil. This shows that the expansion (8) has a quadratic convergence rate and for practical use the first two terms are adequate. The complete farfield potential to this order is given by

$$\varphi(x,y)=\varphi_{lift}+ D\{\frac{x-x_o}{r} +(\gamma+1)[-\frac{(x-x_o)^3}{r^3} +\frac{(x-x_o)^5}{r^4}]\} \tag{10}$$

where $r = (x-x_o)^2 + K(y-y_o)^2$

The three parameters D, x_o, y_o can be determined by computing two solutions for different values of D. From these solutions $\dfrac{\partial \varphi_x}{\partial D}$ can be estimated for the "inner" computed solution and the outer solution (10). From these values and requiring continuous φ_x at the matching boundaries, new corrections $\Delta D, \Delta x_o$ and Δy_o can be computed. The described procedure has quadratic convergence rate in the corrected parameters $(\Delta D, \Delta x_o, \Delta y_o)$ so usually a single correction is enough. More details of this procedure can be found in [8].

THE MULTI-GRID METHOD

The operator L in equation (1) is approximated on a net with spacing h by a discrete operator L_h; such that $L_h(\varphi*)=0$ (11)
To find a $\varphi*$ which satisfies (11) we use a MGM method:
Given a current approximation $\varphi^{(n)}$ to $\varphi*$. We want to find a correction v to $\varphi^{(n)}$ such that

$$v = \varphi* - \varphi^{(n)} \tag{12}$$

This would result in a so-called correction problem

$$\tilde{L}_h(v) = -L_h(\varphi^{(n)}) \tag{13}$$

where $\tilde{L}_h$ is also an approximation to L but might differ from L_h. The correction v can be found, more easily, on a coarse grid H from

$$L_H(v) = -I_H^h \, L_h(\varphi^{(n)}) \tag{14}$$

where I_H^h is an operator which projects the residuals $(-L(\varphi^{(n)}))$ from the fine net h to the coarser one H. Once v is computed it is interpolated to the fine grid. The interpolated correction contains errors which can be represented only by L_h and not by L_H. These errors are equal essentially the Fourier-components with wave-length corresponding to h, while longer wave-length components are almost absent. The errors in the interpolated correction are eliminated

efficiently by a few relaxation sweeps using a proper relaxation operator. The resulting correction after a few relaxation sweeps does not exactly satisfy equation (13), but it serves to compute a new approximation: $\varphi^{(n+1)} = \varphi^{(n)} + v$
The whole procedure can be repeated, using more levels (coarser and coarser grids), until convergence is reached.

The number of levels is usually 5-6 and each new level is constructed by doubling the step size of the previous one.

It should be pointed out that if $\tilde{L}_h = L_h$ then the programming and the computer storage can be made more compact (Brandt's [9] FAS-version). However, taking $\tilde{L}_h \neq L_h$ gives more freedom in choosing a better relaxation operator. A typical example [1] is when L_h is a second order accurate approximation and $\tilde{L}_h$ is a first order one. In such cases $\tilde{L}_h$ is more dissipative, contributing to the overall stability of the iterative method.

The choice of the relaxation operator determines how fast errors with highly oscillating components on the current net can be reduced. This can be determined by local mode analysis, as done in reference [1]. It turned out that in transonic flow computations, line (or point) relaxations may be much less efficient in some regions than in simpler purely elliptic cases. These regions include the shock, leading and trailing edges and possibly some other regions depending upon the local mesh distribution. Furthermore, due to the appearance of the shock the approximation $\varphi^{(n)}$ usually contains larger amplitude error components and this requires more "smoothing" operations.

The difficulties listed above may be overcome by "visiting" more frequently those regions in which the relaxation operator is less efficient. In this way the shock region is swept up to ten times more than a point far away from the profile. The errors which are concentrated to some regions (shocks) are spread out to net points which belong to coarser grids. Relaxations on these levels spread out the errors on the whole field, and reduce their amplitude. This method, applied to the transonic problem has an efficiency quite close to that

of the MG-Poisson solver (Brandt [9]), and it is not too sensitive to increasing shock strength and the extent of the
shock in the computational field.

RESULTS

The problems are solved in the physical domain using a rectangular net with the same spacing (in most cases) in both
directions. The finest net has 81x33 points (for symmetrical
profiles at zero angle of attack). The MG procedure has 5 levels if the results are obtained on the finest net, or 4 levels if the results are computed on the next finest one.

Figure 1 shows the result obtained (K=1.48) using a coarse
grid and vanishing potentials as far-field conditions. Both
first and second order accurate profile boundary conditions
are used. The same problem is recomputed on a finer mesh with
the new far-field conditions (10) and the first order profile
conditions. The pressure distributions on the profile in the
last case and the second-order coarse-grid case are in good
agreement. Figure 2 displays similar results for K=1.34 obtained with other mesh spacings. The higher order method cannot, however, compensate for the coarseness of the net near
the leading edge (of the blunt-nosed profile, such as the
NACA 0012 which is used here), and a treatment like the one
suggested earlier [1] must be used.

The efficiency of the method can be expressed in terms of the
number of work units needed to get a converged solution. A
work unit is defined to be equivalent to the work needed to
carry out one (line) relaxation sweep on the finest net. The
error decay per work unit varies from 0.60 (for pure subsonic
flows) to 0.68 in cases of flows with a shock which extends
into half of the computational domain). Adjustment of the
computational parameters can reduce these figures to about
0.55 as was reported in the past [1]. The total number of
work units to reduce the residual (defined as the term on the
right hand side of equation (13)), by more than 4 orders of
magnitudes is between 18 and 24 the various Mach numbers.

Typical computational times are 10.0 sec for the finest grid
(when the residual decreases from 16 to 6.10^{-4}) and 4.4 sec
on the next finest level (about the same residuals as in the
former case). These times are the total CPU times (as regis-
tered on an IBM 370 machine) needed to execute the whole
program.

We conclude from the numerical experiments that the MGM is
fast and capable to compute transonic flows efficiently. By
using a far-field expression, a second order approximation on
the solid boundary and a possible leading and trailing edge
correction technique, coarse grids may be used. This saves
substantial computational times without loosing too much of
the accuracy.

REFERENCES

1   Fuchs, L.J.: Finite difference methods for plane steady
     inviscid transonic flows. TRITA-GAD-2, 1977

2   Brandt, A.: Multi-level adaptive computations in fluid
     dynamics. AIAA paper No 79-1455, 1979

3   Murman, E.M. and Cole J.P.: Calculation of plane steady
     transonic flows. AIAA Journal No , pp. 114-121, 1971

4   Langley, M.J.: Numerical methods for two-dimensional and
     axisymmetric transonic flows, Aeronautical Research Coun-
     cil, C.P. No 1376, 1977

5   Murman, E.M. and Cole, J.D.: Inviscid drag at transonic
     speeds. AIAA paper No. 74-540, 1974

6   Murman, E.M.: Analysis of embedded shock waves calculated
     by relaxation methods. AIAA Journal Vol. 12, pp 626-633.
     1974

7   Klunker, E.B.: Contribution to methods for calculating
     the flow about thin lifting wings at transonic speeds -
     analytical expressions for the far-field NASA TN D-6530,
     1971

8   Fuchs, L.J.: On the accuracy of numerical solutions of
     transonic Problem, to appear, 1979

9   Brandt, A.: Multi-level adaptive solution to boundary
     value problems. Math. of Comp. vol. 31, pp. 333 - 380,
     1977

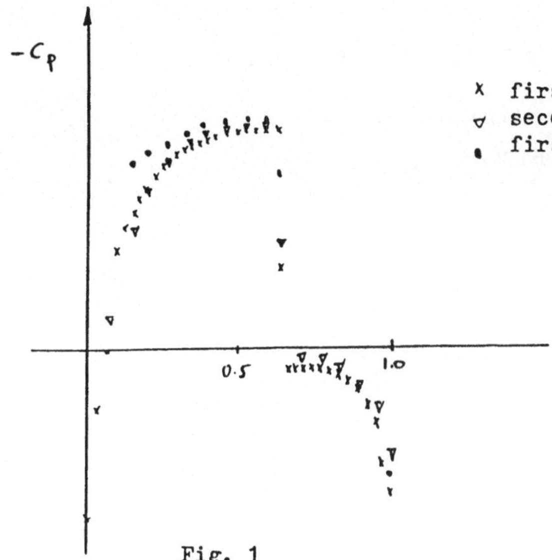

Fig. 1

C_p over a NACA 0012 profile. K=1.48

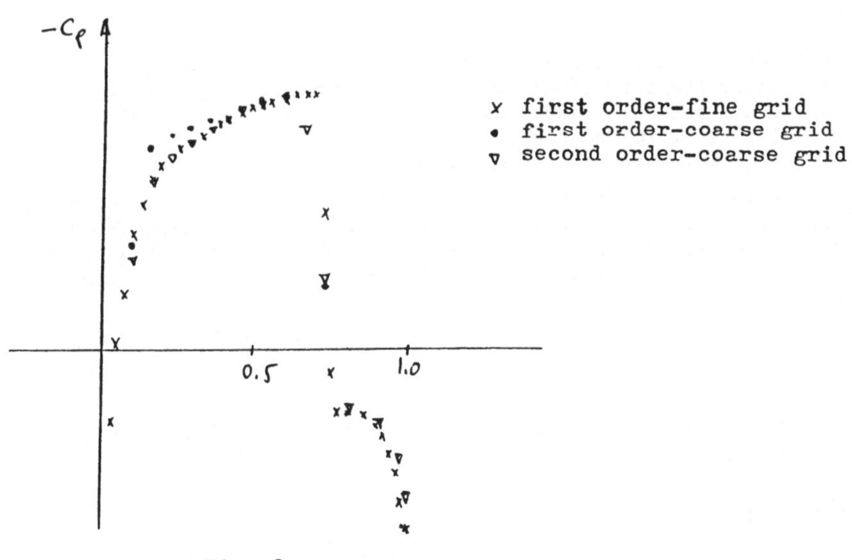

Fig. 2

C_p over a NACA 0012 profile. K=1.34

TRANSONIC FLOW CALCULATIONS WITH FINITE ELEMENTS

H. Deconinck and Ch. Hirsch

VRIJE UNIVERSITEIT BRUSSEL
Dept. Fluid Mechanics
Brussels, Belgium

ABSTRACT

The full potential equation is formulated for transonic flow with an artificial compressibility and discretized with Finite Elements. Bilinear elements are used and the system of equations is solved iteratively with a relaxation method and with an implicit factorized ADI technique. The methods are briefly described and results are discussed for the channel flow problem.

INTRODUCTION

In the past, FE methods have been faced with several difficulties when applied to transonic flow problems. This is mainly connected to the elliptic nature of Galerkin's or any other weighted residual approach to the integral formulation of the equation which then leads to non-positive definiteness of the stiffness matrices in supersonic regions.

Various ways have been presented to circumvent these difficulties [1],[2] [3],[4] of which the method of optimal control developed with the most success [2], [3],[4]. Transonic flow calculations with shocks require the introduction of some form of artificial viscosity (A.V.) in order to damp the non-linear oscillations appearing in the numerical computation. Writing the A.V. under the form of an artificial compressibility [5],[6],[7] allows to write the potential equation under a form which is formally identical to the elliptic form and hence can be treated in a FE formulation as a subsonic problem for which the FE method is extremely well adapted.

This is the way followed in this paper and two methods are presented to solve the discrete equations : the successive line overrelaxation (SLOR) and the ADI or factorized form of the equations. The former method is not new and has been developed previously by Eberle[5] while the factorized form for FE formulations had to be demonstrated.

The basic equations are summarized in the first section while in the second and third section the SLOR-FE and ADI form for FE formulations are discussed.

Results for SLOR and ADI applied with bilinear quadrilateral elements on an arbitrary mesh are presented in section four and five.

1. GOVERNING EQUATIONS

The full potential equation in conservation form is given by :

$$(\rho \ \phi_x)_x + (\rho\phi_y)_y = 0$$

$$\rho = \rho_s \ [1 - \frac{\gamma - 1}{\gamma r T_s} \ (\phi_x^2 + \phi_y^2)]^{\ 1/(\gamma-1)}$$

(1)

where ρ_s and T_s are the stagnation density and temperature ; γ is the ratio
of specific heats and r is the gas constant. Equation (1) expresses mass
conservation in the physical x-y coordinate system for steady and irrotational flows. In order to obtain valid solutions for shock position and
intensity it is essential that mass flow be conserved across a shock. This
is only realised if the equations and its discrete approximation are written in conservative form. The corresponding shock-jump conditions will
then be valid approximations to the Rankine-Hugoniot relations.

The potential flow formulation when conservative is valid for many transonic flow applications, since the error introduced by neglecting the entropy increase due to shocks is small in many cases.

Stable operation in supersonic regions is obtained by introducing artificial viscosity. Again it is necessary that the addition of artificial viscosity should not destroy the conservative form of the equation.

<u>Artificial viscosity</u>

In order to introduce in the potential equation the irreversible character
of transonic flow with shocks, an artificial viscosity term must be added
in the supersonic flow regions. This is usually done in an implicit way
using upwind differencing in the supersonic regions. In classical finite
element methods this technique is not applicable since the discretization
is always of the elliptic central differencing type.

However, the artificial viscosity can be introduced explicitly. Jameson
[8] adds a viscosity term to the equation which can be shown [7] to be e-
quivalent to adding :

$$- [(\nu \, \phi_x \, \rho_x^\leftarrow \, \Delta x)_x + (\nu \, \phi_y \, \rho_y^\leftarrow \, \Delta y)_y] \tag{2}$$

where :

$$\nu = \max (0, 1 - \frac{1}{M^2}) \tag{3}$$

is the switching function setting the artificial viscosity to zero in the
subsonic flow regions, and $\rho_x^\leftarrow$ and $\rho_y^\leftarrow$ are upwind differenced.

Expression (2) is particular interesting since it offers the possibility
to treat the equation as an elliptic one.

The potential equation (1) corrected with the artificial viscosity formulation (2) is given by :

$$(\overset{\sim}{\rho} \, \phi_x)_x + (\overline{\rho} \, \phi_y)_y = 0 \tag{4}$$

$$\overset{\sim}{\rho} = \rho - \nu \, \rho_x^\leftarrow \, \Delta x \tag{5a}$$

$$\overline{\rho} = \rho - \nu \, \rho_y^\leftarrow \, \Delta y \tag{5b}$$

The non linear coefficients $\overline{\rho}$ and $\overset{\sim}{\rho}$ are updated after each solution of the
linearized equation(4) with help of (5), while the equation is solved and
discretized in an elliptic way. So equation (3) could be discretized with
standard Galerkin finite element techniques.

A further simplification can be obtained by replacing equations (5) by [6]:

$$\overline{\rho} - \overset{\sim}{\rho} = \rho - \overset{.}{\nu} \, \rho_s^\leftarrow \, \Delta s \tag{6}$$

with $\rho_s^\leftarrow$ the upwind differencing of ρ along a streamline :

$$\rho^{\leftarrow}_s = \frac{1}{\sqrt{\phi_x^2 + \phi_y^2}} (\phi_x \, \rho^{\leftarrow}_x + \phi_y \, \rho^{\leftarrow}_y) \tag{7}$$

and the potential equation has now again the usual form

$$(\tilde{\rho} \, \phi_x)_x + (\tilde{\rho} \, \phi_y)_y = 0 \tag{8}$$

where the viscosity terms are included in the artificial crompressibility $\tilde{\rho}$.

The form given in equation (6) is used in this paper resulting in an artificial viscocity slightly different from the one introduced by Jameson [8]. The modified conservation law (8) reduces to the original conservation law (1) as the mesh width tends to zero, assuring correct shock capturing. The term $\rho^{\leftarrow}_s \, \Delta s$ is calculated in a straightforward way if one type of coordinate lines is approximately parallel to the flow in supersonic flow regions.

As pointed out by several authors [7],[9], the choice of ν strongly affects the accuracy and stability of solutions to equation (8). The use of the standard definition (3) produces preshock overshoots which can result in numerical instability. The amount of artificial viscosity is therefore increased in high supersonic regions replacing ν by :

$$\nu = \max \left\{ 0, [\, (1 - \frac{1}{M^2}) \, M^n] \right\} \tag{9}$$

where n is chosen by numerical experiment for each case. An alternative definition for ν, introduced by Holst [9] is :

$$\nu = (1 - \frac{\rho}{\rho\star})^{\sigma} \qquad\qquad M > 1 \tag{10}$$
$$\nu = 0 \qquad\qquad M < 1$$

where $\rho\star$ is the sonic value of density and σ has been chosen by numerical experiment to be equal to 6.

2. FINITE ELEMENT SUCCESSIVE LINE OVERRELAXATION METHOD

2.1. FE Discretization

The potential equation (8) with modified density is equivalent to the following weak form :

$$- \int_S \tilde{\rho}(\phi_x W_x + \phi_y W_y) \, dS + \int_{s_n} \tilde{\rho} \, \frac{\partial \phi}{\partial n} \, W ds = 0 \tag{11}$$

for any weightfunction W and where S is the flow domain and s_n the part of the boundary where the Neumann boundary condition $\rho(\partial \phi / \partial n)$ is specified.

In the FE method the potential function is approximated with help of the nodal values ϕ_I and the shape functions N_I corresponding to each node :

$$\phi = \sum_I \phi_I \, N_I \, (x,y) \tag{12}$$

Applying Galerkin method using the shape functions as weightfunction, the discrete equation for iteration (n) is obtained :

$$\sum_J \phi_J^{(n)} \int_S \tilde{\rho}^{(n-1)} \, \nabla N_J \, \nabla N_I \, dS - \int_{s_n} (\tilde{\rho} \, \frac{\partial \phi}{\partial n})^o \, N_I \, ds = 0 \tag{13}$$

Here the non linear equation is linearized as usual using the values of $\tilde{\rho}$

of the previous iteration.

Note that this discretization technique is fully elliptic, the distinction between the supersonic and subsonic points being included in the artificial compressibility expression $\tilde{\rho}$.

A more condensed form for equation (13) is the following :

$$\sum_J \phi_J^{(n)} K_{IJ}^{(n-1)} = f_I \tag{14a}$$

$$K_{IJ}^{(n-1)} = \int_S \tilde{\rho}^{(n-1)} \nabla N_J \, \nabla N_I \, dS \tag{14b}$$

and
$$f_I = \int_{s_n} (\tilde{\rho} \frac{\partial \phi}{\partial n})^0 N_I \, ds \quad \text{for} \quad I \text{ belonging to } s_n$$
$$f_I = 0 \qquad\qquad \text{everywhere else} \tag{14c}$$

In finite difference notation with row and column indices (fig.2) and omitting the iteration index for the stiffness matrix K, one can write equ. (14) as :

$$\sum_{k,l} \phi_{kl}^{(n)} K_{ij}^{kl} = f_{ij} \tag{15}$$

or :
$$\sum_{k,l} \delta\phi_{kl} K_{ij}^{kl} = - R_{ij}^{(n-1)} \tag{16}$$

with :
$$\delta\phi = \phi^{(n)} - \phi^{(n-1)} \tag{17}$$

and the residual $R^{(n-1)}$ is defined by

$$-R^{(n-1)} = f^{(n-1)} - K^{(n-1)} \phi^{(n-1)} \tag{18}$$

It is to be noted that due to the FE approach the boundary conditions are included explicitly in the residual since f contains the contribution from the boundaries, equ. (14c).

2.2. Line overrelaxation solution

In order to solve the system of equ. (16) the line relaxation method is adopted instead of a direct method, as is generally used in subsonic flows, but which diverges in the transonic range. The SLOR method has also been used with success by Eberle [5] in an analog way.

With bilinear quadrilateral elements and shape functions (fig.1), where the (ξ,η) coordinates are the local coordinates of an element :

$$N_I (\xi,\eta) = \frac{1}{4} (1 + \xi\xi_I)(1 + \eta\eta_I) \tag{19}$$

with :
$$\xi_I = \pm 1$$
$$\eta_I = \pm 1$$

and the isoparametric transformation,

$$x(\xi,\eta) = \sum_I x_I N_I (\xi,\eta) \tag{20}$$
$$y(\xi,\eta) = \sum_I y_I N_I (\xi,\eta)$$

allows to handle an arbitrary mesh.

Since the shape functions are locally defined, the matrix elements K_{ij}^{kl} will have non-zero value only for :

$$(i - 1) \leqslant k \leqslant (i + 1) \text{ and } (j - 1) \leqslant \ell \leqslant (j + 1)$$

Therefore along a fixed i column, the system will be tridiagonal in the unknowns $\delta\phi_{i,j-1}$, $\delta\phi_{i,j}$, $\delta\phi_{i,j+1}$.

The SLOR scheme is then

$$\sum_{\ell=j-1}^{j+1} \delta\phi_{i\ell}^{(n)} K_{ij}^{i\ell} = -\omega \left[R_{ij}^{(n-1)} - \sum_{\ell=j-1}^{j+1} K_{ij}^{i-1,\ell} \delta\phi_{i-1,\ell}^{(n)} \right] \tag{21}$$

It is to be noted that the relaxation coefficient ω should be lower than 2 and that due to the isoparametric transformation (20), the method is equally valid for an orthogonal or an arbitrary system.

In order to increase the convergence speed a multigrid technique is applied with four successive meshes going from a very coarse mesh to finer meshes, where at each transition the number of elements is multiplied by four. This technique was shown by Eberle [5] to improve strongly the rate of convergence and is adopted systematically here.

This implies that the number of nodes in the finest mesh on any line should be of the form $k.2^n + 1$; where $(k + 1)$ is the number of nodes on the considered line in the first coarse mesh and n is the number of successive grids.

Application of the SLOR-FE method with an arbitrary mesh is presented in section 4.

3. THE GALERKIN-ADI METHOD

Recently, the introduction of factorized forms for the full potential equation in transonic flows has met with some success [9] when formulated with FD on an arbitrary mesh and with the artificial compressibility form of the basic equation.

It is tempting therefore to investigate the capability of a FE formulation along these lines.

In the mathematical literature, some theoretical formulations can be found for ADI Galerkin methods for elliptic, parabolic or hyperbolic problems on rectangular meshes [10], [11].

In these papers, convergence proofs and error bounds are given, even for some classes of non-linear problems, but no application to practical problems is known to us. Neither has the formulation of an ADI Galerkin method on an arbitrary mesh been derived and proven theoretically. These aspects will be developed and presented in the following sections.

3.1. Galerkin-ADI method on a rectangular mesh

Considering the parabolic equation in time :

$$\phi_t - \vec{\nabla}(\rho \, \vec{\nabla} \phi) = f \tag{22}$$

Douglas and Dupont [10] define a Galerkin approximation of the form,

$$\phi_t = \delta\phi/\tau = (\phi^{(n+1)} - \phi^{(n)})/\tau \tag{23}$$

$$(\frac{\delta\phi}{\tau}, W)' + \lambda(\vec{\nabla}\delta\phi, \vec{\nabla}W) + \lambda^2\tau(\frac{\partial^2\delta\phi}{\partial x\partial y}, \frac{\partial^2 W}{\partial x\partial y}) = (f^n, W) - (\rho^n\vec{\nabla}\phi^n, \vec{\nabla}W) \tag{24}$$

for any weight function $W \in H_0^1$.

If the form function $N_I(x,y)$ factorizes in the form

$$N_I(x,y) = N_i(x) \cdot N_j(y) \tag{25}$$

where I is the node number and i,j the corresponding column and line numbers, and if a Galerkin approximation is applied, namely $W = N$, then the following matrix form is obtained.

$$\left[[M] + \lambda \tau[K] + \lambda^2 \tau^2 K_x \otimes K_y \right]\{\delta\phi\} = - \tau R \tag{26}$$

where R is the residual

$$R = - (f^n, W) + (\rho^n\vec{\nabla}\phi^n, \vec{\nabla}N) \tag{27}$$

$$M = M_x \otimes M_y = \int N_i \cdot(x) N_k(x) \, dx \otimes \int N_j(y) N_\ell(y) \, dy \tag{28}$$

$$K = \int \vec{\nabla}N_I \cdot \vec{\nabla}N_J \, dS \tag{29}$$

$$\text{or}: \quad K = K_x \otimes M_y + K_y \otimes M_x \tag{30}$$

$$\text{with}: \quad K_x = \int \partial_x N_i(x) \, \partial_x N_k(x) \, dx \tag{31}$$

$$K_y = \int \partial y \, N_j(y) \, \partial_y N_\ell(y) \, dy \tag{32}$$

The expression (26) can be factorized as follows :

$$(M_x + \lambda\tau K_x) \otimes (M_y + \lambda\tau K_y)\delta\phi = -\tau R \tag{33}$$

and introducing the intermediate vector g, the following system has to be solved

$$(M_x + \lambda\tau K_x)g = - \tau R \tag{34}$$

$$(M_y + \lambda\tau K_y)\delta\phi = g \tag{35}$$

This form of ADI-Galerkin method corresponds to the approximate operator C in the equation

$$C \, \delta\phi = - \tau R \tag{36}$$

given by :

$$C = 1 - \lambda L + \lambda^2 \partial_{xx} \partial_{yy} \tag{37}$$

where L is the Laplacian.

The same relations could be obtained in the inverse way, namely by applying first a splitting or factorizing of the equations, following Marchuk [12], and then a FE-Galerkin formulation. Writing the operator A as a sum of two terms :

$$A = A_x + A_y \tag{38}$$

or more specifically

$$- A = \vec{\nabla} \rho \vec{\nabla} = \partial_x \rho \partial_x + \partial_y \rho \partial_y \tag{39}$$

a factorized form of the equation is :

$$(1 + \sigma A_x) (1 + \sigma A_y) \delta\phi = - \tau\sigma R \tag{40}$$

which corresponds to the approximate operator

$$C = 1 - \sigma A + \sigma^2 A_x A_y \tag{41}$$

This could be a better approximation then (37), where A is replaced by the Laplacian operator.

Equ. (40) is then splitted as :

$$(1 + \sigma A_x)g = -\tau\sigma R \tag{42}$$

$$(1 + \sigma A_y)\delta\phi = g \tag{43}$$

and applying a FE-Galerkin method, one obtains

$$(M_x + \sigma K_x)g = - \sigma\tau R \tag{44}$$

$$(M_y + \sigma K_y)\delta\phi = g \tag{45}$$

where M_x and M_y are defined by (28) while

$$(K_x)_{ik} = \int \rho^{(n)} \partial_x N_i \cdot \partial_x N_k \ dx \tag{46}$$

$$(K_y)_{j\ell} = \int \rho^{(n)} \partial_y N_j \cdot \partial_y N_\ell \ dy \tag{47}$$

3.2. Application to triangular and bilinear elements

Considering the more general equation

$$(\partial_x p \partial_x + \partial_y q \partial_y) \ \phi = f \tag{48}$$

and the factorized form

$$(1 + \sigma\partial_x p \ \partial_x) \ g = - \sigma\tau \ R \tag{49}$$

$$(1 + \sigma\partial_y q \ \partial_y) \ \delta\phi = g \tag{50}$$

the partial stiffness matrices K_x, K_y become

$$K_x = - \int p\partial_x N_i \ \partial_x N_k \ dx \tag{51}$$

$$K_y = - \int q\partial_y N_j \ \partial_y N_\ell \ dy \tag{52}$$

3.2.1. Triangular elements :

On an arbitrary mesh, only the 6 nodes surrounding (i,j) will contribute to the equation in this point (Fig.3).

However, it turns out in the case of an orthogonal mesh that only the nodes
$(i, j - 1)$, $(i, j + 1)$, $(i - 1, j)$ and $(i + 1, j)$ will contribute and give
for the operator the same expression as the one obtained from a central,
conservative finite difference discretization :

$$(\delta_x \, p^n_{i+1/2} \, \delta_x + \delta_y \, q^n_{j+1/2} \, \delta_y) \, \phi^n \tag{53}$$

where :

$$p_{i+1/2} = \frac{1}{\Delta x^2} \int_{4+5} p \, dS \, , \quad q_{j+1/2} = \frac{1}{\Delta x^2} \int_{3+4} q \, dS \tag{54}$$

The mass matrix, as well as the K_x, K_y matrices are tridiagonal and one obtains for the first equation (x equ.) along line j :

$$\left[\frac{1}{6} (1 \ 4 \ 1) - \frac{\sigma}{\Delta x} \left(p_{i-1/2} \ -(p_{i-1/2} + p_{i+1/2}) \ p_{i+1/2} \right) \right] \begin{Bmatrix} g_{i-1} \\ g_i \\ g_{i+1} \end{Bmatrix} = -\sigma\tau R_i \tag{55}$$

In an FD scheme, the first matrix is diagonal (o 1 o) but the other terms
are identical except that here, the r-h-s of the basic equation ,f, con-
tributes to the residual through the integral :

$$\int f \, N_i(x) \, N_j(y) \, dx \, dy \tag{56}$$

over the hexagon of Fig.3, and that the boundary conditions are automatically included in the residual since :

$$R_J = - \sum_I \phi_I \int_S (p \, \partial_x N_I \cdot \partial_x N_J + q \, \partial_y N_I \cdot \partial_y N_J) dS + \int_{s_n} (p \frac{\partial \phi}{\partial n_x} + q \frac{\partial \phi}{\partial n_y}) \, N_J \, ds$$

$$- \int f \, N_J \, dS \tag{57}$$

over   a hexagon around node J.

For the Laplace equation, one would obtain the equation ($\Delta x = \Delta y = 1$) :

$$\frac{1}{6} (1 \ 4 \ 1) - \sigma(1 \ -2 \ 1) = - \sigma\tau R = -\sigma\tau \begin{bmatrix} & 1 & \\ 1 & -4 & 1 \\ & 1 & \end{bmatrix}$$

which is identical to the FD form except for the mass matrix.

3.2.2. Bilinear elements

In this case, Fig. 2, 8 nodes around (i,j) contribute to the residual and
compared with the triangular elements, the mass matrix remains unchanged
as well as the K_x and K_y matrices while the residuals will give rise to an
integration over the four elements surrounding the node (i,j). Hence the
l-h-s of equation (55) will remain unchanged while the residual will have
another numerical approximation, involving a larger amount of averaging.
This is best illustrated for the Laplace equation where one obtains the
scheme :

$$\frac{1}{6} (1 \ 4 \ 1) - \sigma(1 \ -2 \ 1) = - \frac{\sigma\tau}{6} \begin{bmatrix} 2 & 2 & 2 \\ 2 & -16 & 2 \\ 2 & 2 & 2 \end{bmatrix} = -\frac{\sigma\tau}{3} \left(\begin{bmatrix} & 1 & \\ -4 & & 1 \\ & 1 & \end{bmatrix} + \begin{bmatrix} 1 & & 1 \\ & -4 & \\ 1 & & 1 \end{bmatrix} \right)$$

It can generally be seen that the FE equations lead to a larger amount of
implicitness in the discretized equations since more surrounding mesh
points will contribute to the value of the solution in a given point.
Hence better convergence might be expected compared to the FD form of the
equation.

This is indeed the case, as illustrated on a test problem :

$$\partial_x [(1 + a \phi) \partial_x \phi] + \partial_y [(1 + b \phi) \partial_y \phi] = f \tag{58}$$

in a unit square with f defined such as to obtain

$$\phi = (x - x^2) (y - y^2) e^x \tag{59}$$

as solution.

For an orthogonal mesh, as well as for an arbitrary curvilear mesh, the FE
solution is more accurate than the FD solution after the same number of
iterations. It is known that the FD form, centrally differenced, is only
first order on a non-orthogonal mesh while the FE solution remains second
order for bilinear elements. More specifically, the following values are
obtained for the maximum error on the same mesh (11 x 11)

	FD	FE
orthogonal mesh	.33 E-3	.22 E-3
non-orthogonal mesh	.18 E-2	.45 E-3

These values are obtained after the same number of iterations in both cases, although in the FD cases the corresponding residuals usually have lower values compared to the FE values.

These results show that the FE calculation of the residuals leads to a
higher accuracy than the FD form on an identical mesh and this gain becomes significant on an arbitrary mesh.

3.3. Galerkin-ADI method on an arbitrary mesh

In order to apply the Galerkin-ADI method on an arbitrary mesh one can
either transform the equation from the physical plane to a computational
plane which is rectangular, or work directly on the Galerkin form of the
equation. Both methods lead to the same results, as will be seen in the
following.

As is well known, the basic equation (8), written in tensornotations :

$$\nabla_i \rho \nabla^i \phi = 0 \tag{60}$$

transforms, in a general coordinate transformation :

$$x^i = x^i(\xi^1, \xi^2) \text{ or } \xi^i = \xi^i(x^1, x^2) \tag{61}$$

with metric tensor $g^{i,j}$ defined as :

$$g^{ij} = (J^{\tau}J)^{ij} = \begin{pmatrix} \xi_x^2 + \xi_y^2 & \xi_x \eta_x + \xi_y \eta_y \\ \\ \xi_x \eta_x + \xi_y \eta_y & \eta_x^2 + \eta_y^2 \end{pmatrix} \equiv \begin{pmatrix} A_1 & A_2 \\ \\ A_2 & A_3 \end{pmatrix} \qquad (62)$$

where

$$J^i_j = \frac{\partial \xi^i}{\partial x^j} \qquad (63)$$

is the jacobian matrix of the transformation to [13],[14]

$$\partial_\xi \left[\frac{\rho}{|J|} (A_1 \partial_\xi + A_2 \partial_\eta)\phi \right] + \partial_\eta \left[\frac{\rho}{|J|} (A_2 \partial_\xi + A_3 \partial_\eta)\phi \right] = 0 \qquad (64)$$

or :

$$(\partial_\xi \frac{\rho}{|J|} U + \partial_\eta \frac{\rho}{|J|} V) = 0 \qquad (65)$$

which will be written also as :

$$\left[A_{\xi\xi} + A_{\xi\eta} + A_{\eta\xi} + A_{\eta\eta} \right]\phi = 0 \qquad (66)$$

It is to be noted that this equation is precisely the form which is impli-
citly calculated in the classical finite element approach with isoparametric transformation between the physical and the local coordinates ξ, η
Fig.1 .

Within an element, one has the isoparametric transformation (20) and the
FE representation of equ. (8) is, after partial integration and for Dirichlet boundary conditions, given by the first term of equ. (13).

This equation is numerically calculated as follows [15], with $\vec{\nabla}'$ being the
gradient operation in the ξ,η coordinates :

$$\Sigma \, \phi_I \int \rho \, \vec{\nabla}' \, N_I \, (\xi,\eta) \, (J^{\tau}J) \, \vec{\nabla}' \, N_J(\xi,\eta) \cdot \frac{1}{|J|} \, d\xi d\eta = 0$$

which can be written :

$$\Sigma \, \phi_I \int \frac{\rho}{|J|} (A_1 \partial_\xi N_I + A_2 \partial_\eta N_I) \partial_\xi \, N_J \, d\xi d\eta$$

$$+\Sigma \, \phi_I \int \frac{\rho}{|J|} (A_2 \partial_\xi N_I + A_3 \partial_\eta N_I) \partial_\eta N_J \, d\xi d\eta = 0$$

and with a reverse partial integration :

$$\int [\frac{\partial}{\partial \xi} \frac{\rho}{|J|} (A_1 \phi_\xi + A_2 \phi_\eta) + \frac{\partial}{\partial \eta} \frac{\rho}{|J|} (A_2 \phi_\xi + A_3 \phi_\eta)] \, N_J \, d\xi d\eta = 0$$

This is the formulation of equ. (8) which would be obtained by applying
directly the Galerkin method to equ. (64) in the (ξ,η) plane.

That the form (64) is contained in the standard FE-Galerkin method explains
the flexibility of the FE approach for arbitrary meshes. Moreover, the
type of discretization in FE methods, namely the area average of the residual calculation as discussed in the previous section, allows the FE-

approach to maintain the order of accuracy of the orthogonal mesh also in
the arbitrary mesh calculation. Of course, certain restrictions are imposed in order to maintain this property, which require the mesh not to be
too distorted.

3.3.1. Splitted scheme

The splitting method requires the operator to be written under the form of
a sum of two terms

$$A = A_1 + A_2$$

which contain only one type of derivative.

Writing equ. (64) under the form :

$$\{1 - \sigma(A_{\xi\xi} + A_{\xi\eta} + A_{\eta\xi} + A_{\eta\eta})\delta\phi = -\tau\sigma R \tag{67}$$

and neglecting the mixed operators $A_{\xi\eta}$ and $A_{\eta\xi}$; which corresponds to approximate the metric tensor g^{ij}, equ. (62) by a diagonal matrix one obtains

$$(1 - \sigma A_{\xi\xi})(1 - \sigma A_{\eta\eta})\delta\phi = -\sigma \tau R \tag{68}$$

This form is identical to the AF1 scheme of ref.[7].

Application of the FE-Galerkin approach to this equation follows the same
line as in the previous section.

Explicitely, equ. (68) is written as :

$$[1-\sigma\partial_\xi(\frac{\rho A_1}{|J|} \partial_\xi)][1- \sigma\partial_\eta(\frac{\rho A_3}{|J|} \partial_\eta)] \delta\phi= -\sigma\tau[(\frac{\rho U}{|J|})_\xi^{(n-1)} + (\frac{\rho V}{|J|})_\eta^{(n-1)}] \tag{69}$$

with :

$$\delta\phi = \phi^{(n)} - \phi^{(n-1)}$$

Applying, the FE-Galerkin method one obtains after splitting and for factorizing shape functions in the (ξ,η) plane, the system

$$(M_\xi + \sigma K_\xi)g = - \sigma \tau R \tag{70}$$

$$(M_\eta + \sigma K_\eta)\delta\phi = g \tag{71}$$

where :

$$(K_\xi)_{ik} = -\int \partial_\xi(\frac{\rho A_1}{|J|} \partial_\xi N_i(\xi)) \cdot N_k(\xi) \, d\xi \tag{72}$$

$$(K_\eta)_{j\ell} = -\int \partial_\eta(\frac{\rho A_3}{|J|} \partial_\eta N_j(\eta)) \cdot N_\ell(\eta) \, d\eta \tag{73}$$

and the residual R is given by equ. (18).

Hence, the residual contains automatically the boundary conditions and no
separate treatment has to be introduced. This maintains one of the strong
points of the FE method where the Neumann conditions $\partial\phi/\partial n = 0$ is a natural one and leads to a zero value for the first term in the expression of
the residual.

Numerical results obtained with bilinear elements on arbitrary meshes are
discussed in section 5.

4. RESULTS WITH SLOR-FE

Geometry and boundary conditions

The department being mainly concerned with the calculation of flows in tur-
bomachinery and blade rows, the programs are written for channel flow type
geometries(such as workshop case B), where no cut has to be introduced in
the flow domain. However a single airfoil geometry without lift (case A)
can, due to the symmetry, be transformed in a channel by omitting the lower
half part of the computational domain.

Due to the multigrid technique the proposed meshes are slightly modified
according to the condition mentioned in section 3. Therefore all the calculations are performed on a 73X25 mesh giving 1825 points. In order to respect the proposed geometry on the airfoil (or bump) boundary, the additional
points in the flow direction are located downstream the airfoil (or bump).

The following boundary conditions were adopted for the channel flow (case B):
Dirichlet B C at inlet ($\emptyset$=0), Neumann BC on the walls ($\frac{\partial \phi}{\partial n}$ =0) and Neumann
BC at outlet ($\frac{\partial \phi}{\partial n}$ = U_∞). As to the airfoil, the far field BC appears to influence strongly the calculations : the Dirichlet BC $\emptyset$=U_∞.x gave rise to an
oscillatory behaviour in the far field region and poor convergence, while
the Neumann BC $\frac{\partial \phi}{\partial n}$ = U_∞ cos α (α being the angle between x-axis and normal
direction) proved to be very stable and was finally adopted in all the com-
putations. On the airfoil boundary and symmetry axis simple Neumann BC

$\frac{\partial \phi}{\partial n}$ = 0 is applied.

Channel flow – case B – M_∞ = .85

This is the only case presented in the paper and shows all the features of
the method.

The artificial viscosity with the classical switching function (definition
3) produces a large preshock overshoot, as mentioned before. By multiplying this function with M^n for increasing values of n, the overshoot was
nearly eliminated for the case n=4, showing clearly the need of a stronger
A.V. This was also experienced by Holst[9] and was the reason for the introduction of definition (10). The pressure distribution recovered using this
form is shown in figure 5, with isomach plot figure 6. The distance between the isomachlines is .05. This point needs further examination and
better forms for the switching function might be obtained.

A second remark concerns the convergence speed. By numerical experiment
the optimal value for the relaxation factor was found to be 1.85 in this
case. As convergence norm, the maximum difference of the potential function between two consecutive iterations was used. In order to speed up the
convergence a multigrid technique was applied. After 20 iterations on the
coarse grids the norm was reduced by a factor 10 and the transition was made to the next finer grid. On the final grid, 40 supplementary iterations
were performed, until a stationary situation was reached. During the iterative processus the shock moves downstream and the intensity increases.
Finally the shock is found to be located between .775 and .825, with a
minimal C_p coefficient equal to −.855.

5. RESULTS WITH ADI-FE METHOD

In a first step, the method was applied to the subcritical flow past the
NACA-0012 airfoil with freestream speed M_∞ = .72, α = 0 (case AIi).

The parameters σ and τ were chosen in a standard way [16] and no effort was
done to optimize them. The parameter σ can be considered as an artificial
time step Δt and therefore one stategy for obtaining fast convergence is to
advance time as fast as possible(σ as high as possible). As pointed out in
[16], this is effective for attacking the low-frequency errors but not the
high frequency errors. The best overall approach is to use a σ sequence
containing a spectrum of σ values. The maximal residual history is shown
in figure 4. The oscillatory behaviour is due to the σ sequence containing
5 different values from low to high. The residual is a good estimation for
the high frequency error, but a very bad estimation for the low frequency error
and therefore the lowest value in the oscillation corresponds with the lowest σ-value in the sequence. Although 90 iterations were performed,
the ultimate solution is reached after only 30 iterations with a maximum
residual drop of only two orders. The explanation is that in the ADI- method the whole error spectrum is attacked equally well giving a lower overall error than would be expected from the residual.

The pressure distribution after convergence is shown in figure 7.

Application of this ADI-method to the M_∞ = .80 case diverges : indeed the
ADI operator of equation (41) produces a ϕ_t term which works destabilizing
in supersonic regions. This term should be replaced by a $\phi_{\xi t}$ term [9][10]
giving then the so called AF2 method.

This modification is actually being introduced and no results are yet available.

LITERATURE CITED

1. Hafez, M & Murman,E.M. Recent Developments in Finite Element Analysis for Transonic Flows. Proc. of Advanced Technology Airfoil Research - ATAR - NASA Langley, March(1978).

2. Glowinski,R , Périaux, J , Pironneau,O. Transonic Flow Simulation by the Finite Element Method Via Optimal Control. Second Int. Symp. on Finite Elements in Fluid Flow, Rapallo,(1976).

3. Périaux,J. Résolution de quelques Problèmes non-linéaires en Aérodynamique par des méthodes d'elements Finis et de Moindres Carrés Fonctionnels. Ph. D. Thesis-Univ. Pierre & Marie Curie - Paris VI, (1979).

4. Deconinck,H , Hirsch,Ch. A Finite Element Method Solving the Full Potential Equation with Boundary layer Interaction in Transonic Cascade Flow. AIAA Paper 79-0132, (1979).

5. Eberle,A. Eine Methode Finiter Elemente zur Berechnung der Transsonischen Potential-Stromung um Profile. MBB Bericht UEE 1352(o),(1977).

6. Hafez, M , Murman,E.M. Artificial Compressibility Methods for Numerical Solution of Transonic Full Potential Equation. AIAA Paper 78-1148 (1978).

7. Holst,T.L.,   Ballhaus,W.F. Conservative Implicit Schemes for the Full Potential Equation Applied to Transonic Flows. NASA TM 78469,(1978).

8. Jameson,A. Transonic Potential Flow Calculation using conservation Form. AIAA 2nd Computational Fluid Dynamics Conf. Proceedings,(June 1975),148-155.

9. Holst,T.L. An Implicit Algorithm for the conservative,Transonic Full Potential Equation Using an Arbitrary Mesh. AIAA Paper 78-1113 (1978).

10. Douglas,J & Dupont,T. Alternating Direction Galerkin Methods on Rectangles. Proc. Symp. on Numerical Solution of Partial Differential Equations II - SYNSPADE II, B. Hubbard,Ed. Academic Press, New York, (1971) p. 133-214.

11. Dendy,J. & Fairweather,G. Alternating Direction Galerkin Methods for Parabolic and Hyperbolic Problems on Rectangular Polygones. SIAM J. Numer. Anal. Vol 12, (1975), 144.

12. Marchuk, G.I. Methods of Numerical Mathematics. Springer Verlag,(1975).

13. Viviand, H. Conservative Forms of Gas Dynamic Equations. La Recherche Aérospatiale, 1, (1974), p.65.

14. Vinokur,M. Conservation Equations of Gas Dynamics in Curvilinear Coordinate Systems. J.Comp. Phys., Vol. 14, (1974), p.105-125.

15. Zienkiewicz,O.C. The Finite Element Method, Mc Graw Hill, New York, (1977).

16. W.F. Ballhaus, A. Jameson, J. Albert. Implicit Approximate Factorization Schemes for Steady Transonic Flow Problems. AIAA Journal Vol. 16, N° 6,(June 1978).

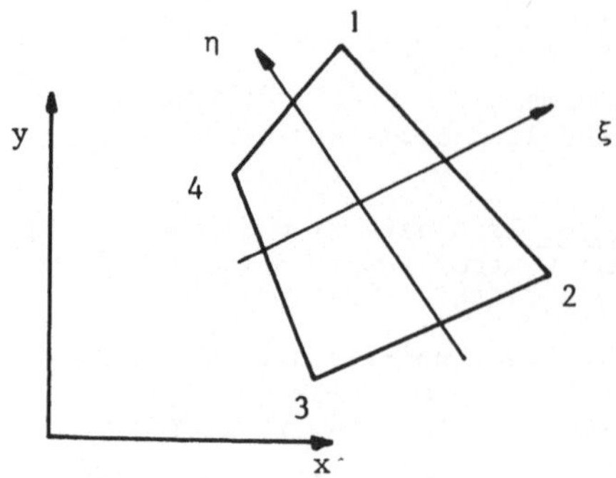
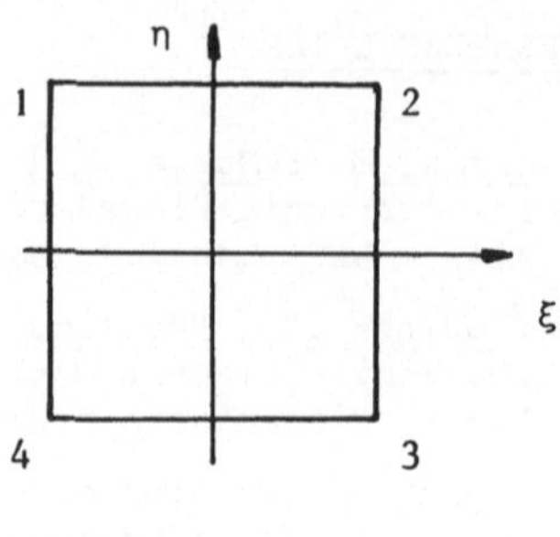

Figure 1

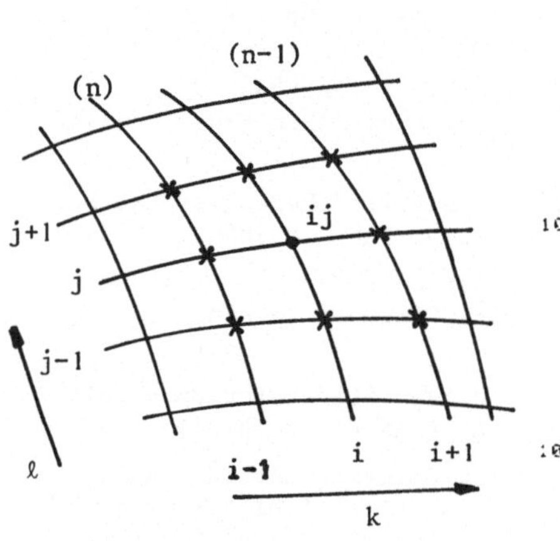

Figure 2

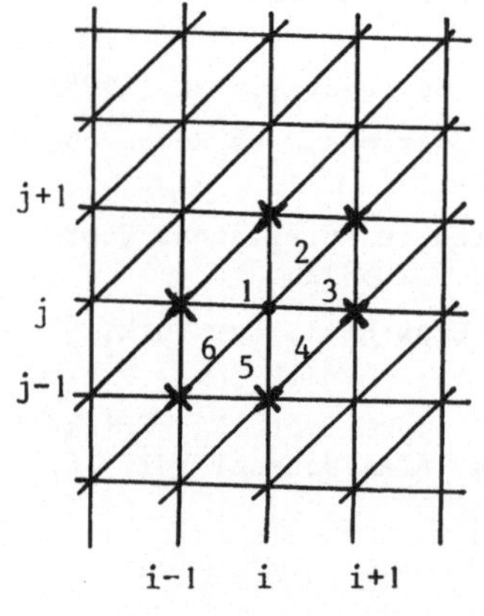

Figure 3

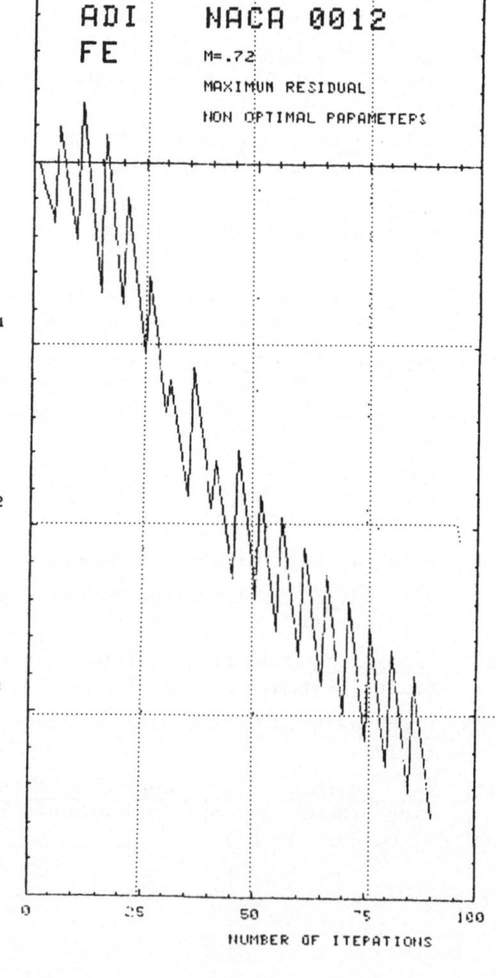

Figure 4

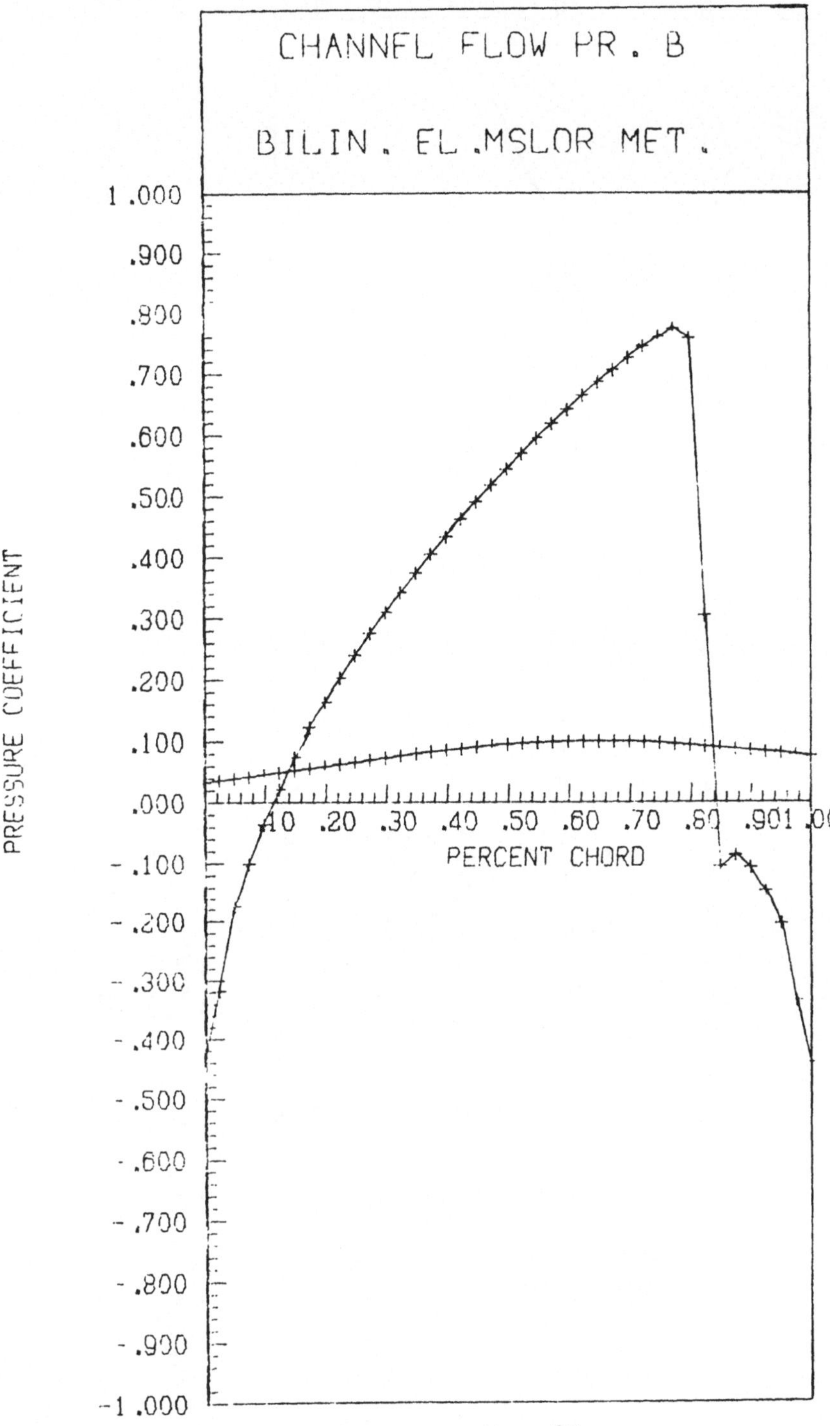

Figure 5. M = .85

CHANNEL FLOW PR . B
INLET ANGLE =   0.00 DEG   OUTLET ANGLE =   0.00 DEG   BILIN.ELEM. MSLOR METH.
INLET MACH NUMBER =   .8500   OUTLET MACH NUMBER =   .8500
ISOMACH LINES

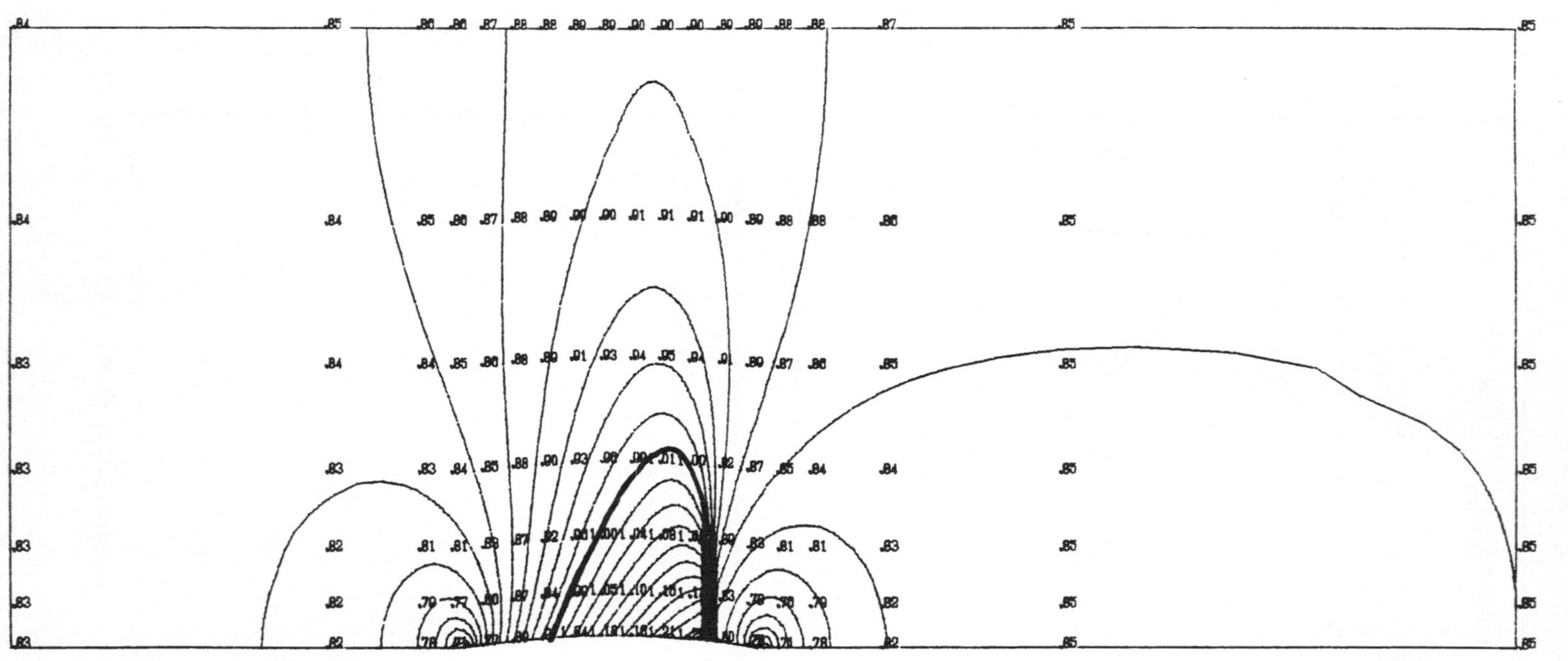

Figure 6.

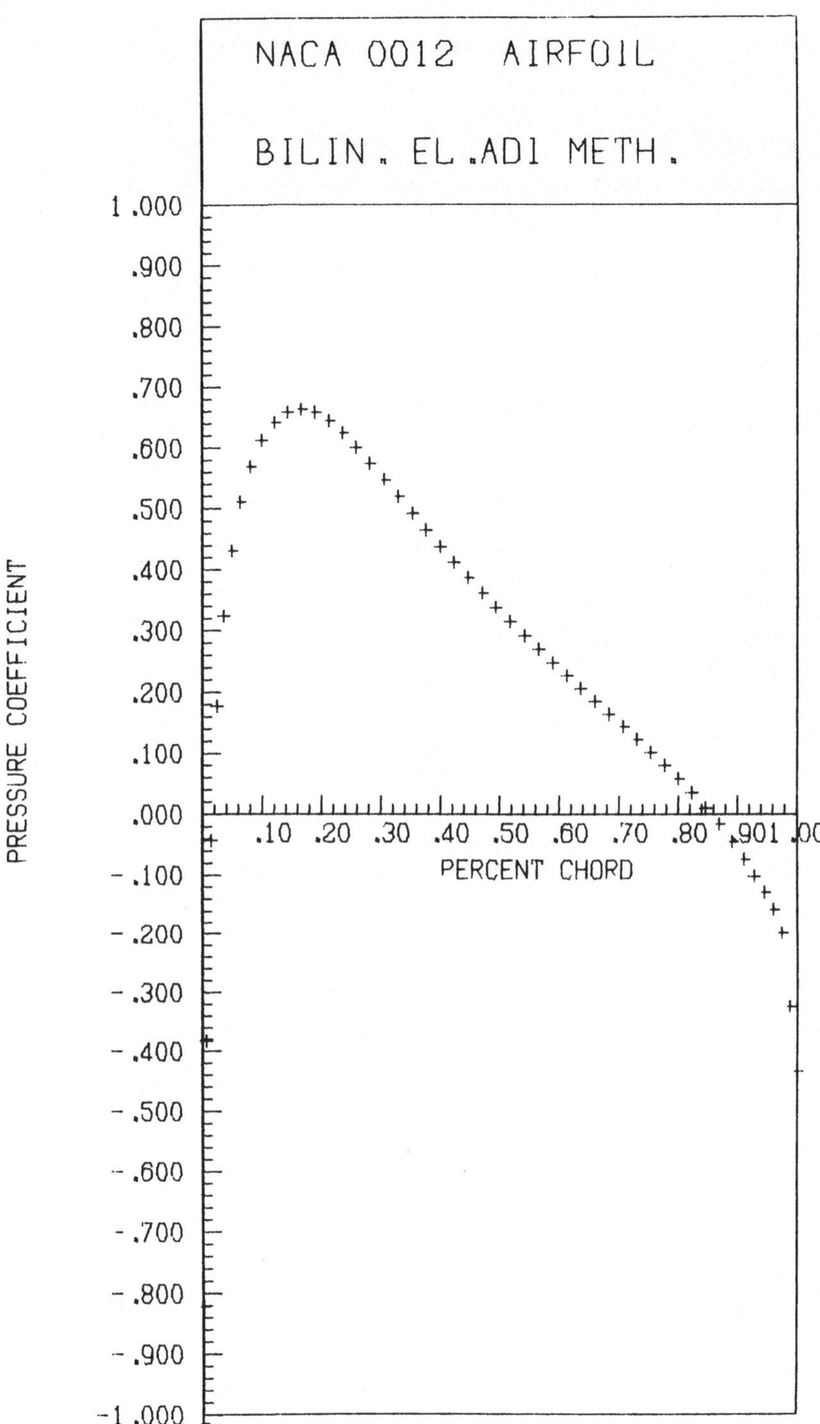

Figure 7 (M = .72 , α = 0).

TRANSONIC FLOW COMPUTATIONS BY A VARIATIONAL PRINCIPLE FINITE ELEMENT METHOD

A. EBERLE

Messerschmitt-Bölkow-Blohm GmbH   FE 122
Postfach 80 11 60   8000 München 80
W.-Germany

INTRODUCTION

In the past quite a lot of procedures have been programmed for the prediction of transonic flow past practically all sorts of configurations occuring in aerodynamic design.

As proved by many examples the present Finite Element Method (FEM) is particularly suited for this purpose due to the ease the codes can be adapted to new problems and the surprisingly little computerwork needed for solving the latter by exploiting successive mesh grid divisions in order to speed up convergence.

It is to these facts why the present method is now in use at several companies as well as research institutions. Researchers rather than engineers tend to emphasize high quality accuracy at the expense of higher computer costs regardless that the latter play a dominant role at industry. It is exactly to this reason why use had to be made of the present FEM for the solution of the compulsory GAMM-Workshop '79 problems despite of the shock-Machnumbers exceeding in several cases the upper bound acceptable for a potential theory approach. Application of the EULER-schemes available at the author's company was precluded due to excessive computer time required.

THEORY

Our starting point is NEWTON's law stating that all forces acting on a fluid particle cancel. For steady irrotational inviscid flow this is expressed by BERNOULLI's equation:

$$dp + \varrho q dq = 0 \tag{1}$$

where p = static pressure
         ϱ = density
         q = velocity

Irrotational flow allows for the introduction of a velocity potential

$$u = \phi_x, \quad w = \phi_z, \quad q = \sqrt{u^2 + w^2}$$

by means of which (1) can be rewritten

$$p_\phi + \varrho q q_\phi = 0 \tag{2}$$

After integration by parts extended over the entire fluid volume we arrive at

$$\iint (p_\phi + \varrho q q_\phi)\, dF = \iint p_\phi\, dF + \oint \varrho\, \vec{q}\vec{n}\, ds - \iint \mathrm{div}\,(\varrho \vec{q})\, dF$$

Due to the continuity equation the third integral on the right side vanishes as well as the boundary integral since the total fluid mass is constant and solid boundaries cannot be penetrated by fluid particles. Noting that the pressure integrals cancel identically we finally arive at

$$\iint \varrho q q_\phi\, dF = 0 \tag{3}$$

NUMERICAL EVALUATION

Since in general no analytic solution of (3) can be found the integral equation is decomposed in as many relations as unknown discrete potential values are to be determined. So $\partial/\partial\phi$ is replaced by

$$\frac{\partial}{\partial\phi} \longrightarrow \frac{\partial}{\partial\phi_i} \qquad i = 1, \ldots N$$

There remains to provide a good interpolation of the flow quantities inside finite intervals in order to form the integrand of (3).

For this purpose the bilinear isoparametric quadrilateral finite element is used. The choice of exactly this element meets the accuracy requirements of aerodynamic engineering. It can be proven analytically that finite elements based for example on full polynomial triangular forms cannot converge to the true solution. The FE representation of the functions needed is

$$(\phi, x, z) = \frac{1}{4} \sum_1^4 (\phi, x, z)_i \cdot (1 \pm \xi)\, (1 \pm \zeta)$$

where the +/- signs are used such that at node i the contributions of the other three nodes vanish.

ARTIFICIAL DENSITY

The AD is a simple device to maintain stability and to allow for the natural evolution of shock waves whenever the local Machnumber exceeds unity. The idea is to replace the stream density ϱq in (3) by a value computed a small distance upstream of the controlpoint such as to allow only for upstream information. So ϱq is replaced by

$$\varrho q \longrightarrow (\varrho q)_H = \varrho q + (\varrho q)_s\, \Delta s$$

which is equivalent to replacing ϱ by the following expression

$$\varrho \to \varrho_H = \varrho \left[1 + \varepsilon \min (0, \frac{1}{\bar{q}^2} - \frac{1}{\bar{a}^2}) \, (q_H^2 - q^2) \right]$$

where a is the speed of sound, ε is a parameter controlling the upwind shift and the bar denotes the average between the upstream element contribution (H) and the control element, i.e.:

$$\bar{q}^2 = q^2 + \varepsilon (q_H^2 - q^2)$$

ε may range between 0.5 (standard) and 1 (total upwinding). For the sake of simplicity neither shock fitting nor a shock operator is entered. So the present procedure is quasiconservative.

The matrix structure for the ϕ_i's is that of an elliptic equation throughout the flowfield and overrelaxation can be applied everywhere.

RESULTS

a)   FEM versus Finite Volume Method

Figure 1 outlines the major differences of the two methods. Inherently the FEM solves the continuity equation using overlapping control surfaces. Applying the trapezoidal rule only one integrand per face has to be formed. As confirmed by numerical experiments computer time of an equivalent FVM is about twice as much. This is also due to the fact that the present method avoids higher derivatives in forming the artificial density.

b)   Mesh grid division

Exploiting this simple device to the extent as shown in figure 2 reduces computer time considerably. A usual airfoil computation needs less than 20 equivalent iterates on the finest grid.

c)   Artificial density

Standard artificial viscosity ($\varepsilon = 0.5$) may lead to unphysical waviness ahead of the shock. It can be suppressed by choosing a higher value for ε at the expense of shock smearing, figure 3.

CONCLUSIONS

The present method exhibits one of the simplest forms of a
FEM-procedure.

Clearly, there exists quite a number of methods superior in
accuracy at the expense of more computer work however.

Industrial experience shows that the choice of an adequate
code for solving a particular flow problem is far away from
being independent of the costs it causes.

REFERENCES

[1] EBERLE A.
     Transonic Flow Computations by Finite Elements:
     Airfoil and Wing Analysis, Airfoil Optimization
     MBB-UFE1428(Ö) / DGLR 78-65   1978

[2] JAMESON A. / CAUGHEY D.A.
     A Finite Volume Method for Transonic Potential
     Flow Computations
     AIAA 3rd Computational Fluid Dynamics Conference
     Albuquerque, N.Mex./June 77   1977

<u>FEM versus Finite Volume Method</u>

FEM Eberle (1978)	FVM Jameson/Caughey (1977)

control   surface   arrange-
ment

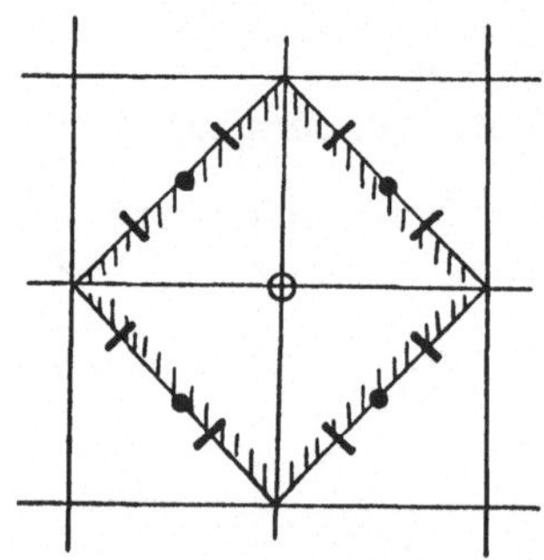

• trapezoidal rule ❘ Gauss quadrature (optional)	Needs 'luming error correc- tion' by shifting from • to ❘

artificial compressibility $\widetilde{g} = g + \mu\,\Delta g_{upstr.}$	artificial viscosity gu retarded in x-direction gv retarded in y-direction gw retarded in z-direction

FIGURE 1

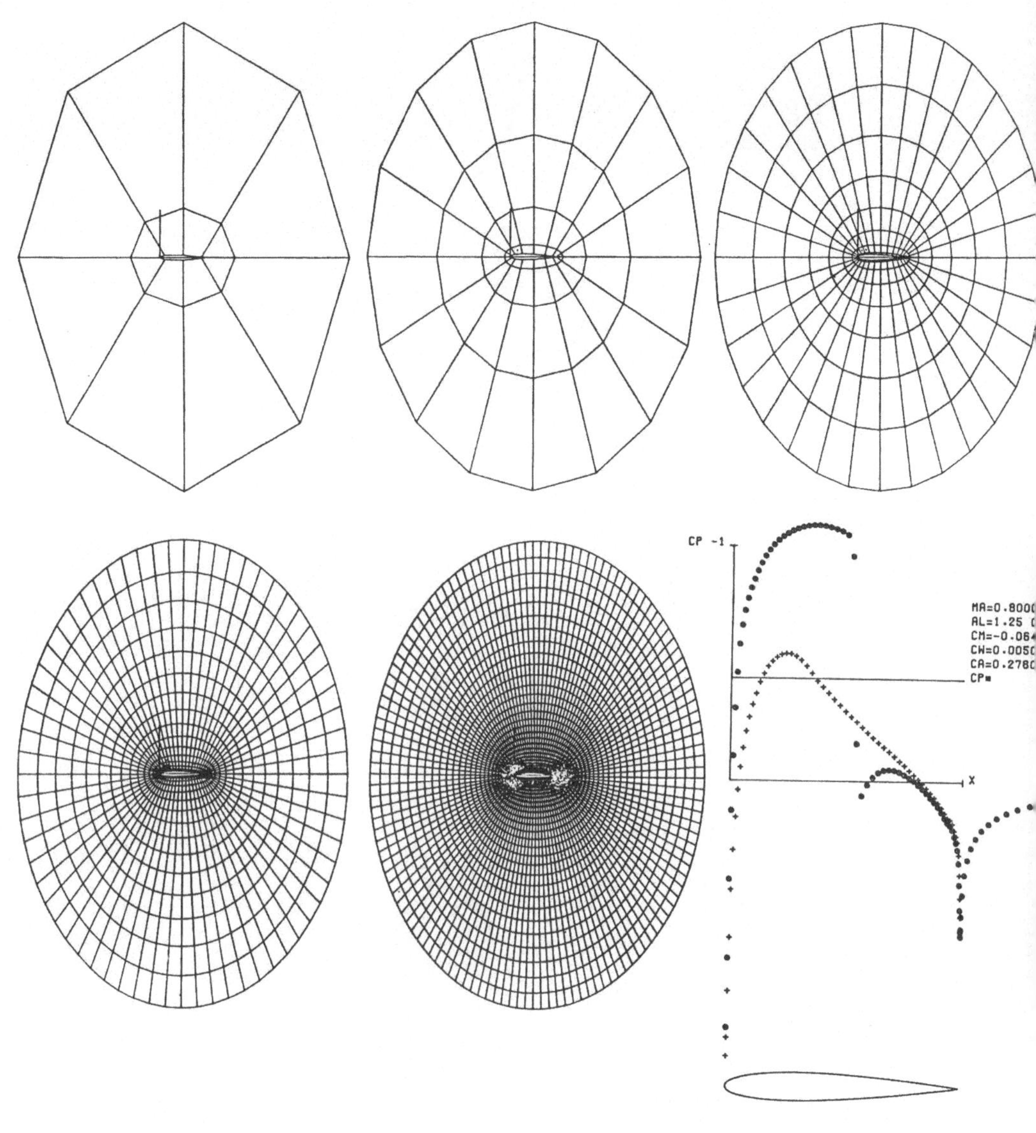

SPEEDING UP CONVERGENCE

FIGURE 2

CP -1
MA=0.8500
AL=1.00 GRAD
CM=-0.1078
CW=0.0156
CA=0.3111
CP*
X
Standard AD
ε = 0.5
CP -1
MA=0.8500
AL=1.00 GRAD
CM=-0.1289
CW=0.0177
CA=0.3404
CP*
X
Strong AD
ε = 1
NACA 0012 AIRFOIL (CLOSED)

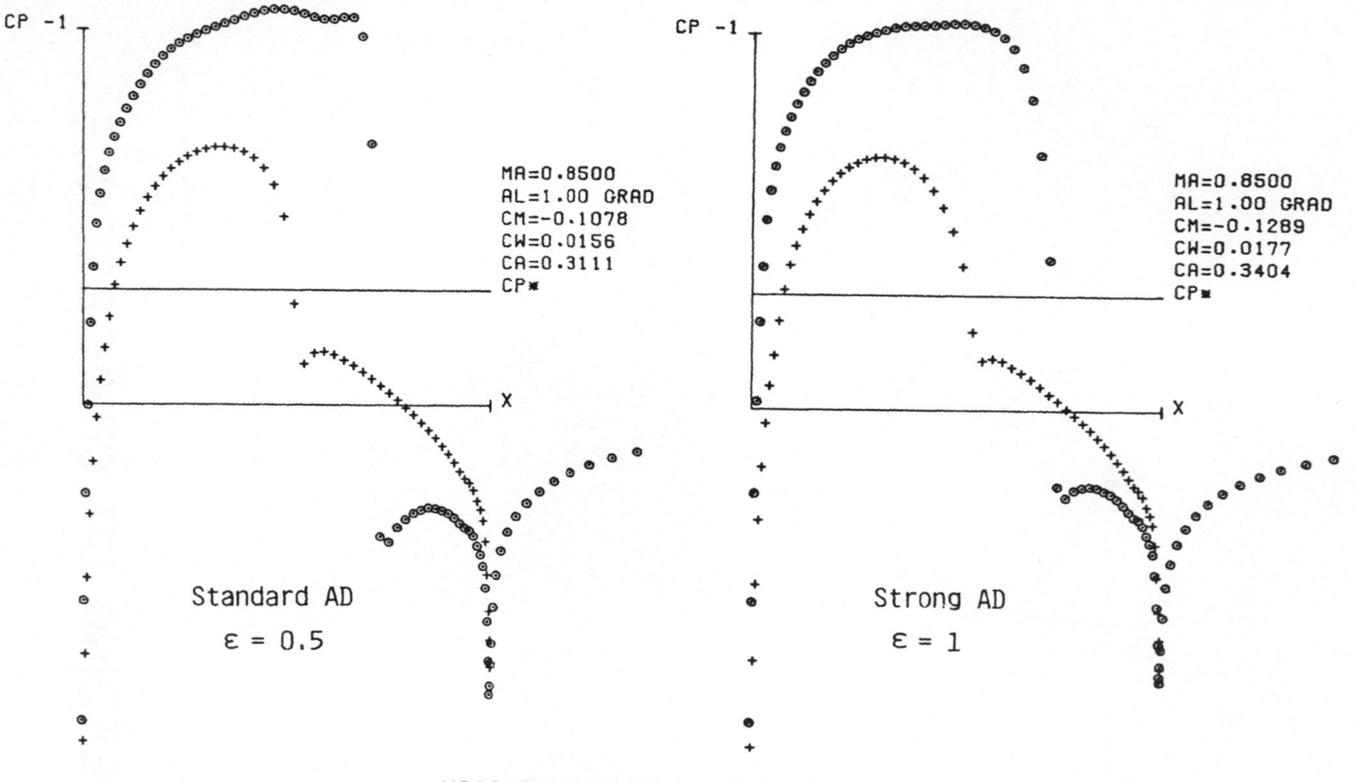

FIGURE 3

A FINITE ELEMENT METHOD FOR COMPUTING TRANSONIC POTENTIAL FLOW

S. A. Jepps

(British Aerospace Aircraft Group, Warton)

The finite element method has gained widespread acceptance
as a technique for solving elliptic partial differential
equations, where its ability to handle quite general
geometries has proved an advantage over finite differences.
For the steady flow of an inviscid fluid, the governing
equations are elliptic as long as the flow speed is everywhere
subsonic. In such cases the effects of compressibility are
generally small, and results of adequate accuracy for
engineering purposes can be obtained by applying approximate
corrections to the solution obtained for incompressible
flow. However, incompressible flow can be calculated using
the so-called panel method or boundary integral equation
approach. Because it reduces the problem from one of solving
for an entire field to that of solving for quantities on the
boundary, the latter approach is more efficient than either
finite elements or finite differences. This fact has
prevented the widespread use of finite elements for purely
subsonic flow. Attention has therefore recently focused on
the possibility of using the finite element method to treat
flows which are not purely subsonic, and which thus contain
regions in which the governing equation is hyperbolic. This
paper describes one such attempt which makes use of the
'artificial compressibility' concept developed in ref. 1.

The governing equation for inviscid potential flow is

$$\mathrm{div}\ (\rho\ \mathrm{grad}\ \phi) = 0 \qquad\qquad (1)$$

where ϕ is the velocity potential and ρ the dimensionless
fluid density. The density is related to the fluid velocity
V by the relation

$$\rho = \left\{ 1 + \frac{M_\infty^2}{5} (1 - V^2) \right\}^{5/2} \qquad\qquad (2)$$

assuming a ratio of specific heats $\gamma = 1.4$ where M_∞ is the
free stream Mach number. In the present method the nonlinear
system of equations represented by (1) and (2) is solved
iteratively by assuming a fixed distribution of ρ at each
iteration and solving (1) for ϕ. The resulting ϕ distribution
is then substituted in (2) to give a revised estimate for ρ,
which is re-used in (1). With this technique a linear,
elliptic partial differential equation has to be solved at
each step. As might be expected, the above iteration only
converges if the flow is everywhere subsonic. However, the
artificial compressibility concept, by introducing a
streamwise shift in the density distribution, enables the
iteration to converge even when the flow includes locally
supersonic regions. Moreover, it ensures that the physically
correct solution is obtained, ie. one which excludes expansion

shocks. The details of the way in which this shift is applied
will be described later. For the present we consider the
problem of solving (1) for ϕ with a fixed distribution of ρ.

Since equation (1) is elliptic in ϕ, it is possible to use
standard finite element techniques for solving elliptic
partial differential equations. In common with ref.2, we
use first-order quadrilateral isoparametric elements. Each
element is mapped from the physical x, y plane into the
local ξ,η computing plane by the transformation

$$x = \sum_{i=1}^{4} x_i \, N_i \, (\xi,\eta), \qquad\qquad y = \sum_{i=1}^{4} y_i \, N_i \, (\xi,\eta) \qquad (3)$$

where the N_i are pyramid functions defined by

$$N_i \, (\xi,\eta) = (\tfrac{1}{2} + 2 \, \xi_i\xi) \, (\tfrac{1}{2} + 2 \, \eta_i\eta) \qquad (4)$$

where each of the ξ_i, η_i is equal to $+ \tfrac{1}{2}$. The distribution
of potential ϕ within the element is then approximated by

$$\phi = \sum_{j=1}^{4} \phi_j \, N_j \qquad (5)$$

where the ϕ_j are the nodal values of potential. Using a
Galerkin approach to discretise equation (1) (ref. 3), the
ϕ_j are related by a system of linear equations whose matrix
is made up of elements of the form

$$a_{ij} = \int_{-\frac{1}{2}}^{\frac{1}{2}} \int_{-\frac{1}{2}}^{\frac{1}{2}} \rho \, (\text{grad } N_i \, . \, \text{grad } N_j) \, J \, d\xi d\eta \qquad (6)$$

where J is the Jacobean of the transformation given by (3).
In the present implementation ρ is assumed to be constant
within each element and the integrals in (6) are evaluated
by Gaussian integration using two points in each direction.

The resulting individual element matrices are then assembled,
using the classical matrix assembly concept of finite element
theory, to yield a global matrix for the entire flow domain.
This system of equations is then solved by a relaxation
process. For lifting aerofoils the Kutta condition at the
trailing edge is satisfied by calculating two solutions
corresponding to zero and unit circulation respectively.
These two solutions are then combined at each iteration in
such a way that the resultant distribution of ϕ satifies the
Kutta condition.

We note in passing that two points is probably the minimum
that can be used for the Gaussian integration. When
applied to Laplace's equation on a grid of unit squares, the
above technique results in a discretisation which involves

the eight points surrounding the point in question. However,
if an attempt is made to estimate the integral in (6) by
sampling the integrand at a single Gauss point, the resulting
discretisation involves only the four neighbouring points
lying on the diagonals through the point in question.

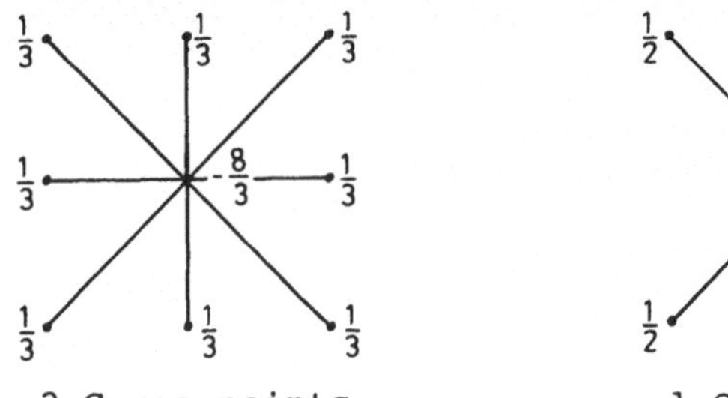

2 Gauss points   1 Gauss point

The use of such a diagonal molecule decouples alternate grid
points, resulting in a system which is very susceptible
to high frequency ripples.

The technique used to apply the streamwise density shift is
as follows: It is assumed that equation (1) has been solved
during the previous iteration, yielding values of the
potential ϕ at all the nodes of the computational grid.
Because the gradient of the potential is discontinuous at the
element boundaries, it is not possible to ascribe a unique
value to the velocity at each node for use in equation (2).
Instead, we compute the limit of the velocity as the node
is approached from within each of the finite elements which
meet at that node . The arithmetic mean of these values is
then. taken to be the velocity at the node. Equation (2) is
then invoked to compute nodal values of the density. At nodes
where the flow is currently computed to be locally supersonic
this value of the density is then modified as follows:

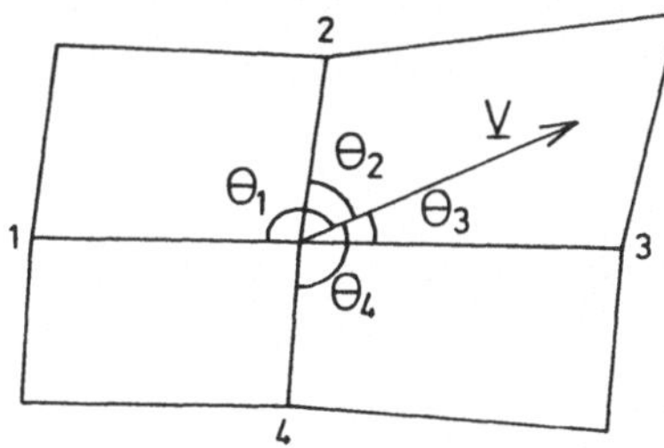

First, that one of the four surrounding nodes which is closest
to being directly upstream from the node in question is
selected, i.e. that node for which θ is greatest (see diagram
above). Denoting this node by i, the modified density $\tilde{\rho}_o$
at the point O is then computed as

$$\tilde{\rho}_o = (1 - \mu)\rho_o + \mu\rho_i \tag{7}$$

where μ is a quantity which corresponds to the 'artificial viscosity' which is used in certain other numerical methods. Finally, the four nodal density values at the corners of an element are averaged to yield the constant density value within the element.

There is some uncertainty as to the optimum form to be used for the artificial viscosity μ. Early techniques for computing transonic flow employed finite differences, with centred difference formulae in the subsonic region and backward differences in the supersonic region. It can be shown that switching to backward differences in the supersonic region is equivalent to introducing an artificial viscosity given by

$$\mu = \max\left\{ 0, \ 1 - \frac{1}{M^2} \right\}$$

where M is the local Mach number. It might be argued that this amount of artificial viscosity should be used in the finite element formulation. Experience (see Fig. 1) shows, however, that considerably more artificial viscosity is needed in the finite element method to yield good results. An empirical factor ϵ is therefore included in the expression for μ. However, it is found that using a high value of ϵ can restrict the maximum local Mach number for which the procedure converges. This is presumably due to the value of μ exceeding unity, in which case equation (7) represents an extrapolation rather than an interpolation. It is thus advisable to enforce the restriction $\mu \leqslant 1$. The final expression for μ is thus

$$\mu = \begin{cases} 0 & \text{if} \quad M \leqslant 1 \\[2ex] \epsilon\left(1 - {}^{1}/M^2\right) & \text{if } 1 < M < \left(\frac{\epsilon}{\epsilon-1}\right)^{\frac{1}{2}} \\[2ex] 1 & \text{if} \quad M \geqslant \left(\frac{\epsilon}{\epsilon-1}\right)^{\frac{1}{2}} \end{cases} \qquad (9)$$

Because the finite element method is flexible regarding the form of computing grid, there is some room for manoeuvre regarding the means used to set up the grid for an aerofoil problem. In practice we have opted for a simplified version of Thompson's algorithm (ref. 4), in which the Laplace equation is solved for the x,y coordinates of the grid points. The grid extends from the aerofoil contour outward to a circular far-field boundary. At the far-field boundary the solution is matched with an approximate far-field solution consisting of the fields of a point doublet and vortex in Prandtl-Glauert space. An example of a computing grid is shown in Fig. 2.

To summarise, a finite element method has been developed for
solving the equations of two-dimensional compressible
potential flow. The artificial compressibility concept is
applied in order to obtain a converged solution for flows
involving embedded supersonic regions and shock waves. The
use of finite elements promises the ability to handle flow
domains which are geometrically or topologically complex.

References

1)   M. Hafez, J. South and E. Murman

     Artificial Compressibility Methods for Numerical Solution
     of Transonic Full Potential Equation.

     Paper presented at AIAA 11th Fluid and Plasma Dynamics
     Conference, Seattle, July 1978.

2)   A. Eberle

     Eine Methode finiter Elemente zur Berechnung der
     transsonischen Potential-Strömung um Profile.

          MBB Bericht Nr UFE 1352 (Ö), Sept. 1977.

3)   S. A. Jepps

     Application of the Finite Element Method to Aerodynamics

          BAe (Warton) Report No. Ae/A/602   March 1979

4)   J. F. Thompson, F. C. Thames and C. W. Mastin

     Automatic Numerical Generation of Body-Fitted Curvilinear
     Coordinate Systems for Fields Containing Any Number of
     Arbitrary Two-Dimensional Bodies.

          Journal of Computational Physics 15,299 (1974)

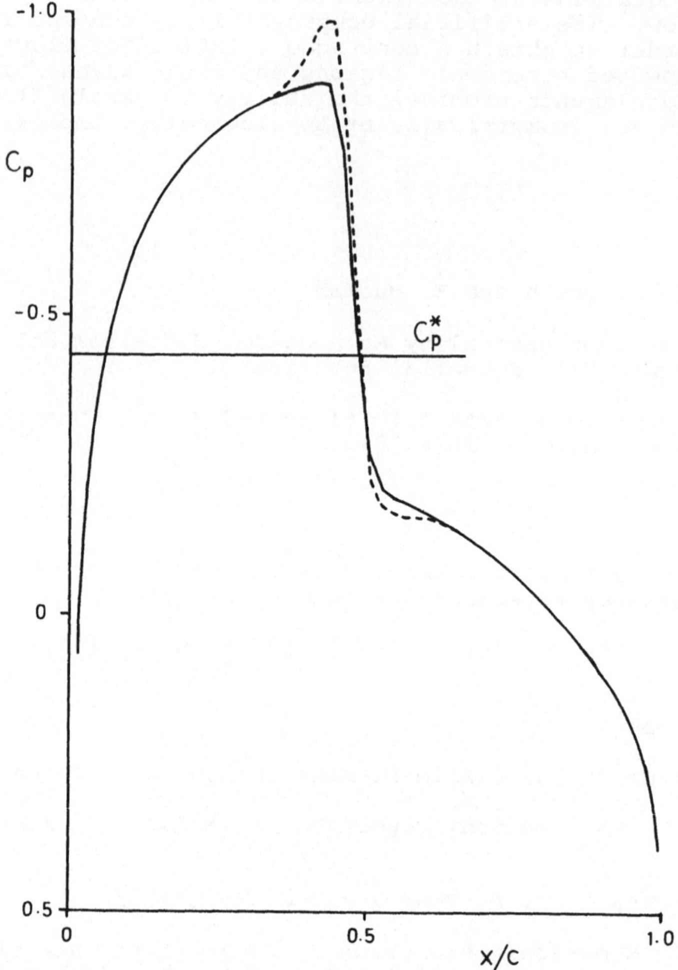

Fig. 1   Pressure distribution on NACA0012 aerofoil at

$\alpha = 0$, M = 0.8

——— ε = 3
- - - - - ε = 2
ε = 1 - No convergence

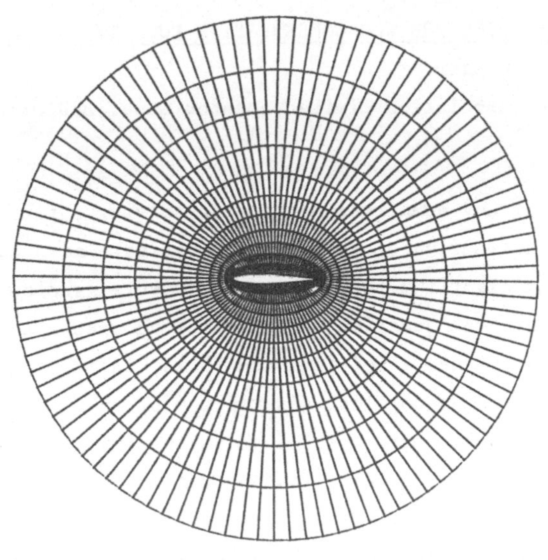

Fig. 2   Computing grid for the NACA0012 aerofoil.

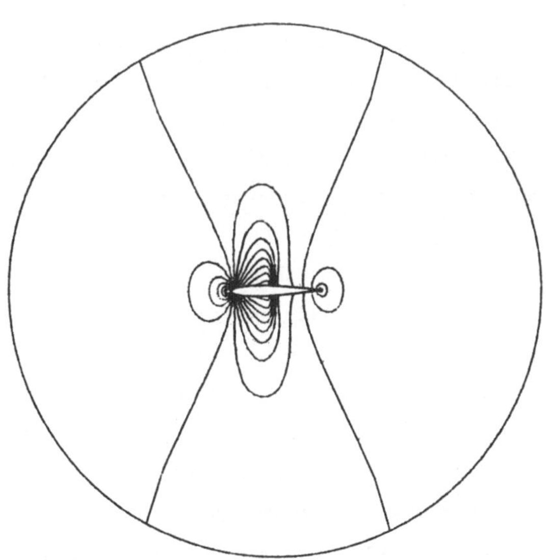

Fig. 3   Mach number contours for NACA0012 aerofoil
at $\alpha = 0$, $M = 0.8$

<u>FLOW CALCULATIONS USING THE NON-CONSERVATIVE POTENTIAL EQUATION</u>

by T J Baker and M P Carr

Aircraft Research Association Limited, Manton Lane, Bedford, England

<u>Introduction</u>

The success of any numerical method for solving a fluid flow problem depends on the construction of a suitable coordinate mesh and the use of a stable, convergent algorithm for solving the discretised set of equations. In a similar manner, the question of numerical accuracy falls into two essentially separate areas. For inviscid flows, it is accepted that the Euler equations provide a complete description of the flowfield and that the adoption of an approximate model such as the full potential equation must necessarily introduce certain inaccuracies. The comparisons presented at this Workshop should provide some insight into the relative accuracy of the various approximate mathematical models used to describe transonic flows.

The second and particularly difficult question concerns the effect of the truncation errors introduced by the discretisation of the governing differential equations. In principle this problem can be reduced to manageable proportions by using a sufficiently fine mesh. Just how fine the mesh should be is not always known, and in any case, the number of mesh points available for most practical computations is usually dictated by considerations of computer storage and run times. In this paper we hope to touch on some of the mesh related errors and to indicate how changes in the mesh can influence the numerical solution to a considerable extent. First, however, we shall outline the main features of the mathematical model and numerical method we have used.

All the computations on the Workshop test cases have been carried out using the full potential equation and, as far as possible, we have attempted to achieve a common approach to the different problems. The use of the potential equation to model a fluid flow rests on the assumption that the flow is everywhere isentropic. Although this assumption is only strictly valid for subcritical cases, useful results can be obtained for transonic flows provided shock strengths are not too large. A potential flow calculation cannot conserve both mass and momentum across a shock wave and a certain ambiguity therefore exists depending on which type of numerical differencing is used to solve the flow equation.[1]

In this paper we consider the full potential equation in its non-

conservative form. The differential equation is first discretised and the
resulting set of finite difference equations is then solved by an iterative
algorithm. Second order accurate central difference formulae are used to
approximate all first derivatives and all second derivatives when the flow
is locally subsonic. In supersonic regions a combination of centred
differencing and first order accurate upwind differencing is employed in
order to produce the right amount of dissipation required to capture any
shock waves. It follows that, in regions of supersonic flow, the
truncation error is formally of first order and hence one would expect this
loss of accuracy to be reflected in the computed results. This topic will
be discussed later in connection with the airfoil computations when
comparisons between different mesh sizes will be presented.

The algorithm chosen to solve the set of finite difference equations
for all test cases is an approximate factorisation scheme known as AF3.[2]
This scheme has been found to converge more rapidly and to be generally
more reliable than the conventional line relaxation (SLOR) method. The
convergence rate of the approximate factorisation method is so fast that it
is possible to allow each case to converge to the limit set by round-off
error. Provided the round-off error is sufficiently small, the accuracy of
the numerical solution will then be determined entirely by the truncation
error of the difference formulae used to approximate the governing equation.
The computer used for the calculations is a PRIME 400 and on this machine
the accuracy of single precision arithmetic only extends to seven
significant figures. In practice this means that the maximum absolute
value of the residual can only be reduced to about 10^{-3} before round-off
error sets in. A few sample computations using double precision arithmetic,
however, indicate that the single precision limit is adequate to define
lift and drag coefficients correct to four decimal places.

A Unified Approach

Our original objective had been the construction of a general computer
code that would accept an arbitrary set of mesh points, then compute the
required transformation derivatives and finally use the AF3 algorithm to
solve the resulting set of difference equations. With this approach the
only part of the program specific to any particular flow problem would be
a single subroutine which handled the application of the boundary
conditions. It was intended that this code could be used for all the test
cases and hence form a common basis from which one could examine the
accuracy of the computed results.

We assume that the mesh is given by a set of ordinate pairs

$$\{x_{i,j}, y_{i,j} \mid i = 1, \ldots, IL, \; j = 1, \ldots JL\}$$

and that it is possible to define a new pair of independent variables ξ and η such that ξ lines pass through the sets of points

$$\{x_{i,j}, y_{i,j} \mid i = 1, \ldots, IL\}$$

for each fixed value of j. Similarly, each η line corresponds to a fixed value of i. In order to obtain the quantities

$$\frac{\partial x}{\partial \xi}, \; \frac{\partial x}{\partial \eta}, \; \frac{\partial y}{\partial \xi}, \; \frac{\partial y}{\partial \eta}$$

second order accurate finite difference formulae of the form

$$\frac{\partial x}{\partial \xi}\bigg|_{i,j} = \frac{x_{i+1,j} - x_{i-1,j}}{2\Delta \xi}$$

were used. These transformation derivatives were then combined together in the usual way to form the metric tensor components and a finite difference approximation to the full potential equation was thus obtained. Some of the terms that appear in the potential equation involve derivatives of the metric tensor components with respect to ξ and η. Once again second order accurate finite difference formulae were used.

The computer code was first tried out on a few simple shapes such as a circle and ellipses in which the set of mesh points had been previously generated from a body normal type transformation. The results obtained were in very good agreement with the corresponding results from a code that solved the potential equation expressed exactly in body normal coordinates (ie the transformed equation obtained by the appropriate analytic transformation). These comparisons suggested that the rather crude finite difference approximation used to obtain the transformation derivatives in the general computer code was adequate at least for well behaved orthogonal meshes. A similar exercise using a sheared type of mesh over a parabolic arc confirmed that the general code could perform well for non-orthogonal meshes. It was also felt that these comparisons strongly suggested that no program bugs were present.

Channel Flow Problem

The computer code was then applied to Problem B of the Workshop test cases and some interesting results were obtained. The flow tangency condition on the upper and lower walls of the channel leads to Neumann type boundary conditions. At the upstream and downstream ends of the channel the flow is required to be uniform and at a Mach number of 0.85. The coordinate mesh has 72 points in the streamwise direction of which 49

points occupy an interval that extends from a position just upstream to a
point just downstream of the circular bump. The remaining 23 points are
split between the upstream and downstream sections resulting in a mesh that
is dense over the interval occupied by the bump but which is highly
stretched in the upstream and downstream regions. With this set of mesh
points, the computer code prediction was clearly wrong with spikes
appearing in the pressure distribution at the two positions where the mesh
changes from a dense uniform spacing to the highly stretched spacing.

In order to improve the flow prediction some further calculations were
carried out using progressively finer meshes in the upstream and downstream
sections. Thus by placing an extra 10 points in both the upstream and
downstream sections, a mesh of 92 x 21 points was obtained. Similarly by
placing an extra 15 and then 30 points in both sections, meshes of 102 x 21
and 132 x 21 points were produced. In all these examples the number and
hence density of mesh points in the central region was left unchanged. It
follows that the effect of increasing the total number of mesh points in
the streamwise direction led to a more gentle stretching of the mesh in the
upstream and downstream sections and hence a smoother variation of the
transformation derivatives. The results of this exercise are presented in
Figure 1 which shows the lower wall pressure distribution computed using
the three finer meshes. An examination of the mass flux through the channel
reveals a sudden change in the mass flux at the mesh changeover points.
For the finest mesh (ie 132 x 21 points) the mass flux error at the mesh
changeover points is about 0.4% which is less than the mass increase of
0.8% that occurs on passing through the shock. For other mesh sizes
considered here, the mass flux error that occurred at the mesh changeover
points was larger than the mass increase introduced by the shock. It
therefore seems reasonable to consider the result obtained on the finest
mesh (solid line in Figure 1) as the best that we can achieve for this
problem using a non-conservative potential method.

The use of fairly crude finite difference formulae to obtain the
transformation derivatives was probably responsible for some of the
difficulty experienced with the more highly stretched meshes.
Unfortunately within the time available to do these computations, it was
not possible to examine the effect of using higher order differencing or
curve fitting to obtain more accurate approximations. It is, however,
worth noting that similar effects have been observed with highly stretched
meshes even when the exact analytical form is used to evaluate the

transformation derivatives. In other words, rapid variations in the distribution of mesh points can have the effect of introducing forcing terms that lead to anomalous results.

This is unfortunate since the application of potential flow methods to general 3D flow problems will require an economical distribution of mesh points and fairly rapid variations in mesh spacing over some regions of the flowfield would appear to be inevitable. A more sophisticated treatment of the numerical differencing such as the use of finite volume techniques[3] might alleviate the problem and it is possible that developments along these lines will be needed before satisfactory solutions to complicated 3D potential flow problems can be achieved.

<u>Airfoil Flow Problems</u>

The application of the general computer code to the airfoil problems was less successful and a typical result is shown in Figure 2. This example is for the NACA 0012 airfoil at a freestream Mach number of 0.72 and zero degrees of incidence. The solid line represents what we consider to be the correct result obtained using the circle plane mesh. The dashed line shows the prediction we obtained when our general computed code was used with the standard set of mesh points provided for this problem. In view of our simplified treatment for extracting transformation derivatives from the set of mesh points, it is perhaps not surprising that the general computer code failed to produce sensible results for this mesh. Within the time available it was not feasible to examine this problem further and so it was decided to compute all the airfoil test cases using the circle plane mesh.

This mesh which was first introduced by Sells[4] is well suited to the airfoil problem. Under this transformation the region exterior to the airfoil is mapped onto the interior of the unit circle. The relaxation methods of Garabedian and Korn[5] and Jameson[6] both use this mesh and our application using an approximate factorisation technique is described in Ref.2. The circle plane mesh varies smoothly and we would not expect to encounter any problems of the mesh stretching type. It follows that the mesh related errors should be due entirely to the truncation error of our difference formulae for the potential derivatives.

In order to gain some insight into the magnitude of the truncation errors we have computed each airfoil case using first a mesh of 106 x 28 points which is roughly comparable to the size of mesh proposed for these problems, and then using a mesh of 210 x 56 points.

For all the subcritical cases no plotable difference between the coarse
and fine mesh pressure distributions was evident. However differences of
up to 0.001 in lift coefficient were apparent showing that this measure is
a very sensitive indicator of the solution accuracy. The drag coefficient
obtained with the coarse mesh typically showed a thrust of one or two
counts; on the fine mesh a zero drag count was predicted.

The difference formulae used when the flow is locally supersonic are
only first order accurate and the effect of this increase in truncation
error is apparent in the comparisons between the coarse and fine mesh
computations. A typical result is shown in Figure 3 which compares the
coarse and fine mesh computed pressure distributions for the KORN 1 airfoil
at a freestream Mach number of 0.75 and one degree of incidence. The fine
mesh (solid line) and coarse mesh (dashed line) pressure distributions are
clearly different. This is also reflected in the lift coefficients which
differ by 0.0065 and the five counts difference in drag coefficient. One
does not expect to obtain accurate predictions of the absolute wave drag
from a potential flow calculation. However it seems reasonable to look for
relative differences between slightly different flow conditions and to
expect accurate predictions of the relative differences in wave drag. It
is therefore rather disturbing that the coarse and fine mesh results shown
in Figure 3 should be so different. For flow computations over complicated
3D shapes the number of mesh points available for any one region is
severely restricted. One would therefore anticipate that inaccuracies due
to truncation errors could be a major consideration in such problems. One
obvious improvement would be the use of second order accurate difference
formulae throughout the flowfield for supercritical as well as subcritical
cases. Whether it is possible to find stable second order accurate
difference formulae for supersonic flow regions that will capture strong
shocks remains an open question.

<u>Conclusion</u>

Although the assumption of isentropic flow is only strictly valid for
subcritical cases, the potential equation does provide a useful model for
inviscid transonic flows provided shock strengths are not too large. The
development of stable and rapidly convergent algorithms for solving the
potential equation has promoted its widespread use and accurate numerical
solutions of the potential equation can be obtained if the coordinate mesh
is carefully constructed. Most mesh points should be placed where large
flow accelerations are expected and the point distribution should vary

smoothly between dense and sparse regions of the mesh.

The flow computations that we have carried out on the Workshop test cases show that rapid or sudden changes in the distribution of mesh points can lead to erroneous results. In order to relax this constraint on the choice of coordinate mesh, there is a need for improved difference formulae that are less sensitive to changes in mesh point distribution. The use of first order accurate upwind difference formulae in supersonic regions is another feature that restricts the numerical accuracy of potential flow calculations and the development of stable higher order accurate difference formulae for supersonic regions would be highly desirable.

<u>References</u>

1.   Jameson A, 'Numerical Computation of Transonic Flows with Shock Waves', IUTAM Symposium Transsonicum II, Göttingen, September 1975.

2.   Baker T J, 'Approximate Factorisation Methods for the Non-Conservative Potential Equation', to be published.

3.   Jameson A and Caughey D A, 'A Finite Volume Method for Transonic Potential Flow Calculations', AIAA Paper 77-635, 1977.

4.   Sells C C L, 'Plane Subcritical Flow past a Lifting Aerofoil', Proc Roy Soc, London, Vol.A308, 1968, pp.377-401.

5.   Bauer F, Garabedian P and Korn D, 'Supercritical Wing Sections', Vol.1, Springer, 1972.

6.   Jameson A, 'Transonic Flow Calculations for Airfoils and Bodies of Revolution', Grumman Aerodynamics Report 390-71-1, 1971.

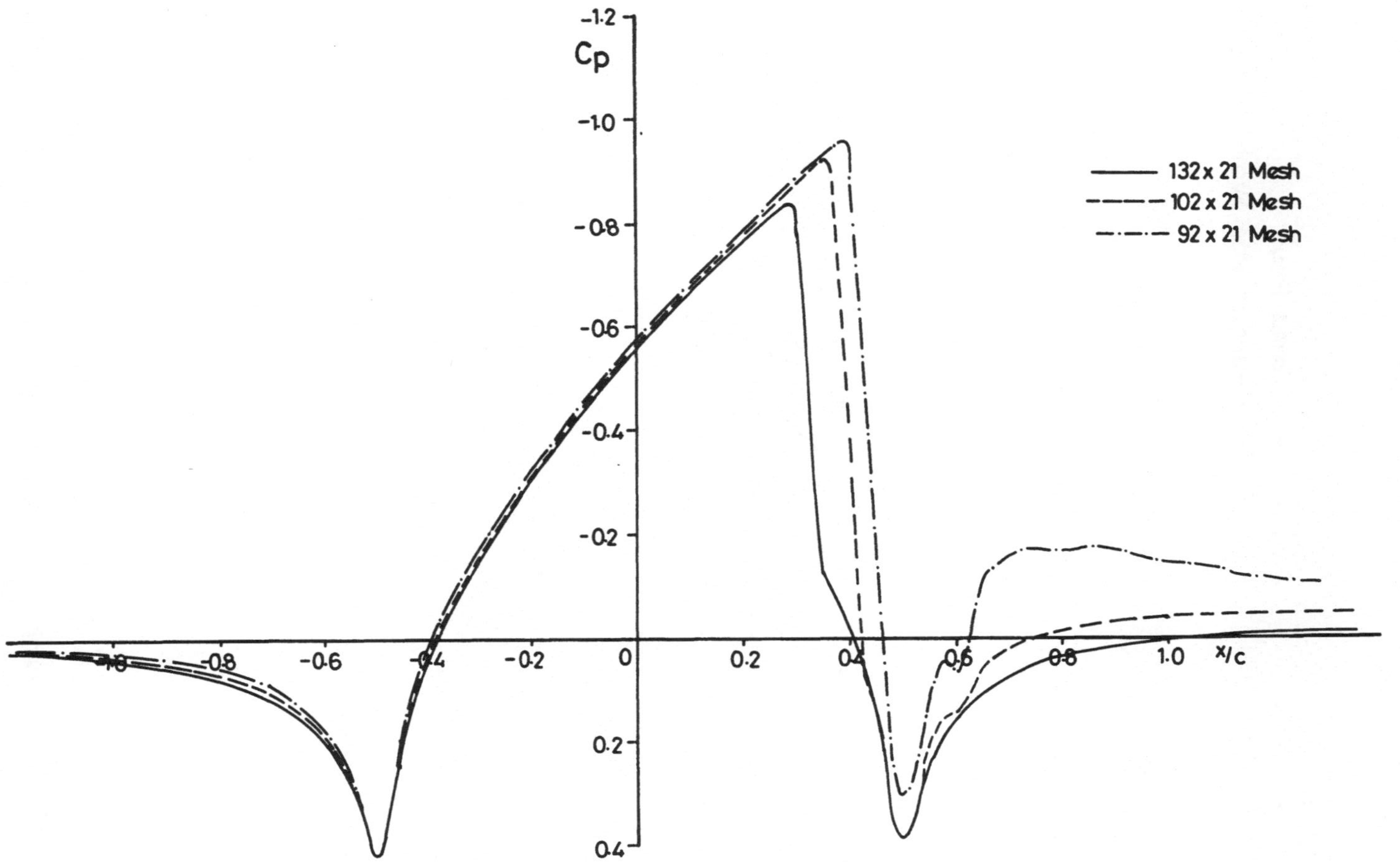

FIG. 1 Theoretical Pressure Distributions along the Channel Lower Wall $M_\infty = 0.85$

105

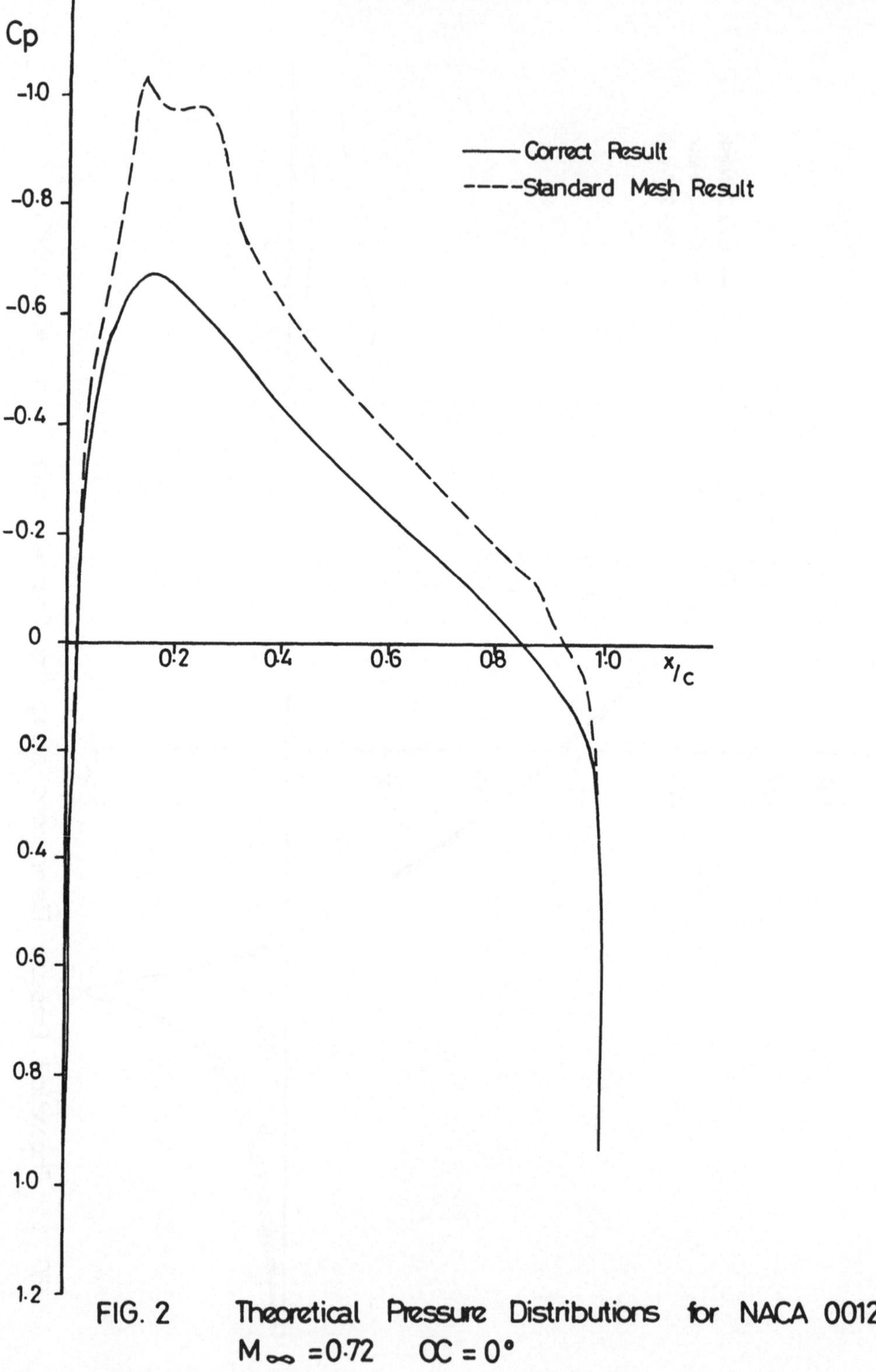

FIG. 2   Theoretical Pressure Distributions for NACA 0012
$M_\infty = 0.72$ $\alpha = 0°$

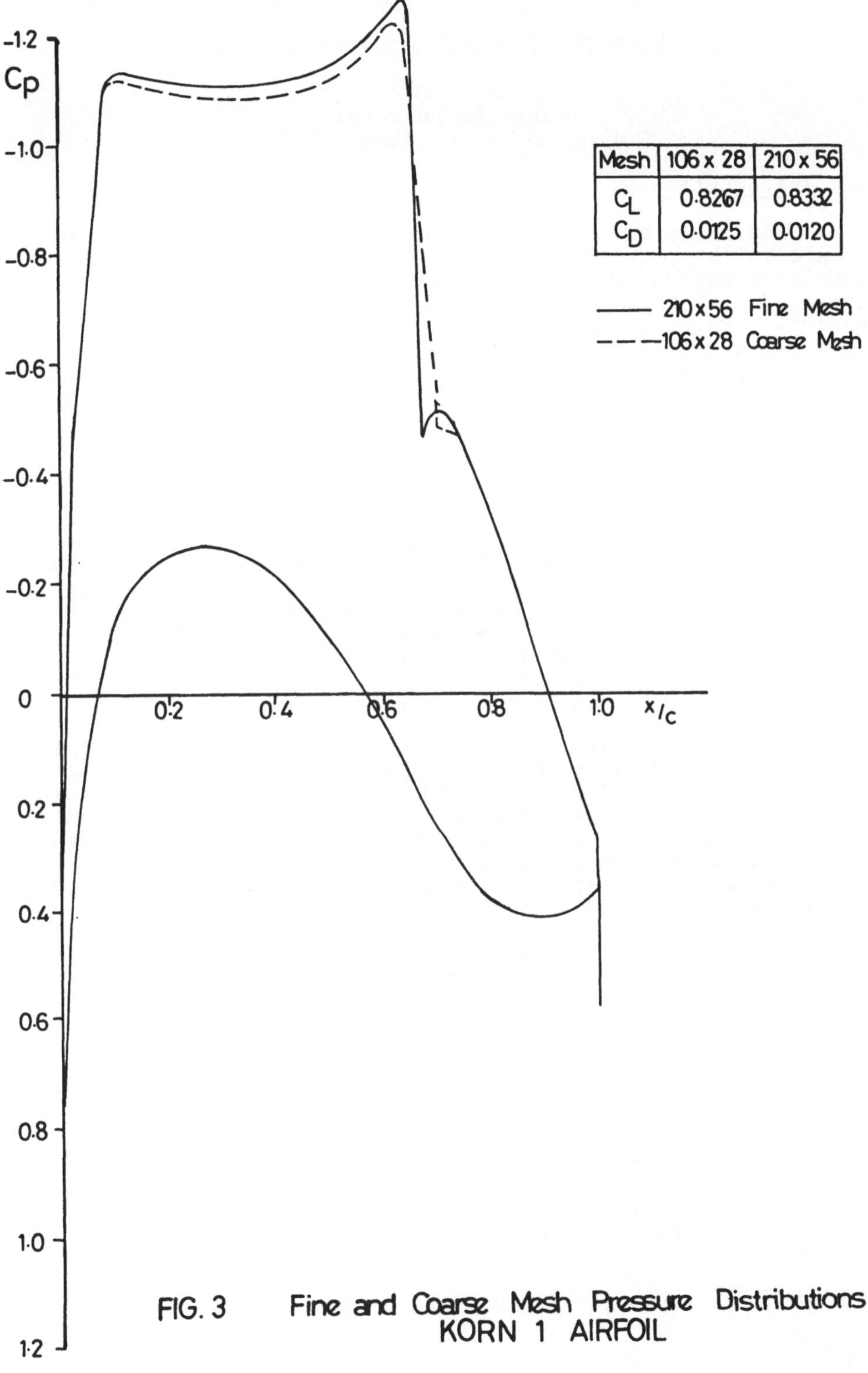

Mesh	106 x 28	210 x 56
C_L	0.8267	0.8332
C_D	0.0125	0.0120

FIG. 3   Fine and Coarse Mesh Pressure Distributions
KORN 1 AIRFOIL

TEST PROBLEMS FOR INVISCID TRANSONIC FLOW

Leland A. Carlson[*]
Texas A&M University
College Station, Texas, 77843

I. Introduction

This paper will briefly discuss some of the results obtained in the process of solving the test problems for the
GAMM Workshop on "Numerical Methods for the Computation of
Inviscid Transonic Flow with Shock Waves" with the TRANDES
program. Briefly, this method,[1-5] utilizes the full, inviscid, perturbation-potential flow equation in a Cartesian grid
system that is stretched to infinity. This equation is represented by a non-conservative system of finite difference
equations that includes at supersonic points a rotated difference scheme and is solved by column relaxation. The solution
usually starts from a zero perturbation potential on a very
coarse grid (typically 13 x 7) followed by several grid halv-
ings until a final solution is obtained on a fine grid (97x49).
Occasionally, for cases having high local Mach numbers, the
solution must be started on the coarse grid (25 x 13). Since
the airfoil does not coincide with the grid points, the surface boundary conditions are represented as two-term Taylor
series about dummy points inside the airfoil. On the outer
boundaries, the exact infinity conditions are used. This
method can, if desired, include the effects of weak viscous
interaction or be used in the design mode. [2-6]

All of the results presented at the workshop and in this
paper were obtained at the rate of 10^4 pts/sec on an Amdahl
470/V6 using a FORTG compiler and single precision arithmetic
(less than 7 significant digits). A typical run took 4-8 minutes, although good engineering results were obtained in one
minute on the medium grid; and convergence was obtained on the
medium grid to at least a maximum cyclic perturbation change
of less than 5E-5. On the fine grid, the $\Delta\phi$max was usually
larger due to significant digit error in the far-field. It
should be noted that the workshop cases were run 1.5-5.0 times
longer than usual due to the availability of computer time
from TAMU, and that the TRANDES program has never been optimized for time.

In order to maintain the goal of common discretization,
the grid stretchings were setup so that the number of airfoil
points, number of wake points, Δx and Δy at the trailing edge,
and the location of the last finite vertical grid column
matched the suggested grids as closely as possible. For cases
involving large supersonic zones, the y-grid was extended so
that the last finite horizontal grid line was subsonic. Otherwise, the rotated difference scheme might have used undefined
values.

---

[*] Professor, Aerospace Engineering Department

Finally, this paper will not attempt to discuss in detail all the workshop problems, which are completely presented in Ref. 6. Instead, it will discuss some of the difficulties and problem areas associated with the application of the Cartesian grid formulation to the test cases in order to aid those individuals considering the usage of a Cartesian type of formulation.

II. Comments on Results

a. NACA 0012

These cases were all computed using 98 points on the airfoil, 24 points in the wake, $\Delta x_{te} = \Delta y_{te} = 0.012$, and $X_{IMAX1} = 7.0$. In the supersonic zones, damping was added via a parameter EPSS, which ranged from 0.4 ($M = 0.8, \alpha = 0°$) to 1.0 ($M_\infty = 95$, $\alpha = 0°$). The subcritical results agreed well with those of Lock[7] for C_p and aerodynamic coefficients.

At present, a fully conservative version of TRANDES is under development, and it is believed it would be of interest to see the difference in results for conservative and non-conservative formulations. Consequently, some preliminary conservative medium grid results are compared to those from TRANDES on Fig. 1. Note the difference in C_p distributions and shock location. These differences should be kept in mind when comparing the present workshop results with those obtained using fully conservative schemes.

For the $M_\infty = 0.95$, $\alpha = 0°$ case two grids were investigated. One was the same as above (98 points on airfoil) and one closely matched the FFA suggested grid (50 pts on airfoil, 36 in wake $\Delta X_{te} = 0.03$, $\Delta Y_{te} = 0.025$, $X_{IMAX1} = 12$, $Y_{JMAX1} = 23.3$) Considerable difficulty was encountered due to the presence of supersonic points in the wake on the furthest horizontal grid line. (Y_{JMAX1}). In general the wake solution converged slowly, was sensitive to the vertical spacing and fineness, and is probably not correct. Conversely, the flow near the airfoil converged rapidly and exhibited no sensitivity to grid structure, except at the trailing edge. There, the coarse grid smoothed the shock across the trailing edge and created the frequently observed ski-ramp shape just upstream of the shock wave. For both grids, the solutions contained a super-sonic-subsonic shock near the trailing edge, a small subsonic pocket at the trailing edge, and a supersonic-supersonic oblique shock off the airfoil. In the wake, a weak normal shock was located 1.3 chord lengths aft of the TE, but its position was probably not converged. This lack, however, did not seem to affect the results over the airfoil.

Finally, the Cartesian formulation places very few points in the leading-trailing edge regions; and, thus, the accuracy of the wave drag, which is computed by integration, is questionable. While several correction methods have been suggested [3,5], none were applied here. Nevertheless, the CD's for the test cases appear to show correct trends. (I.E. For C_L = 0.0 --M_∞ = 0.72, CD = .0004; M_∞ = 0.8, CD = 0.0100; M = 0.85, CD = 0.0381; M_∞ = 0.95, CD = 0.0989)

b. <u>Bump in Channel</u>

This problem was solved by treating the bump as a symmetrical airfoil in a solid wall wind-tunnel. The solution used 41 points on the upper surface, 16 in the wake, 21 pts vertically from the centerline to the wall, ΔX_{te} = 0.0251, ΔY_{te} = 0.03, and ϕ_y = 0 on the channel walls. For the upstream/downstream infinity conditions two approaches were tried. The first used the asymptotic form derived by Murman[8] while the second had ϕ_x = 0 imposed and allowed the solution to float. The results on the bump and between it and the wall were identical. However, some slight differences were observed upstream and downstream, with the floating solution showing less blockage type influence. While the floating approach is easier to implement, it is more prone to significant digit errors due to the magnitude of the floating potential ($O(1)$ instead of $O(0.1)$).

Finally, several other upstream/downstream boundary conditions were imposed, such as ϕ=0; and all yielded essentially the same results. Since the channel case was coded rapidly for the workshop, this lack of sensitivity may be due to either coding errors or a special feature of the test case.

c. <u>RAE 2822</u>

This problem was solved using the same grid as for the NACA 0012 cases. During the solutions, it was discovered that the subcritical results were sensitive to the location of the first and last points on the airfoil, which are normally at 0.01c and 0.99c (X4 = 0.49). The initial results (solid line, Fig. 2) indicated a possible peak in C_{pu} near the leading edge. To resolve this peak, the first points were moved to 0.005c and 0.995c (X4 = 0.495), and slightly different results were obtained as shown on Fig. 2. Interestingly, EPSS had to be increased to 1.0 for this second case due to the appearance of supersonic points on the coarser grinds. Obviously, the Cartesian grid placement sometimes affects the airfoil effective α; and it is probably best to correlate results versus C_L rather than α.

d. <u>CAST 7</u>

Again the basic NACA 0012 grid was used. Here, in both the subcritical and supercritical cases, oscillations were observed in C_{pl} at 0.03c and in C_{pu} at 0.92c. Since the surface slopes were also oscillatory in these areas, it is believed this behavior was due to the sparsity of the given coordinates and the use of spline fits to determine the computational ordinates and slopes. Also, in both cases, the aft C_{p_l} bucket turned around at the TE and approached stagnation. This behavior is reasonable due to the large TE angle (12.5°). In actuality, viscous effects would mitigate this trend.

e. <u>KORN 1</u>

The basic NACA 0012 grid was also used for this case. For the design point, the resultant C_p distribution was close to the theoretical hodograph values and was almost shockless. Again some sensitivity to the grid size and X4 value was observed, as can be seen in Fig. 3. Also, a double precision calculation (16 digits) was made and essentially yielded the same results as with single precision (7 digits). However, for the 16 digit case the $\Delta\phi$max on the fine grid was steadily decreasing instead of oscillatory and smaller (6E-5 vs. 2E-4).

III. Conclusions

Except for the NACA 0012, $M_\infty = 0.95$ case, all test problems were solved straight-forwardly and appeared to be converged or close to convergence. The only difficulty was some sensitivity to grid placement, which is typical of the Cartesian formulation. Again, note that all results were obtained in single precision (less than 7 significant digits) on an Amdahl 470/V6.

IV. Acknowledgments

This work was primarily supported by NASA Grant NSG 1174 and partially supported by the Texas Engineering Experiment Station. The author expresses his appreciation to Robert Simmons, TAMU Aerospace Engineering Department, for his assistance in the preparation of the figures.

V. References

1. Carlson, L.A., "Transonic Airfoil Analysis and Design Using Cartesian Coordinates," J. of Arcft, Vol. 13, May 1976 pp. 349-356.
2. Carlson, L.A., "Transonic Airfoil Flowfield Analysis Using Cartesian Coordinates," NASA CR-2577, Aug. 1975.
3. Carlson, L.A., "TRANDES: A Fortran Program for Transonic Airfoil Analysis and Design," NASA CR-2821, June 1977.
4. Carlson, L.A., "Inverse Transonic Airfoil Design Including Viscous Interaction," NASA CP-2001, Nov. 1976, pp 1387-1395.
5. Carlson, L.A. and Rocholl, B.M., "Application of Direct Inverse Techniques to Airfoil Analysis and Design," NASA CP-2045, Pt. 1, pp 55-72.
6. Carlson, L.A., "Test Problems for Inviscid Transonic Flow," Texas Engr. Expt. Station Rept. TAMRF-3224-7904, 1979.
7. Lock, R.C., "Test Cases for Numerical Methods in Two Dimensional Transonic Flow," AGARD Rept R-575-70, 1970.
8. Murman, E.M., "Computation of Wall Effects on Ventilated Transonic Wind Tunnels," AIAA Paper No. 72-1007, Sept. 1972.

<u>Summary of Test Problem Results</u>

Problem	Airfoil	M_∞	α	CL	CD	CMC4	EPSS	Comments
A-I (i)	NACA 0012	0.72	0.0°	0.0	0.0004	0.0	----	
(ii)	"	0.63	2.0°	0.339	-0.0002	-0.002	----	
II(i)	"	0.80	0.0°	0.0	0.0100	0.0	0.4	
(ii)	"	0.85	0.0°	0.0	0.0381	0.0	0.6	
(iii)	"	0.95	0.0°	0.0	0.0958	0.0	1.0	
(iv)	"	0.80	1.25°	0.321	0.0199	-0.035	0.6	
(v)	"	0.85	1.00°	0.283	0.0444	-0.075	0.8	
B	Bump	0.85	0.0°	0.0	0.0240	0.0	0.6	Murman Asymptotic
	"	"	"	0.0	0.0236	0.0	0.4	Floater
C-I	RAE 2822	0.676	1.0°	0.551	0.0047	-0.106	0.4	X4 = 0.49
	"	"	"	0.574	-0.0043	-0.109	1.0	X4 = 0.495
II	"	0.75	3.0°	1.154	0.0583	-0.210	0.8	
D-I	CAST 7	0.70	-1.0°	0.436	0.0065	-0.138	----	
II	"	0.76	0.5°	0.898	0.0259	-0.210	0.8	
G-I	KORN 1	0.75	0.115°	0.622	0.0012	-0.160	0.4	X4 = .4925, Single Precision
	"	"	"	0.609	0.0074	-0.148	0.4	X4 = .49, FFA Grid
	"	"	"	0.606	0.0058	-0.147	0.4	X4 = .49, St'd Grid
	"	"	"	0.624	0.0012	-0.150	0.4	X4 = .4925, Double Precision
G-II	"	"	1.0°	0.833	0.0181	-0.171	0.6	

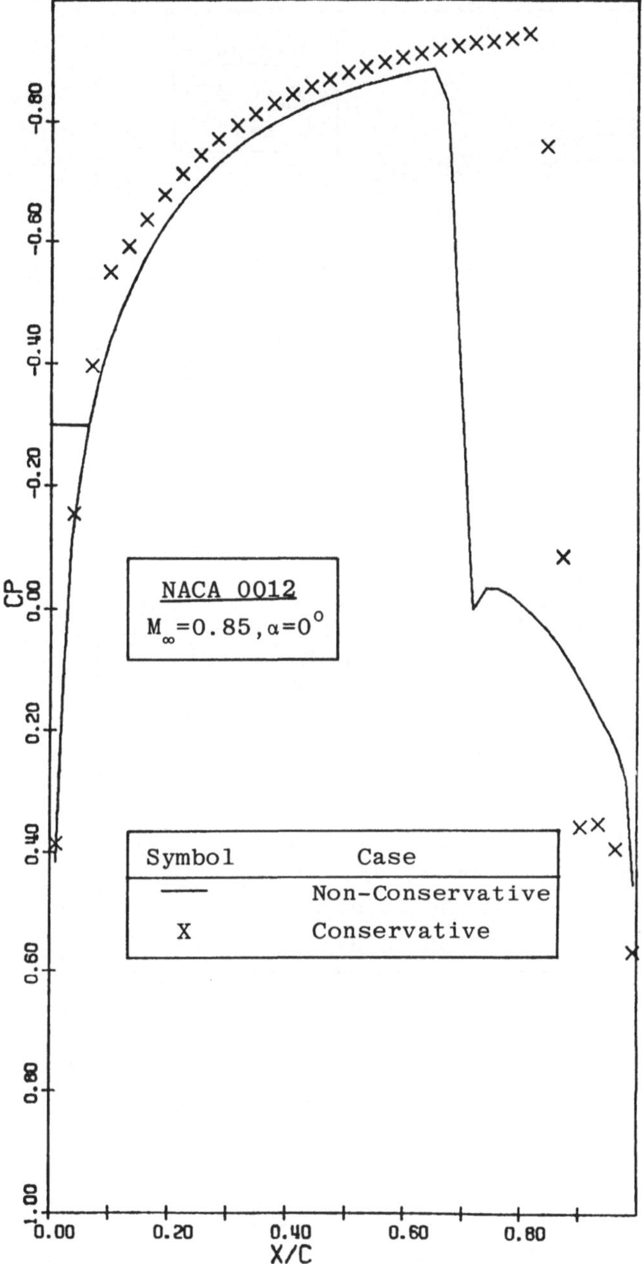

Figure 1: Non-Conservative versus Conservative Results

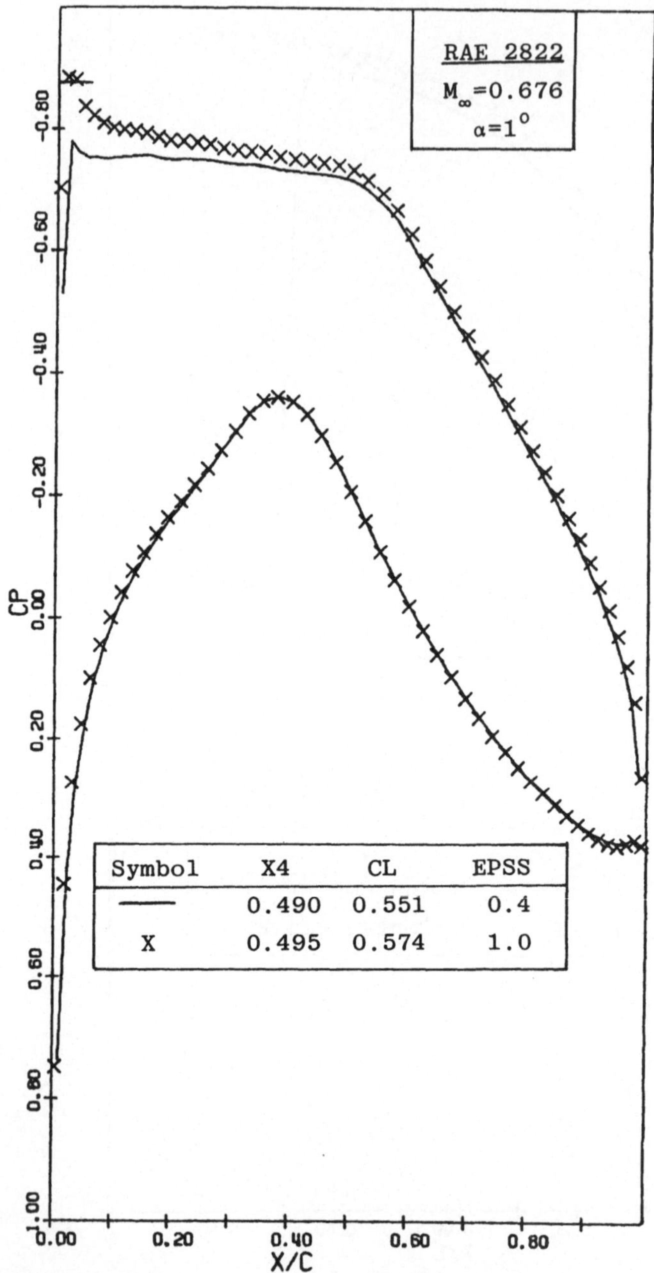

Figure 2: The Effect of Grid Placement on C_p

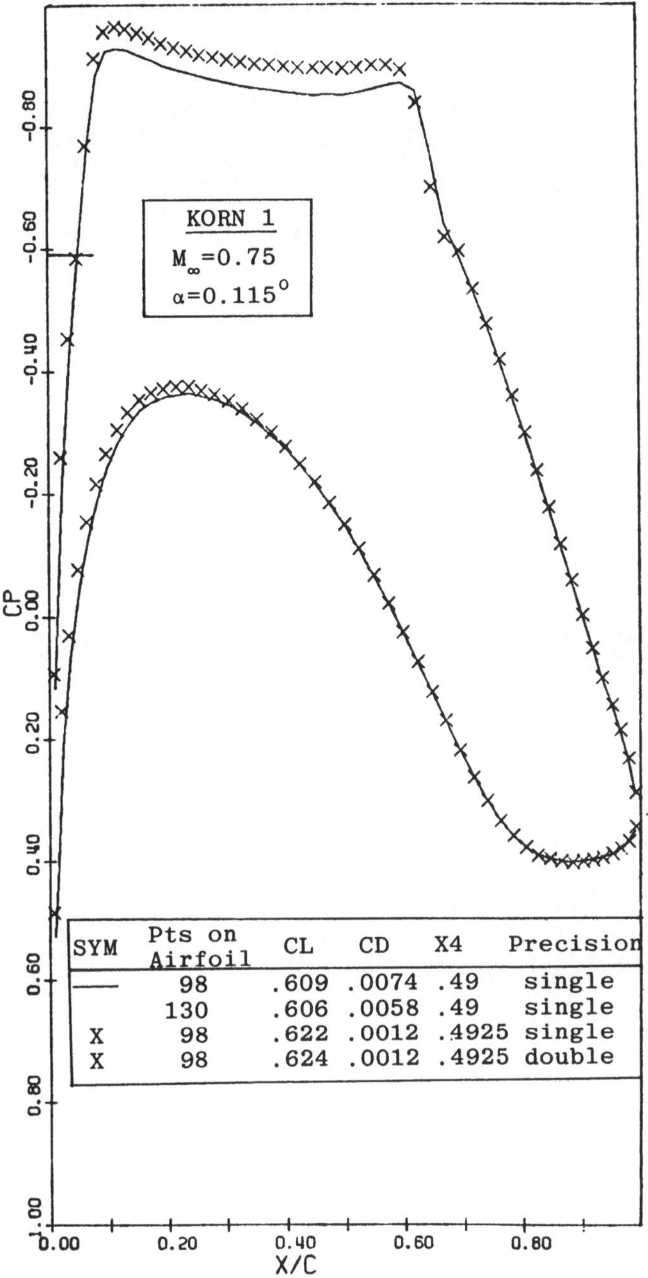

SYM	Pts on Airfoil	CL	CD	X4	Precision
──	98	.609	.0074	.49	single
	130	.606	.0058	.49	single
X	98	.622	.0012	.4925	single
X	98	.624	.0012	.4925	double

Figure 3: Results for the KORN 1 Airfoil

A MODIFICATION TO THE METHOD OF GARABEDIAN AND KORN*

by

R. C. Lock

Aerodynamics Dept, Royal Aircraft Establishment,
Farnborough, Hampshire, England

SUMMARY

A modification is described to the method of Garabedian
and Korn for calculating the potential flow round aerofoils.
When the flow is supercritical and shock waves are present, the
modification enables a solution to be obtained intermediate
between the two extremes - non-conservative (N-C) and conservative (C) - of the existing method. A parameter λ (between
0 and 1) is introduced which leads to a N-C solution of $\lambda = 0$
and to a C solution if $\lambda = 1$, while taking an intermediate
value allows the solution to be adjusted so that the pressure
jump across the shock wave is a reasonable approximation to the
true physical (Rankine-Hugoniot) jump. In this way it is hoped
to achieve an overall solution which is closer to a solution of
the full Euler equations while retaining the computational
speed of the Garabedian and Korn method.

* *This paper was written for the GAMM Workshop on Numerical methods for the
computation of inviscid transonic flow with shock waves, held at FFA,
Stockholm, September 1979*

1   INTRODUCTION

        Ever since its publication in 1971, the method of
Garabedian and Korn (1,2) for calculating the compressible
inviscid flow round aerofoils has been widely used both because
of its accuracy and computational speed. There are no doubts
on either of these points for a subcritical flow; but when the
flow is supercritical and shock waves of appreciable strength
are present, the problem arises that a method of this type, in
which the flow is assumed to be isentropic and irrotational,
cannot give a correct solution to the full (Euler) equations of
motion. In particular, the increase in entropy across the
shock wave is not included, and the rotational flow to which
this gives rise cannot be allowed for. In the past, however,
it was impossible to determine the size of the errors occurring
in potential flow solutions, because of the dearth of accurate
solutions to the Euler equations for examples of practical
aerofoils; and one of the main aims of the present Workshop is
to provide such solutions against which the accuracy of potential flow, and other relatively simple methods, can be
assessed.

        In the case of the potential flow methods, the problem is
complicated by the lack of uniqueness of solutions (when shock
waves are present). Published versions of the Garabedian and
Korn method (2,3) employ finite-difference schemes which yield
results at two extremes; the original version (2) used a
'non-conservative' (N-C) scheme which produces shock waves such
that the downstream pressure is only just subsonic; while the
latest version (3) uses Jameson's 'quasi-conservative' (Q-C)
scheme for which the shock pressure jump is a good approximation to the 'isentropic shock' relation - and thus appreciably
greater than the true physical (Rankine-Hugoniot) jump when the
shock strength is large. In going from one extreme to the
other the shock thus becomes much stronger, and it also moves
back along the chord, to an extent which may be quite large.

        For these reasons, it is to be expected that neither
extreme is likely to represent a good approximation to the true
solution, either as regards shock strength or position. So it
seems desirable to devise a numerical scheme which will yield
intermediate solutions in a controlled manner. Such a scheme
is described below; it is based on a simple empirical modification to the Murman-Jameson 'shock-point' operator involving an
arbitrary parameter λ (between 0 and 1), such that $\lambda = 0$
yields a N-C solution, $\lambda = 1$ a Q-C solution. Then λ can be
adjusted until the shock jump takes the required value; this
naturally produces an intermediate shock position, and it is
hoped that this will be nearer the true position than with
either of the two extremes ($\lambda = 0$ and $\lambda = 1$).

2   MODIFICATION TO THE NUMERICAL METHOD

2.1   The modified shock point operator

        In the computing plane, inside the unit circle to which
the region exterior to the aerofoil is transformed, coordinates

(θ, r) and corresponding velocity components (u,v) are used. Then the equation for the velocity potential Φ has second order terms

$$(a^2 - u^2)\Phi_{\theta\theta} - 2uvr\Phi_{r\theta} + (a^2 - v^2)r^2\Phi_{rr} \quad , \qquad (1)$$

where a is the local velocity of sound.

Using suffices i,j to denote values at $\theta = i\Delta\theta$, $r = 1 - j\Delta r$, the central difference expression to the first term in (1) is

$$A_{ij}(\Phi_{\theta\theta})_{ij} \quad ,$$

where $A_{ij} = (a^2 - u^2)_{ij}$

and

$$\left(\Phi_{\theta\theta}\right)_{ij} = \frac{\Phi_{i-1,j} - 2\Phi_{ij} + \Phi_{i+1,j}}{(\Delta\theta)^2} \quad . \qquad (2)$$

In the original (N-C) formulation by Garabedian and Korn (1,2), this central difference expression is used whenever the point (i,j) is subsonic, and replaced by the 'upwind' expression $A_{ij}\left(\Phi_{\theta\theta}\right)_{i-1,j}$ when (i,j) is supersonic. A version of the 'quasi-conservative' method of Jameson (3,4) may be obtained in the following way:

An artificial viscosity term is added to (1), namely

$$- P_{ij} + P_{i-1,j} \qquad (3)$$

where $P_{ij} = \mu_{ij}\left(\Phi_{\theta\theta}\right)_{ij}$

and

$$\mu_{ij} = \begin{cases} A_{ij} & (A_{ij} < 0) \\ 0 & (A_{ij} > 0) \end{cases} .$$

[Note that the switch in μ_{ij} does not occur at precisely the sonic points, which are given by $u^2 + v^2 = a^2$; but since normally $v \ll u$ the difference should be insignificant.]

The effect of this term is as follows: If both (i,j) and
$(i - 1,j)$ are subsonic points $(A > 0)$, then the central difference expression $A_{ij}(\Phi_{\theta\theta})_{ij}$ is retained. If both points are
supersonic $(A < 0)$, then $A_{ij}(\Phi_{\theta\theta})_{ij}$ is replaced by the upwind
expression*

$$A_{i-1,j}\left(\Phi_{\theta\theta}\right)_{i-1,j} \quad .$$

Now at the crucial 'shock point', the first subsonic point
downstream of a shock wave where $A_{ij} > 0$ but $A_{i-1,j} < 0$, an
expression is produced which is the sum of central and upwind
expressions, namely

$$A_{ij}\left(\Phi_{\theta\theta}\right)_{ij} + A_{i-1,j}\left(\Phi_{\theta\theta}\right)_{i-1,j} \quad . \tag{4}$$

This is Murman's shock point operator (5). Now the corresponding expression at this point in the original (N-C) method would
be $A_{ij}\left(\Phi_{\theta\theta}\right)_{ij}$. It is therefore natural to control the degree
to which mass is conserved across the shock by multiplying the
second term in (4) by an arbitrary parameter, λ , leading to

$$A_{ij}\left(\Phi_{\theta\theta}\right)_{ij} + \lambda A_{i-1,j}\left(\Phi_{\theta\theta}\right)_{i-1,j} \tag{5}$$

where $0 \leqslant \lambda \leqslant 1$.

    Clearly $\lambda = 0$ gives a N-C scheme**, $\lambda = 1$ a Q-C
scheme; while intermediate values of λ naturally give intermediate values for both the position of and the pressure jump
across the shock (see section 3 for examples). This scheme may
be called 'partially conservative' (P-C).

2.2   Second order terms

    The finite-difference scheme described above is only of
first order accuracy in the supersonic region, and our experience suggests that the truncation errors will usually be
larger with the present method than was the case with the

---

* Note that this differs from the corresponding term in
   Garabedian and Korn's original method, which is $A_{ij}\left(\Phi_{\theta\theta}\right)_{i-1,j}$.

** Not exactly equivalent to the original Garabedian and Korn
   method - see previous footnote.

original method of Garabedian and Korn, because both factors
in the leading term of (1) are now taken at an upwind point,
as opposed to only the second. It is therefore important to
add suitable second-order terms in the supersonic region. This
may be done by replacing P_{ij} by

$$P_{ij} = \mu_{ij}\left\{\left(\Phi_{\theta\theta}\right)_{ij} - \varepsilon\left(\Phi_{\theta\theta}\right)_{i-1,j}\right\} \tag{6}$$

where ε is an 'artifical viscosity' parameter.

In the supersonic region, this leads to the following finite-
difference expression for the first term of (1)

$$A_{i-1,j}\left(\Phi_{\theta\theta}\right)_{i-1,j} + \varepsilon\left[A_{ij}\left(\Phi_{\theta\theta}\right)_{i-1,j} - A_{i-1,j}\left(\Phi_{\theta\theta}\right)_{i-2,j}\right] . \tag{7}$$

This expression (7) can be considered as an approximation to

$$A_{i-1,j}\left(\Phi_{\theta\theta}\right)_{i-1,j} + \varepsilon\Delta\theta\left\{\frac{\partial}{\partial\theta}\left(A\Phi_{\theta\theta}\right)\right\}_{i-1,j} \quad ,$$

and hence to $\left(A\Phi_{\theta\theta}\right)_{i-1+\varepsilon,j}$. Thus as $\varepsilon \to 1$ the expression
for $A\Phi_{\theta\theta}$ becomes more and more nearly second order accurate.
In practice we have found that the value $\varepsilon = 0.8$ can normally be used without causing convergence problems.

Computing times with the present method are essentially
the same as with the standard G and K method. Using the usual
160 × 30 'fine' grid, 300 iterations are normally sufficient
to bring the residuals down to a satisfactorily low level
($<10^{-4}$) , and on a DEC KL10 computer this takes about 8 minutes (equivalent to about 4 minutes on a CDC 6600).

3   SOME APPLICATIONS OF THE METHOD

Although the results of calculations by this method for
the GAMM Workshop examples are being presented separately, it
seems desirable at this stage to show a few typical examples,
and in particular to discuss the choice of the parameter λ .
From previous experience (6) it has been suggested that 'λ
should be about 0.5 or slightly less', but the evidence on
which this was based was slight and the present GAMM exercise
will provide much more information on this point.

As a first example we show in Fig 1 the effect of vary-
ing both ε and λ for the NACA 0012 aerofoil at $M_\infty = 0.75$,
$\alpha = 2^o$. Note the sensitivity of the shock position to the
value of the artificial viscosity parameter ε ; increasing ε
from 0 to 0.8 can move the shock downstream by up to 5% chord.

120

But the effect of varying λ is much greater; as λ increases from 0 to 1 the shock pressure jump increases markedly and the shock moves downstream about 20% chord. The shock pressure jump with $\lambda = 1$ is in good agreement with the theoretical value for an 'isentropic shock' (7); and that with $\lambda = 0.5$ is slightly greater than the value for a real (Rankine-Hugoniot) shock. Bearing in mind the existence of the Oswatitsch/Zierep (8) singularity at the foot of a normal shock-wave, it is desirable that the pressure jump across the 'shock' observed in the numerical solution should be somewhat less than the R-H value. So from this and other evidence a value 0.3 has been chosen for λ in the examples presented at this Workshop.

One example where this leads to very good agreement with a solution of the Euler equations by Sells (9) is shown in Fig 2, referring to the NACA 0012 aerofoil at $M_\infty = 0.85$, $\alpha = 0$. The potential flow solution with $\lambda = 0.3$ agrees almost perfectly with Sells' results, both as regards the position of the shock wave (about $x/c = 0.73$) and the pressure rise through it, which in both cases is slightly less than the R-H value - a plausible discrepancy for the reason given above. The drag coefficients (0.0400 by Sells' method and 0.0404 by the modified G and K method) are also in excellent agreement. Note that increasing λ from 0 through 0.3 to 0.7* causes the shock position to move aft from $x/c = 0.67$ through 0.73 to 0.80, with a corresponding increase in pressure rise from a value which is much too low to one which is much too high.

This excellent agreement with a solution of the Euler equations is unfortunately not reproduced in a lifting case. Fig 3 shows solutions for the NACA 0012 aerofoil at $M_\infty = 0.8$, $\alpha = 1.25^\circ$. It appears that, compared with the solution by Sells' method, the partially conservative (P-C) solution with $\lambda = 0.3$ predicts both the position and pressure jump for the shock wave on the upper surface quite well - the shock positions are $x/c = 0.62$ and 0.66 respectively. But the suction level ahead of the shock is higher in the P-C solution, while the peak suction on the lower surface is appreciably less than in Sells' solution. These factors, combined with a greater negative loading in Sells' solution near the trailing edge, combine to produce a relatively large discrepancy in lift coefficient - 0.31 by Sells' method and 0.40 by the G and K method. In fact the solution with $\lambda = 0$ gives $C_L = 0.31$, and the lower surface pressure distribution agrees well with Sells' prediction; but on the upper surface the shock is too far forward and the pressure jump too low.

That the discrepancy is almost entirely due to the difference in lift coefficient can be seen from Fig 4, where Sells' solution to the Euler equations for $M_\infty = 0.8$, $\alpha = 1.25^\circ$ is compared with a P-C solution (with $\lambda = 0.3$) at (roughly) the same lift coefficient, which is obtained at $\alpha = 1.0^\circ$. Here we see reasonable agreement in nearly all

---

* It was not possible to obtain a converged solution for $\lambda > 0.7$.

respects, and in particular the predicted strengths and positions of the shock waves on both surfaces are almost the same for both methods.

The reason for this apparent overestimation of lift by the modified G and K method, when λ is adjusted to give the correct shock pressure jump, is not yet clear. It may perhaps be associated with the vorticity which is formed behind the shock in the real flow, which is not of course predicted by a potential flow method.

REFERENCES

1   P.R. Garabedian   Analysis of transonic aerofoils.
    D.G. Korn   Comm. Pure App. Math., Vol 24, pp 841-851
    (1971)

2   F. Bauer   Supercritical wing sections.
    P.R. Garabedian   (Lecture notes in economics and mathematical systems, No 66)
    D.G. Korn   Springer-Verlag (1972)

3   F. Bauer   Supercritical wing sections II.
    P.R. Garabedian   (Lecture notes in economics and mathematical systems, No 108)
    D.G. Korn
    A. Jameson   Springer-Verlag (1975)

4   A. Jameson   Numerical computation of transonic flows with shock waves.
    Symposium Transsonicum II, p 384, Springer-Verlag (1975)

5   E.M. Murman   Analysis of embedded shock waves calculated by relaxation methods.
    Proc. AIAA Computational Fluid Dynamics Conference, Palm Springs, July 1973

6   M.R. Collyer   Improvements to the viscous Garabedian
    R.C. Lock   and Korn method (VGK) for calculating transonic flow past an aerofoil.
    RAE Technical Report 78039 (1978)

7   J.L. Steger   Shock waves and drag in the numerical
    B.S. Baldwin   calculation of isentropic transonic flow.
    NASA TND-6997 (1972)

8   K. Oswatitsch   Das problem des senkrechten stosses an
    J. Zierep   einen gekrümmten wand.
    ZAMM 40, p 143 (1960)

9   C.C.L. Sells   Numerical solutions of the Euler equations for transonic flow past a lifting aerofoil.
    Paper to be presented at GAMM Workshop, Stockholm, September 1979

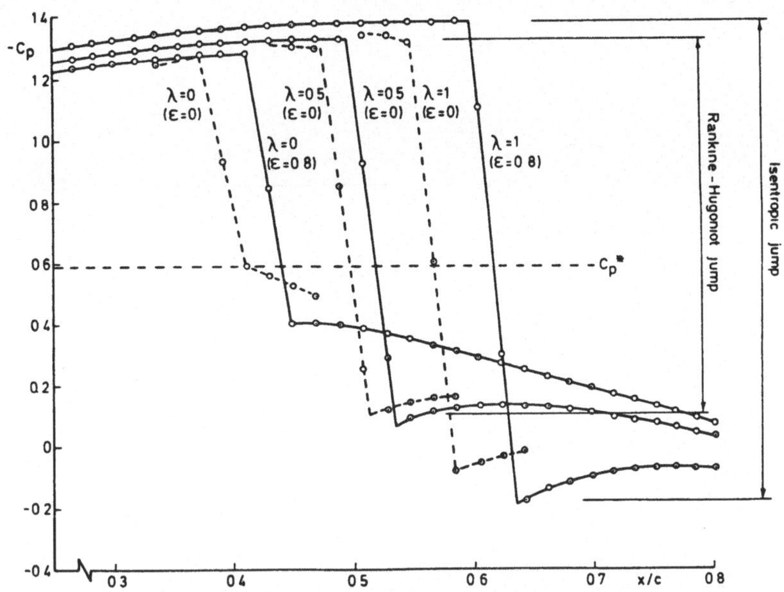

Fig 1   NACA 0012. A comparison of inviscid solutions at
$M_\infty = 0.75$, $\alpha = 2^0$

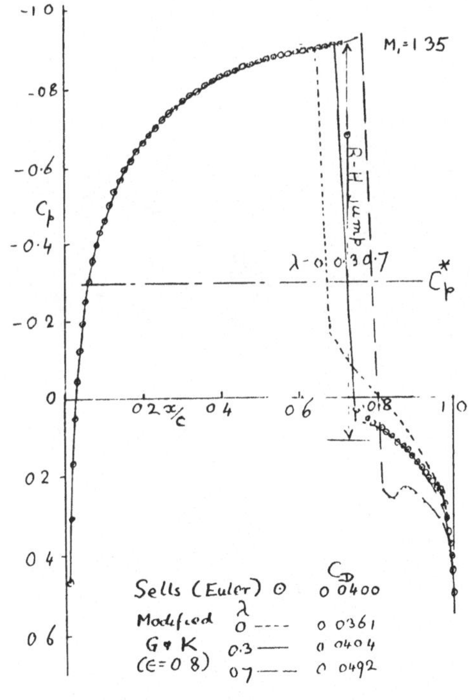

NACA 0012 $M_\infty = 0.85$ $\alpha = 0$

Fig 2   GAMM Ex AII(ii)

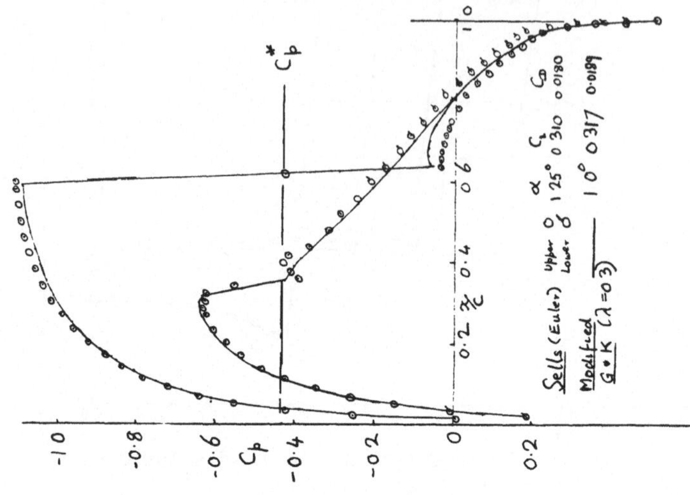
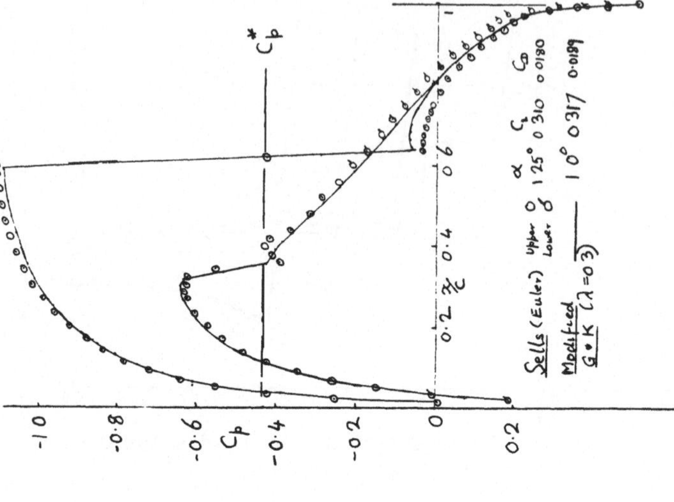

NACA 0012 M_∞ = 0.8 $C_L \simeq$ 0.31

Fig 4

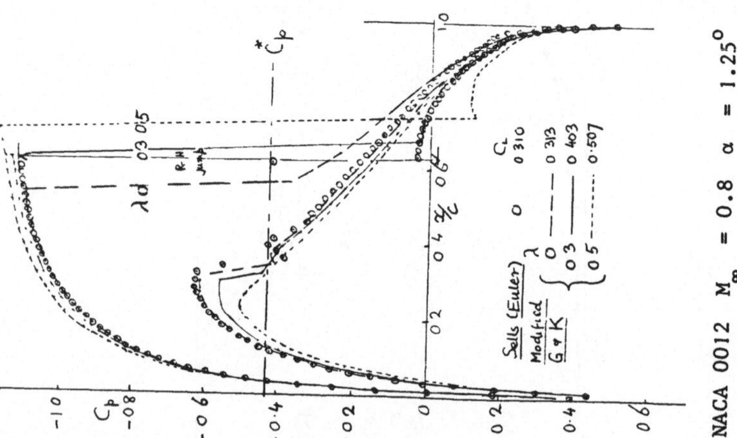

NACA 0012 M_∞ = 0.8 α = 1.25°

Fig 3   GAMM Ex AII(v)

A PHYSICALLY CONSISTENT TIME-DEPENDENT METHOD FOR THE SOLUTION OF THE EULER
EQUATIONS IN TRANSONIC FLOW (°)

Luca Zannetti (*), Guido Colasurdo (*), Luciano Fornasier (**),
Maurizio Pandolfi (*)

PRELIMINARY

The central theme of this GAMM Workshop is the comparison of the respective
performance of various computing procedures in current use today for the numerical solution of inviscid steady transonic flow.
The present paper will be then located partially out of this theme because:

- The computing procedure here presented is fairly new and many aspects
  still need deeper investigations, to understand several peculiarities.
  Therefore, we may say that the method is far to be considered as "in
  current use today".

- The performance, as regards the computing time, is not exciting. In fact
  the approach we propose is based on a time-dependent technique and no
  accelerating procedures are used, so the computing time is strictly related to the corresponding physical time needed in the physical transient
  to achieve a steady flow configuration.

- The codes we are using at the moment are quite "rough". They have been
  written in order to investigate a new methodology and to work out numerical experiments with several options as regards the computational procedure. No attempts have been done so far to achieve any kind of optimization of the codes. So, even because of this fact, the computing times,
  we get presently, should be not considered in giving a figure of merit
  to the proposed method.

- In the light of the previous two points, we have been not able to perform
  computations with the standard mesh sizes, which have been suggested for
  the comparison of the different results presented at the Workshop. In
  fact the proposed higher number of computational points would have brought
  the computing times to a too much high level for our "capabilities",
  whereas the results obtained with the rough meshes we have used, seem to
  be already satisfactory.

However there are other peculiarities of the proposed method which appear
interesting and it seems reasonable to present them at the Workshop, in a
direct comparison with other well tested and reliable computational procedures:

- The mesh size we are using is very rough. If the comparison of our results
  matches fairly well those got by other well experimented and optimized procedures, we may hope to be optimistic for the future of this method where
  the formulation of the flow equations and the numerical algorithm allow
  for few computational points for describing correctly complicated flows.

(°) The present research has been supported by the "Consiglio Nazionale
     delle Ricerche" (Contract N. 115.6799 78 02427.07).

(*) Istituto di Macchine e Motori per Aeromobili, Politecnico di Torino,
     Torino, Italy.

(**) Combat Aircraft Group, Aeritalia, Società Aerospaziale Italiana p.A.,
     Torino, Italy.

- The present approach based on the "true" time dependent technique enables
  us to predict unsteady flows (for example flutter problems), withouth any
  limitation with respect to those methods which are confined to the particular class of the steady flows.

- The limitation of some methods to isentropic flow restricts the use of
  the relative codes to low supersonic Mach numbers. Even if we present
  here only one case where the gradients of entropy due to the shock are
  taken into account, we may reasonably hope that the rotational flow investigation can be generalized easily to all the problems where entropy variations are not negligible.

After these preliminary remarks, we can now proceed to a short review of the
method proposed. It is based on the time dependent approach and on the integration in time, through finite differences, of a particular formulation of
the hyperbolic partial differential equations which describe the compressible inviscid unsteady flow. The reader may refer to $/\overline{\ 1,2,3\ }/$ to get infor-
mations on the steps done in the past according to this line and to $/\overline{\ 4\ }/$
where the basic ideas have been discussed and several examples worked out
and reported. A more detailed description of the particular methodology,
here briefly presented, may be found in $/\overline{\ 5\ }/$.

FEW WORDS ON THE METHOD

The method we use wants to be consistent with the physical nature of the
problem we are dealing with: the description of a wave propagation phenomenon whose mathematical model is given by a hyperbolic set of equations.
From a general point of view the solution of these equations through finite
differences is achieved by integrating in time, at a given point P, the local time-derivatives of the flow properties. These are provided by the governing equations, once the space derivatives are approximated through finite differences.
From a physical point of view these time-derivatives are the result of the
interaction of the signals carried on the bicharacteristics which converge
on the given point P, at a given time. It is interesting to note that such
"signal" are represented by particular combinations of the primitive dependent variables (velocity components and pressure) which in the simple case
of the 1D flow are well known as the "Riemann variables". The variables defined by these combinations appear to be the "most significant properties"
of the unsteady flows we are dealing with.
We want our method to be consistent, in a numerical procedure, with such a
feature of the physical phenomenon, as much as possible. Therefore the governing equations are recast in order to put in evidence these new variables
and to retain only their derivatives along bicharacteristics. The space components of such derivatives are then evaluated by one-side differences, according to the direction of the propagation of the "signals" along the bicharacteristics. A further combination of their time components defines the
local time derivatives of the primitive variables (compnents of the velocity and pressure).

THE EQUATIONS

The full compressible inviscid unsteady flow equations are written in the
particular, but notrestrictive, assumption of isentropic flow:

$$P_t + \overline{q} \cdot \nabla P + \gamma \nabla \cdot \overline{q} = 0 \qquad \text{(continuity)}$$

$$\overline{q}_t + (\overline{q} \cdot \nabla)\overline{q} + T \nabla P = 0 \qquad \text{(momentum)} \tag{1}$$

Let us consider in the (x,y,t) frame the characteristic conoid that leaves
the point P (Fig. 1). The line "m" is the intersection of the characteristic
surface touching the cone on the bicharacteristic "1" with the t = const.
space like surface.
Compatibility equations on characteristic surfaces can be obtained by combining Eqs. 1:

$$[\vec{q}_t + (\vec{q}\cdot\nabla)\vec{q} + T\nabla P]\cdot\vec{\xi} + \frac{a}{\gamma}[P_t + \vec{q}\,\nabla P + \gamma\,\nabla\cdot\vec{q}] = 0 \qquad (2)$$

where $\vec{\xi}$ is the unit vector normal to $\bar{m}$ and running on the constant time
plane.
Eq. 2 may be now written as:

$$\left[\frac{d}{dt}(\vec{q}\cdot\vec{\xi})\right]_\ell + \frac{a}{\gamma}\left[\frac{d}{dt}P\right]_\ell + a\frac{\partial}{\partial m}(\vec{q}\cdot\bar{m}) = 0 \qquad (3)$$

where:

$$\left[\frac{d}{dt}(\ )\right]_\ell = \frac{\partial}{\partial t}(\ ) + (u + a\cos\delta)\frac{\partial}{\partial x}(\ ) + (v + a\sin\delta)\frac{\partial}{\partial y}(\ )$$

$$\frac{\partial}{\partial m} = -\sin\delta\,\frac{\partial}{\partial x}(\ ) + \cos\delta\,\frac{\partial}{\partial y}(\ ) \qquad (4)$$

By putting:

$$Q = \vec{q}\cdot\vec{\xi} + \frac{2}{\gamma-1}\,a$$

Eq. 3 becomes:

$$\left[\frac{d}{dt}Q\right]_\ell = -a\frac{\partial}{\partial m}(\vec{q}\cdot\bar{m}) \qquad (5)$$

Eq. 5 represents the compatibility equation over a general characteristic
surface which is defined by the bicharacteristic "1" and by the line "m".
Following the suggestion given in $\underline{/\,6\,\underline{/}}$, it is convenient to select the four
bicharacteristics (1_1, 1_2, 1_3, 1_4) defined by $\delta_1 = 0$, $\delta_2 = \pi$, $\delta_3 = \pi/2$ and $\delta_4 = \frac{3}{2}\pi$.
The corresponding compatibility equations are:

$$\left[\frac{d}{dt}C\right]_{\ell_1} = -a\,v_y \qquad\qquad (\delta_1 = 0)$$

$$\left[\frac{d}{dt}D\right]_{\ell_2} = -a\,v_y \qquad\qquad (\delta_2 = \pi)$$

$$\left[\frac{d}{dt}E\right]_{\ell_3} = -a\,u_x \qquad\qquad (\delta_3 = \pi/2) \qquad (6)$$

$$\left[\frac{d}{dt}F\right]_{\ell_4} = -a\,u_x \qquad\qquad (\delta_4 = \frac{3}{2}\pi)$$

where the particular values of Q over these four bicharacteristics are:

$$C = \frac{2}{\gamma-1}\, a + u \qquad ; \qquad D = \frac{2}{\gamma-1}\, a - u$$

$$E = \frac{2}{\gamma-1}\, a + v \qquad ; \qquad F = \frac{2}{\gamma-1}\, a - v \tag{7}$$

Three linear combinations of Eqs. 6 and the continuity equation , provide the time derivatives, at the point P, of the primitive variables:

$$u_t = -\frac{1}{2}\left[(u+a)C_x + v C_y - (u-a)D_x - v D_y\right]$$

$$v_t = -\frac{1}{2}\left[u E_x + (v+a)E_y - u F_x - (v-a)F_y\right] \tag{8}$$

$$P_t = -\frac{\gamma}{2a}\left[(u+a)C_x + (u-a)D_x + (v+a)E_y + (v-a)F_y\right]$$

It may be noticed that all the space derivatives in Eqs. 8 refer to the variables C,D,E,F and that each of them represents the space component of the derivative along the corresponding bicharacteristic. All the derivatives along the space lines "m" have been dropped out.
We may then recognize that the space derivatives in Eqs. 8 should be evaluated, in a discreting procedure, as one-side differences, according to the sign of the components of the propagation velocity along the corresponding bicharacteristic:

$$
\begin{array}{lllll}
(u + a) & , & v & \text{for} \quad C_x \text{ and } C_y & \text{along } l_1 \\
(u - a) & , & v & \text{for} \quad D_x \text{ and } D_y & \text{along } l_2 \\
(v + a) & , & u & \text{for} \quad E_y \text{ and } E_x & \text{along } l_3 \\
(v - a) & , & u & \text{for} \quad F_y \text{ and } F_x & \text{along } l_4
\end{array}
$$

The integration in time of Eqs. 8 allows for the prediction of the unsteady flow.
This analysis may be extended easily to the more general case of curvilinear coordinates if transformation of coordinates and mapping are needed.
The reader may refer to $/\,5\,/$ where the analysis has been developed more extensively and the physical interpretation of the numerical methodology is given.

NUMERICAL PROCEDURE

The integration is based on a two levels predictor-corrector scheme, quite similar to the well known suggested many years ago by MacCormack. However, by taking in mind what has been said previously, the space derivatives in Eqs. 8 will be evaluated with one-side differences. In order to ensure the second order of accuracy, the one-side differences are evaluated according to a 2 or 3 points discretization, as indicated in $/\,3,4\,/$.
Eqs. 8 will be integrated in time and the flow properties P, u, v will be then described during a transient until a steady flow configuration is achieved.
We will report in the following on two series of numerical experiments concerning:
  - the external 2D flow past the NACA 0012 airfoil.
  - the internal 2D flow through a parallel channel having a circular arc bump with the 4.2% thickness.

For each of them we will indicate the transformation of coordinates and the mapping used, the boundary conditions and the kind of transient.

RESULTS

The results on the flow past the NACA 0012 airfoil have been achieved on the base of the grid pattern which is obtained through the conformal mapping of the region external to the airfoil (up to   infinity) on to the region inside the unit circle. The grid we have used is shown in Fig. 2, it presents 60 intervals all   around the airfoil and 15 intervals between the profile and   infinity.
The governing equations (Eqs. 1) have been written with respect to the orthogonal curvilinear coordinates shown in Fig. 2. The integration in time has been then carried out on a set of equations which differ from Eqs. 8 only for some modifications which account for the modulus of the derivative of the map function.
As regards the boundary conditions, the computational points representing the infinity are considered as unperturbed by the airfoil. On the profile, the velocity normal to the wall is set equal to zero and the tangential component is computed as the interior points, whereas the pressure is computed by combining the compatibility equations on the bicharacteristics running along the wall with the one along the normal, taken twice $/$ 5 $/$.
The trailing edge point has been here computed very roughly, by assuming there the instantaneous average of the values extrapolated from the upstream neighboring wall points on the two sides of the profile.
A better approach should account for the vortex sheet left downstream of the profile during a transient. The analysis reported in $/$ 7 $/$ gives indications on this matter and several numerical examples we have worked out, not yet published, give   us confidence for a more correct and clear trailing edge treatment in the next future.
The transient is very simple: the profile is assumed to start from the rest condition and to move up to the prescribed steady velocity with a constant acceleration.
We have performed computations for four cases and the results are reported in Fig. 3,4,5,6.
From the supercritical cases (Figs. 5,6)   may be seen the powerful capability of the method to "capture" the shock in only one mesh interval, even if the present formulation is not conservative. The reader may refer to $/$ 4,5 $/$ for the discussion on this remarkable feature of the method.
It is also worthwhile to mention that the total number of computational points is not imposed by the requirement of avoiding spurious oscillations around the shock as well as its spreading over a too   large physical width. However it was the description of the strong expansion around the leading edge which has prevented the use of a rougher mesh size.
A selfexplaining description of the transient which leads finally to the steady flow configuration is reported in the sequence of Figs. 7. Here the isobars have been plotted at several time steps and illustrates the wave interaction around the airfoil during the transient, not yet extinguished.
The second set of results refers to the internal flow and has been achieved on the base of the grid reported in Fig. 8. The transformation of the coordinates is given by simple stretchings both in x and y in order to cluster computational points near the trailing edge of the bump.
The computations have been performed with 6 intervals along y and 28 or only 14 along x.
The governing equations (Eqs. 1) written in the (x,y) cartesian frame are transformed into non-orthogonal computational coordinates.
The derivatives of the variables C,D,E,F with respect to these transformed coordinates have been evaluated as one-side differences, according to the

sign of the corresponding contravariant component of the propagation velocity along the bicharacteristic.
We distinguish here two kinds of boundaries. The permeable ones represent
the inlet and outlet of the channel and have been treated according to the
algorithm shown in /̲ 8 ̲/, which has been proved to work successfully in a
large variety of problems. At the upper and lower walls of the channel (impermeable boundaries), the tangency condition has been imposed and the algorithm follows the same line as for the previous case about the airfoils.
No explicit treatment has been done here at the stagnation points, leading
and trailing edges, which are imbedded in mesh intervals (Fig. 8).
Few words now about the treatment of the shock. It has been here considered
as a discontinuity (double value points) as it may be seen from the results
reported in the Figs. 9,10,11. The governing equations and the compatibility ones have been modified in order to account for the gradients in entropy.
Then the shock has been defined as regards its position and strength by setting double value points over it.
By extrapolating from upstream and downstream the suitable flow properties
at the shock points, it is possible to compute at each time step the speed
of the shock, according to the Rankine-Hugoniot conditions. The shock moves
then, during the transient, according to the waves which reach it from the
surroundings regions. The explicit treatment of the shock allows for the
correct computation of the entropy downstream of it. The present approach
is rather primitive and needs further implementations, mainly for a better
two-dimensional treatment; however it is very simple as regards the coding
and presents many of the advantages obtained with the correct but complicated "true" shock fitting technique.
The steady flow configuration has been computed as the final result of the
following transient. The starting conditions are represented by gas at rest
and with static pressure equal to the total pressure of the upstream boundary conditions (unit value). At the exit of the channel a diaphragm separates
the internal region from a discharging capacity where the pressure level is
set equal to .6235. Suddenly the diaphragm is removed, so that an expansion
wave travels upstream, interfere with the inlet and creates the incoming
flow. After a certain number of paths of waves along the channel a steady
solution is achieved.
The numerical results are reported in Figs. 9,10,11. The pressure coefficient
on the lower wall is shown in Fig. 9. The black circles and the connecting
solid line represent the case with 28 intervals along x, whereas the triangles, where visible, refer to the 14 intervals computation. In both the
cases the shock location is almost the same. The effect of the mesh size on
y has been investigated by doubling the number of intervals across the
channel (12 instead of 6) and the results were practically unchanged.
Because of the non explicit computation at the leading and trailing edges,
the dotted lines connecting the corresponding theoretical total pressure
with the values of the neighboring points is somehow ficticious.
The isobars pattern of such a flow is reported in Fig. 10 (28 intervals in
x); the number labeling the curves refers to the ratio of the static pressure to the total upstream pressure.
The isoentropy lines are shown in Fig. 11 and represent also the streamlines
downstream of the shock. The number labeling the curves refers to the entropy normalized with respect to the constant volume specific heat, which in
turn is related to the pressure and density as:

$$\frac{p/\rho^{\gamma}}{p_{\infty}/\rho_{\infty}^{\gamma}} - 1 = \exp(S/c_{v}) - 1$$

<u>NOMENCLATURE</u>

a	speed of sound	x,y	cartesian coordinates
p	pressure	P	logarithm of the pressure
q	flow velocity	S	entropy
t	time	T	temperature
u,v	flow velocity components	γ	ratio of the specific heats

All the quantities are normalized with respect to reference values:
l_{ref} , P_{ref} , T_{ref} , q_{ref} = $\sqrt{RT}_{ref}$, t_{ref} = l_{ref}/q_{ref}.

<u>REFERENCES</u>

1 Gordon, P., "The Diagonal Form of Quasi-linear Hyperbolic Systems as a Basis for Difference Equations". General Electric, Final Report NOL Contract No. N60921-7164, 1968.

2 Courant, R., Isaacson, E. and Rees, M., "On the Solution of Nonlinear Hyperbolic Differential Equations by Finite Differences". Comm. on Pure and Appl. Math., Vol. 5, 1952.

3 Pandolfi, M. and Zannetti, L., "Some Tests on Finite Difference Algorithms for Computing Boundaries in Hyperbolic Flows". GAMM Workshop on Boundary Algorithms for Multidimensional Inviscid Hyperbolic Flows, Notes on Numerical Fluid Mechanics, vol. 1, Vieweg, 1978.

4 Moretti, G., "The λ-scheme". Computers and Fluids, vol. 7, No. 3, September 1979.

5 Zannetti, L. and Colasurdo, G., "A Finite Difference Method Based on Bicharacteristics for Solving Multidimensional Hyperbolic Flows". Istituto di Macchine e Motori per Aeromobili, Politecnico di Torino, Report N. PP217, 1979.

6 Butler, D.S., "The Numerical Solution of Hyperbolic Systems of Partial Differential Equations in Three Independent Variables". Proceedings of the Royal Society, Series A, April 1960.

7 Zannetti, L., "A Time Dependent Method to Solve the Inverse Problem for Internal Flow". AIAA Paper 79-0013, 1979.

8 Pandolfi, M. and Zannetti, L., "Some Permeable Boundaries in Multidimensional Unsteady Flows". Lecture Notes in Physics, vol. 90, Springer Verlag, 1979.

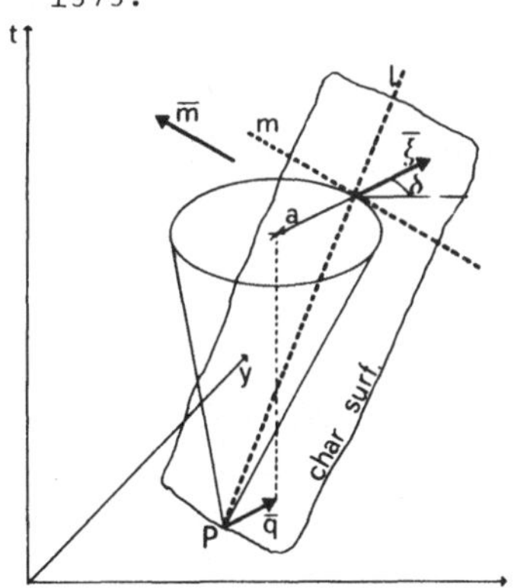

Fig. 1

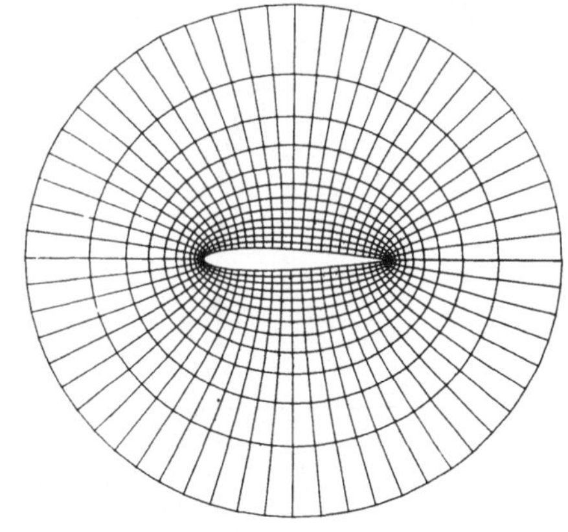

Fig. 2 - Mapping around the NACA 0012 airfoil.

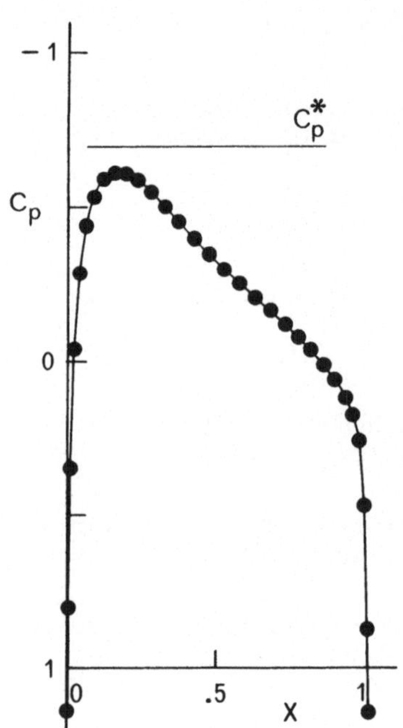

Fig. 3 - NACA 0012.
(M_∞ = .72 , α = 0°)

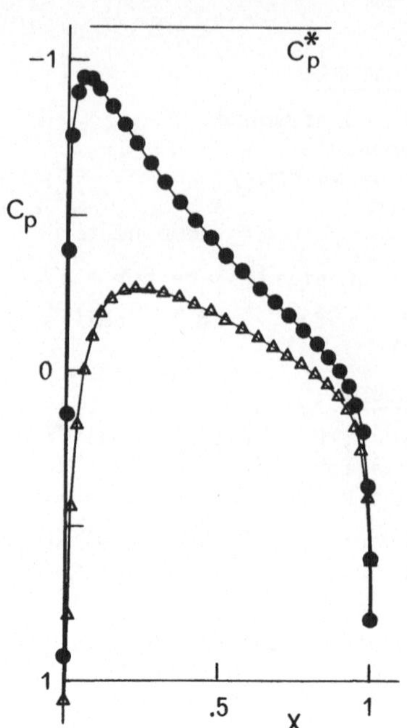

Fig. 4 - NACA 0012.
(M_∞ = .63 , α = 2°)

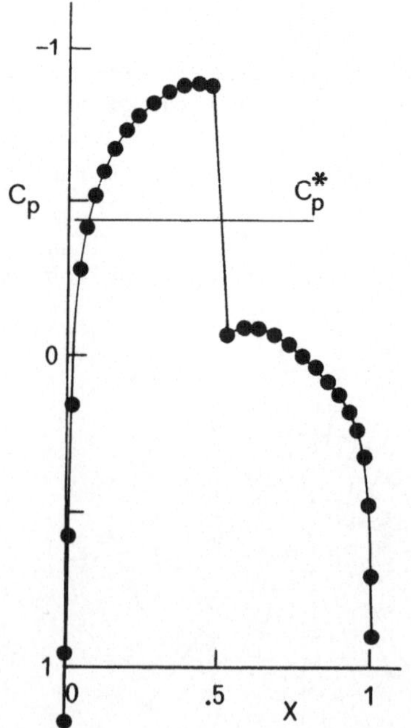

Fig. 5 - NACA 0012.
(M_∞ = .80 , α = 0°)

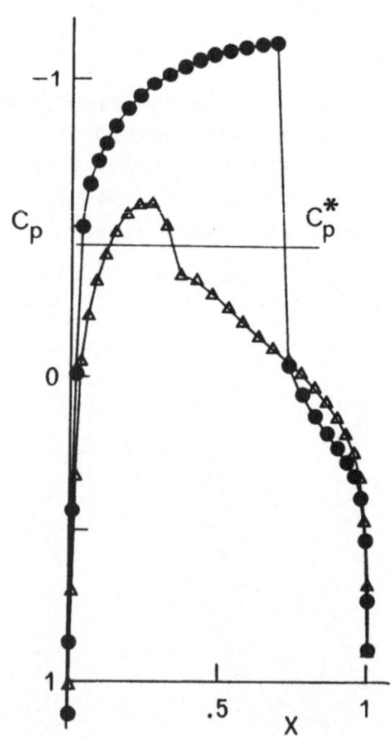

Fig. 6 - NACA 0012.
(M_∞ = .80 , α = 1.25°)

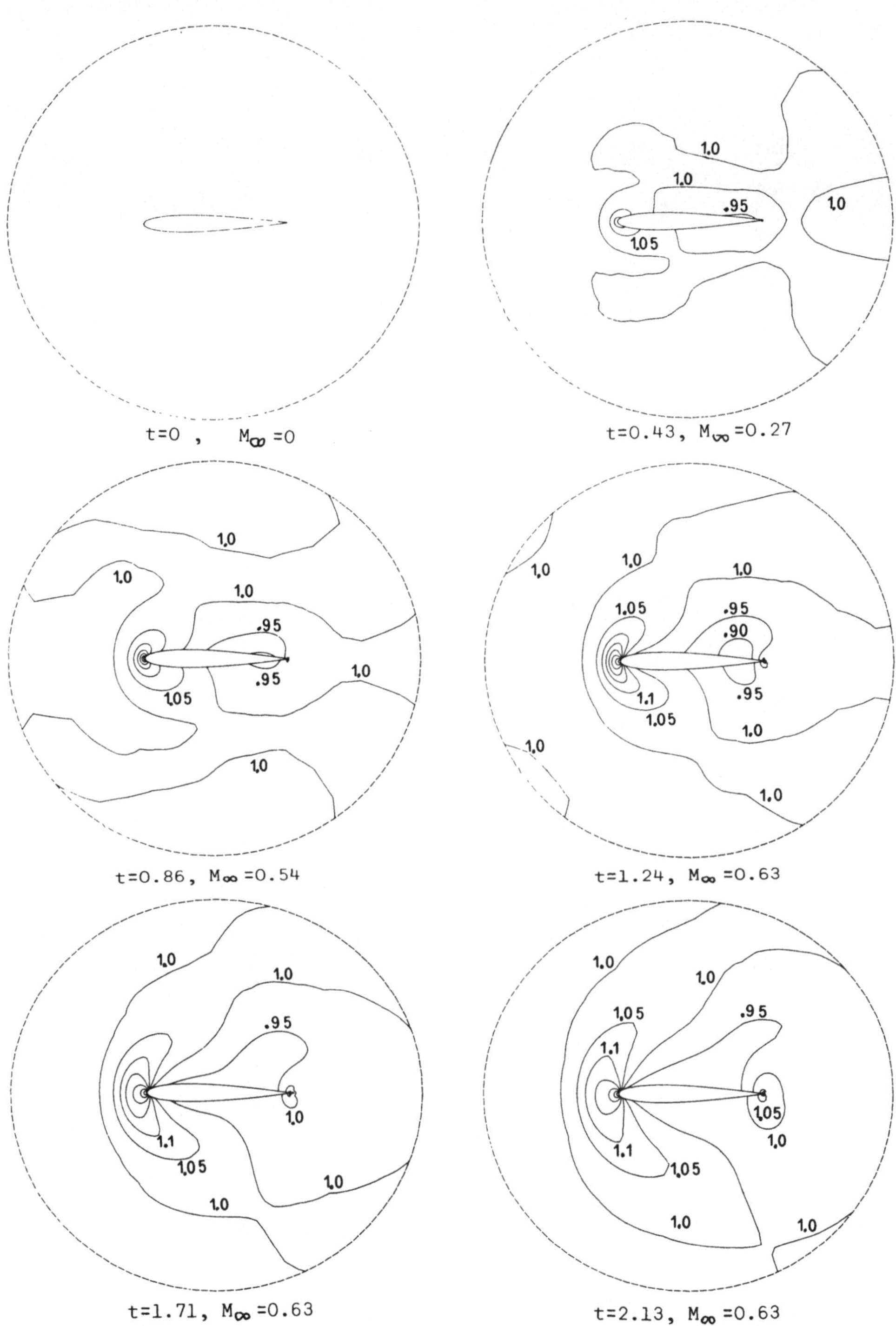

Fig. 7 - Isobars around the NACA 0012 airfoil during the transient (M_∞ = .63 , α = 2°).

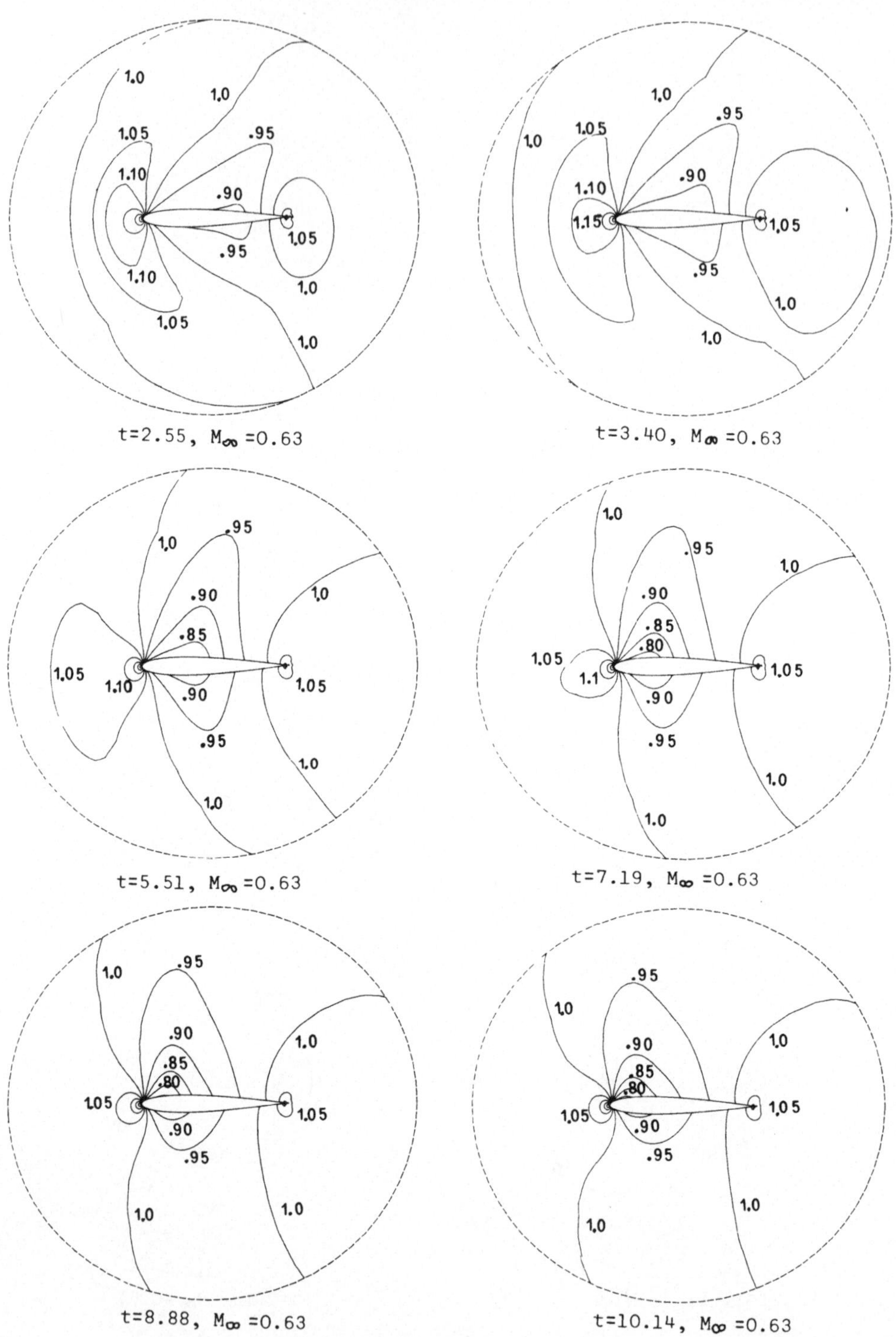

Fig. 7 - Isobars around the NACA 0012 airfoil during the transient ($M_\infty = .63$, $\alpha = 2°$).

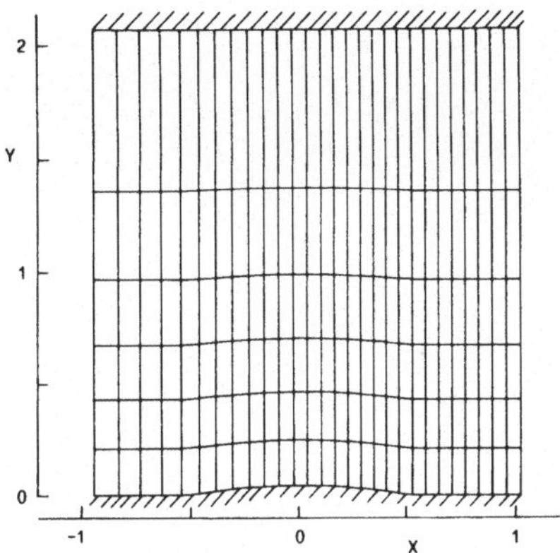

Fig. 8 - Grid used on the
internal 2D flow

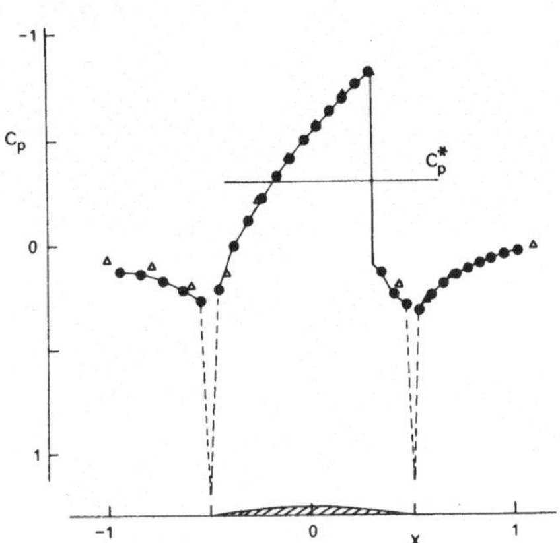

Fig. 9 - Internal 2D flow.
Pressure coefficient at
the lower wall.

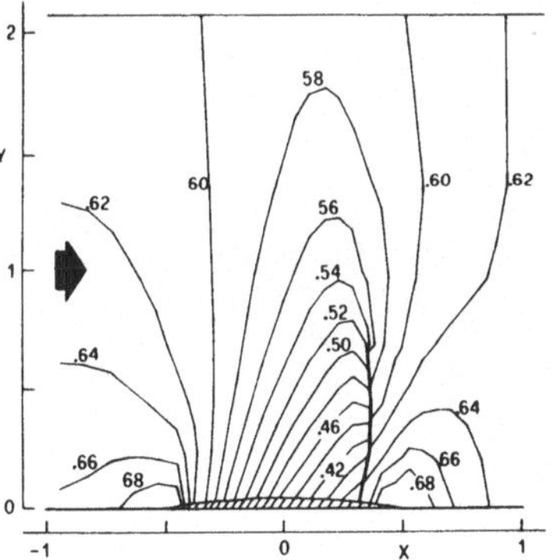

Fig. 10 - Internal 2D flow.
Constant pressure lines.

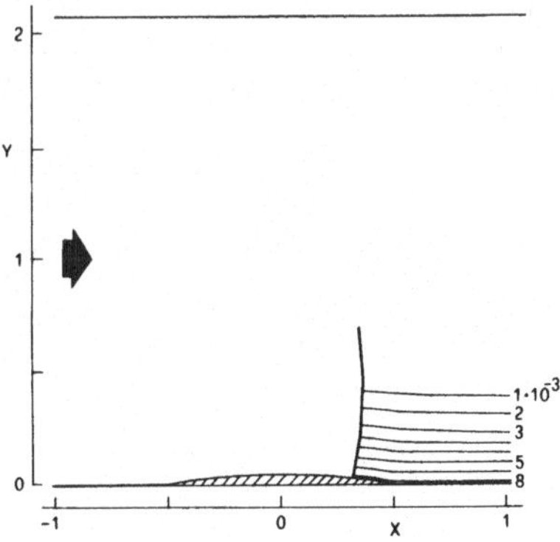

Fig. 11 - Internal 2D flow.
Constant entropy lines.

NUMERICAL SOLUTIONS OF THE EULER EQUATIONS
FOR STEADY TRANSONIC FLOW
PAST A LIFTING AEROFOIL

by

C. C. L. Sells
Royal Aircraft Establishment, Farnborough, Hampshire, England

The well-known method of MacCormack is easily programmed, can be extended to two- and three-dimensional problems, and will capture shocks. However, it does have some bad features. The predictor equation assumes that the characteristics are always in one preselected direction at each point, though the corrector stage often cancels the resulting error. Near shocks, computed solutions frequently contain 'wiggles', spaced regularly along the mesh and clearly induced by the method. The results can depend greatly on the value of the local Courant-Friedrichs-Lewy (CFL) number. Also, the method cannot sort out incoming and outgoing waves at computational boundaries in subsonic flow; thus signals are reflected and wander about, unable to escape, and even when dissipation terms are added convergence to a steady flow is slowed down. It would be worthwhile to develop a method with the advantages of MacCormack's method but without the drawbacks, and Phil Roe at RAE Bedford has shown how to obtain a very effective method for one-dimensional problems; the author at RAE Farnborough has borrowed many of his techniques and applied them to the problem of the title, and fuller details will be given in Ref 1 in due course.

Roe's method takes account of characteristic directions and also splits flux vectors into corresponding eigenvectors, so for each new set of governing equations these must be calculated. We show briefly how this is done for the unsteady one-dimensional Euler equations. Suppose there is a discontinuity in the flow, propagating with speed s ; let the discontinuity separate two states with suffices 1 and 2. Write $\Delta\rho = \rho_1 - \rho_2$, etc. Then

$$s\Delta\rho \;=\; \Delta(\rho u) \;=\; \Delta m \qquad\qquad (\text{with } m = \rho u)$$

$$s\Delta m \;=\; \Delta(p + \rho u^2) \;=\; \Delta(p + mu)$$

$$s\Delta e \;=\; \Delta[(e + p)u] \;,$$

with
$$e \;=\; \frac{p}{\gamma - 1} + \tfrac{1}{2}\rho u^2 \;=\; \frac{p}{\gamma - 1} + \tfrac{1}{2}mu \;.$$

Expressing all product differences using averaged values
$\bar{\rho} = \frac{1}{2}(\rho_1 + \rho_2)$ etc, we have

$$\Delta m = \bar{\rho}\Delta u + \bar{u}\Delta\rho$$

$$\Delta e = \frac{1}{\gamma - 1}\Delta\rho + \tfrac{1}{2}\bar{m}\Delta u + \tfrac{1}{2}\bar{u}\Delta m$$

$$(s - \bar{u})\Delta\rho = \bar{\rho}\Delta u$$

$$(s - \bar{u})\Delta m = \Delta p + \bar{m}\Delta u$$

$$(s - \bar{u})\Delta e = (\bar{e} + \bar{p})\Delta u + \bar{u}\Delta p \quad .$$

These are five linear homogeneous equations for five unknowns,
and so determine an eigenvalue problem for s ; the three
eigenvalues come out as the mean speed $\bar{u}$ and the mean speed
of sound waves propagating with and against the flow, as we
expect. As the differences go to zero the eigenvectors tend
toward those of the Euler equations, so we can use them to
treat the very small disturbances considered as the result of
discretizating a smooth flow field. Also, if the states
correspond to a discontinuity such as a shock, which does not
go to zero with the mesh size, then the splitting will recover
that shock, and all other eigenvectors will vanish. If two or
more discontinuities momentarily coalesce, the splitting will
be somewhat in error compared with the exact solution of the
Riemann problem, but this corresponds to an isolated point in
(x, t) space, such points forming a subset of measure zero,
and in a numerical method the overall error is likely to be
small.

    We now show how we take account of characteristic direc-
tions. At time level n , the interaction between cells j
and j + 1 at predictor level in a plain MacCormack scheme to
solve the equation $\partial U/\partial t + \partial F/\partial x = 0$ might send the whole of
a flux difference $\Delta\underline{F}$ to cell j + 1 :

$$\underline{U}_{j+1}^{n+\frac{1}{2}} = \underline{U}_{j+1}^{n} - \frac{\Delta t}{\Delta x}\left(\underline{F}_{j+1}^{n} - \underline{F}_{j}^{n}\right) \quad .$$

In Roe's modification, if an eigenvalue is positive measured
from j to j + 1 , then the appropriate eigenfunction is
added to cell j + 1 ; otherwise it is added to cell j . It
is noteworthy that, if the characteristic changes direction
locally, a particular cell may receive two, one or no
increments.

    To understand how the corrector is modified, we first
interpret MacCormack's scheme as a three-stage operator.

For the scalar equation $\partial u/\partial t + \partial f/\partial x = 0$, a plain
MacCormack scheme can be written thus:

Advance predictor: $u_j^{n+\frac{1}{2}} - u_j^n = -\sigma\left(f_j^n - f_{j-1}^n\right)$ $(\sigma = \Delta t/\Delta x)$

Retreat predictor: $u_j^{n+\frac{1}{2}} - u_j^* = -\sigma\left(f_{j+1}^{n+\frac{1}{2}} - f_j^{n+\frac{1}{2}}\right)$

Corrector: $u_j^{n+1} = u_j^{n+\frac{1}{2}} - \tfrac{1}{2}\sigma\left[\left(f_{j+1}^{n+\frac{1}{2}} - f_j^{n+\frac{1}{2}}\right) - \left(f_j^n - f_{j-1}^n\right)\right]$.

In the retreat predictor, conditions at the new time
level are used to pseudo-predict the solution at the old time
level, the characteristic direction being maintained so that the
domain of dependence is the opposite of that for the advance
predictor. When f is a non-linear function of u , the
retreat value u_j^* is not the same as the initial value u_j^n ,
and in the corrector stage their difference is used to complete
a second-order algorithm. We do not write this difference in
terms of u_j^* , or even compute u_j^* , but we write it in terms
of the flux increments. When this interpretation is
generalized to take proper account of characteristic directions,
the square bracket is understood as the sum of any and all
increments received.

To alleviate the 'wiggles' which bedevil the plain
MacCormack scheme, we test the contribution of each component
of each eigenvector at the corrector stage, to see if it would
induce an oscillation locally in the computed solution. If,
and only if, the test proves positive, then the square bracket
in the corrector stage is shifted one cell in the direction of
the corresponding characteristic velocity. For example, if
this velocity is in the direction of j increasing, the square
bracket is added to u_{j+1} rather than to u_j . As at the predictor stage, a cell is allowed to receive more than one incre-
ment, conservation still being assured.

Steger and Warming[2] also use an eigensplitting scheme
together with an upwind-corrector MacCormack algorithm equivalent to applying our corrector shift at every point; they
alternate this algorithm with the plain MacCormack algorithm,
and in shock regions their results appear to show only a few
small wiggles. However, the present method seems even more
effective for that purpose.

Boundary conditions are easily handled. A solid wall is
simulated by an image cell, carrying the same values of density
and pressure as the adjacent real cell, but the velocity is
reversed. Then eigenfunctions propagating into the image cell
from the wall interface are ignored. The rationale for this

approach is that a smooth, locally odd function is certain to
vanish at the origin.

We are now ready to extend the method to two-dimensional
flow past an aerofoil. We surround the aerofoil with a computing mesh of quadrilateral cells, so that rows of cells run
right round the aerofoil and columns of cells go out from the
aerofoil to the computational boundary with the free-stream.
The Euler equations for unsteady flow are written in vector
conservation form with four components; we can modify our one-
dimensional subroutine to include a discontinuity in slip
velocity along a cell interface, corresponding to the fourth
component, and then we use a time-split finite-volume formulation and set up two one-dimensional subroutines or operators,
one considering all the row interfaces and the other all the
column interfaces.

The boundary condition at a solid surface is handled as
in the one-dimensional case, with the addition of a component
of velocity along the interface, which is matched in the image
cell. If the surface is curved, the pressure in the image cell
is modified using a difference form of the steady momentum
equation across a streamline. The outer boundary condition is
treated similarly; we imagine a cell carrying free-stream
values, or values modified by adding a 'compressible vortex',
and again ignore flux increments travelling outwards. Thus
the boundary absorbs the major part of an outward-travelling
signal and only a small residual signal is reflected back,
depending on the discretization error. This provides a
mechanism for self-damping which is otherwise difficult to
achieve at outer boundaries in the plain MacCormack method.

Like other explicit time-marching schemes, this method
has a difficulty of long computing times. For good flow definition near the trailing-edge, we need many small cells there,
especially in lifting flows. Hence the allowable timestep is
very small. In a true time-dependent method the solution must
advance by equal timesteps everywhere, even though the cells
in the mid-chord region and outboard of the aerofoil would
admit a much greater timestep. Hence the runtimes for subsonic flow tend to be impracticably long, especially when the
local fluid speed is close to the local sound speed so that
upstream propagation is reduced to a crawl. For a typical
case, a 6 per cent circular-arc aerofoil at $M_\infty = 0.8$, with

a 40 × 12 mesh, a sound wave would need about 10000 timesteps
to get from the trailing-edge to the leading-edge - and the
flow would still have to evolve further.

For the moment, then, we abandon true time-dependence
and rest content with a quasi-time-dependent method tending
toward the final steady-state. For the early stages of the
run, we associate each mesh cell with its own (local) timestep
and multiply the flux increments destined for that cell by this
local timestep instead of keeping the same timestep everywhere.
Thus all cells are working at unit CFL number in at least one
direction, and disturbances propagate much more quickly
through the mesh. Since the results are not sensitive to CFL

number provided it is less than one, this device will only affect regions where the local timestep is changing rapidly from cell to cell, which means the leading-edge in practice; in case the difference is substantial, after the computation with local timesteps has settled down, we do some more calculations with uniform timestep to finish - this is usually the longest part of a run.

In this method, the computation of drag as a contour integral of the pressure is unreliable; a small difference of large numbers is required, and since (among other possible sources of discretization error) the steady-flow entropy is not computed quite constant along streamlines, the large numbers tend to be insufficiently accurate. So we also evaluate the wave drag; if a shock is found on either or both surfaces, a column of cells downstream of the shock gives a control surface S_2 , and we can integrate along S_2 .

We assume that the flow is steady, or nearly so. We take a 'wake surface' S_w normal to the flow, and far downstream, so that the wake curvature is negligible. Then the drag D is

$$D = \int_{S_w} \rho_w u_w (U_\infty - u_w) dS_w$$

where ρ_w and u_w are density and velocity at S_w . Now consider a streamtube running from the surface S_2 to the wake surface S_w . Denote flow quantities at S_2 by subscript 2. Entropy is conserved along the streamtube, hence

$$ENF_2 = \frac{p_2/p_\infty}{(\rho_2/\rho_\infty)^\gamma} = \frac{1}{(\rho_w/\rho_\infty)^\gamma} \ .$$

Total enthalpy is conserved everywhere:

$$H_0 = \frac{\gamma}{\gamma - 1} \frac{p_\infty}{\rho_w} + \tfrac{1}{2} u_w^2 = \frac{\gamma}{\gamma - 1} \frac{p_\infty}{\rho_\infty} + \tfrac{1}{2} U_\infty^2 \ .$$

Eliminating ρ_w , we have u_w in terms of conditions on S_2 :

$$u_w = \left[2H_0 - \frac{2\gamma}{\gamma - 1} (ENF_2)^{1/\gamma} \frac{p_\infty}{\rho_\infty} \right]^{\frac{1}{2}} \ .$$

Also, mass is conserved along the streamtube; thus, letting u_{n2} be the normal component of velocity,

$$\rho_w u_w \delta S_w = \rho_2 u_{n2} \delta S_2 \quad .$$

Hence, replacing the integral by a trapezoidal sum,

$$D \doteqdot \sum_{S_2} (U_\infty - u_w)\rho_2 u_{n2} \delta S_2 \quad .$$

Finally, if c is the aerofoil chord, we have the drag coefficient

$$C_D = \frac{D}{\left(\tfrac{1}{2}\rho_\infty U_\infty^2 c\right)} \quad .$$

This technique has a defect: for weak shocks the discretization error in entropy is of the same order as the entropy jump through a shock. Fortunately, in practice the jump in the entropy function ENF is of the same order as the value ENF* that we would expect to get in an error-free computation for the same observed local Mach number:

$$ENF^* \doteqdot ENF_2 - ENF_1 + 1.0$$

where ENF_1 is a suitable value upstream of the shock. This expression for ENF* is then used in place of ENF_2 when calculating u_w .

REFERENCES

1   C.C.L. Sells: Solution of Euler equations for transonic flow past a lifting aerofoil. RAE Technical Report (to be published)

2   J.L. Steger and R.F. Warming: Flux vector splitting of the inviscid gasdynamic equations with application to finite difference methods. NASA Technical Memorandum 78605 (1978)

$$\text{FINITE-VOLUME METHODS \quad FOR THE SOLUTION OF EULER EQUATIONS}^{*}$$

by A. Lerat[*] and J. Sidès

Office National d'Etudes et de Recherches Aérospatiales (ONERA)
92320 Châtillon (France)

---

1 - <u>INTRODUCTION</u>

This contribution to the GAMM Workshop on Numerical Methods for the Computation of Inviscid Transonic Flow with Shock Waves is concerned with finite-volume methods to solve a pseudo-unsteady system deduced from the unsteady Euler equations by using the condition of constant total enthalpy. This simplification is consistent with the steady-state solution in the present case of iso-energetic flows. The reason for the choice of finite-volume methods is their property of being exactly in conservation form so that in a balance of computed mass, momentum or energy in a domain made of mesh cells, the interior numerical fluxes cancel out two by two.

In this paper, we shall shortly review two finite-volume methods we have used. The first one is the classical method developed at NASA from the Mac Cormack scheme (see /1/-/3/) applied here without time-splitting and with two different treatments of boundary conditions. The second one is based on a new predictor-corrector scheme requiring knowledge of only seven points like the Mac Cormack scheme. Since a finite-volume method can be defined only from the approximation formula of the fluxes, we first describe the general form of the method in §2 and then we give the numerical fluxes corresponding to the Mac Cormack scheme and to the new scheme in §3 and 4. Numerical results are presented for three test problems of the Workshop : the internal flow with shock through a channel having a "bump" on the lower wall and two external supercritical flows over the NACA 0012 airfoil.

2 - <u>INTEGRAL CONSERVATION LAWS AND FINITE-VOLUME METHODS</u>

Inviscid transonic flows are governed by the Euler equations. These equations are originally stated as a system of conservation laws written in the integral form

$$\iint_{\Omega} w\, dx\, dy \Bigg|_{t_1}^{t_2} = - \int_{t_1}^{t_2} \int_{\Gamma} \left(f(w)\, dy - g(w)\, dx \right) dt \tag{1}$$

for any time interval (t_1, t_2) and any bounded domain Ω with boundary Γ fixed in an absolute orthogonal cartesian frame $(0 ; x, y)$. Being interested only in the steady-state solution, we assume that the total enthalpy is constant and we consider the pseudo-unsteady system (1) with

$$w = \begin{bmatrix} \rho \\ \rho u \\ \rho v \end{bmatrix}, \qquad f(w) = \begin{bmatrix} \rho u \\ \rho u^2 + p \\ \rho u v \end{bmatrix}, \qquad g(w) = \begin{bmatrix} \rho v \\ \rho u v \\ \rho v^2 + p \end{bmatrix},$$

for density ρ , pressure p, x-and y- velocity components u and v.

---

[*] Work performed with the financial support of DRET

[*] Ecole Nationale Supérieure d'Arts et Métiers, 75640 Paris Cédex 13 and
Laboratoire de Mécanique théorique, Université Paris VI - Consultant at ONERA.

The energy equation becomes

$$\frac{\gamma}{\gamma-1}\frac{p}{\rho} + \frac{1}{2}\left(u^2 + v^2\right) = H_\infty \, ,$$

where H_∞ is the freestream total enthalpy and the fluid has been supposed to be a polytropic gas with a ratio of specific heats $\gamma = 1.4$.

Let us now divide the time axis into intervals (t^n, t^{n+1}) and the flow domain into a number of cells $\Omega_{i,j}$ delimited by a curvilinear mesh and denote by $\Gamma_{i,j}$ the boundary of $\Omega_{i,j}$ (see Fig. 1). By introducing the mean value of w in cell $\Omega_{i,j}$ of area $S_{i,j}$

$$w_{i,j}^n = \frac{1}{S_{i,j}} \iint_{\Omega_{i,j}} w\, dx\, dy \bigg|_{t=t^n}$$

and also the mean value of the flux $\int f\, dy - g\, dx$ across $\Gamma_{i+1/2,j} = \Gamma_{i,j} \cap \Gamma_{i+1,j}$ and $\Gamma_{i,j+1/2} = \Gamma_{i,j} \cap \Gamma_{i,j+1}$ respectively, during a time step $\Delta t = t^{n+1} - t^n$

$$F_{i+1/2,j}^{n+1/2} = \frac{1}{\Delta t} \int_{t^n}^{t^{n+1}} \int_{\Gamma_{i+1/2,j}} \left(f(w)\, dy - g(w)\, dx \right) dt \tag{2}$$

$$G_{i,j+1/2}^{n+1/2} = \frac{1}{\Delta t} \int_{t^n}^{t^{n+1}} \int_{\Gamma_{i,j+1/2}} \left(f(w)\, dy - g(w)\, dx \right) dt \, , \tag{3}$$

the exact system (1) can be expressed as

$$w_{i,j}^{n+1} = w_{i,j}^n - \frac{\Delta t}{S_{i,j}} \left(F_{i+1/2,j}^{n+1/2} - F_{i-1/2,j}^{n+1/2} + G_{i,j+1/2}^{n+1/2} - G_{i,j-1/2}^{n+1/2} \right) \tag{4}$$

for any interval (t^n, t^{n+1}) and any mesh cell $\Omega_{i,j}$.

An explicit numerical method for system (1) is said to be a "finite-volume method" if it can be written as

$$w_{i,j}^{n+1} = w_{i,j}^n - \frac{\Delta t}{S_{i,j}} \left(\mathcal{F}_{i+1/2,j}^{n+1/2} - \mathcal{F}_{i-1/2,j}^{n+1/2} + \mathcal{G}_{i,j+1/2}^{n+1/2} - \mathcal{G}_{i,j-1/2}^{n+1/2} \right) \tag{5}$$

where $\mathcal{F}_{i+1/2,j}^{n+1/2}$ and $\mathcal{G}_{i,j+1/2}^{n+1/2}$ are consistent approximations of (2) and (3) depending on values $w_{k,\ell}^n$ at time t^n in cells $\Omega_{k,\ell}$ close to $\Gamma_{i+1/2,j}$ and $\Gamma_{i,j+1/2}$, respectively.

Such a method could be called a "method in conservation form" since (5) is a direct extension of the definition given by Lax and Wendroff /4/ for a one-dimensional scheme in conservation form (see /5/).

3 - FINITE-VOLUME METHOD BASED ON MAC CORMACK SCHEME

3.1 Computation of interior fluxes. The finite-volume method based on the second-order accurate scheme of Mac Cormack can be written in the form (5) with

$$\mathcal{F}_{i+1/2,j}^{n+1/2} = \frac{1}{2} \left(f_{i+1,j}^n + \tilde{f}_{i,j}^{n+1} \right) \Delta y_{i+1/2,j} - \frac{1}{2} \left(g_{i+1,j}^n + \tilde{g}_{i,j}^{n+1} \right) \Delta x_{i+1/2,j}$$

and a similar expression for $\mathcal{G}^{n+1/2}_{i,j+1/2}$, where $(\Delta x, \Delta y)_{i+1/2,j}$ are the projections of $\Gamma_{i+1/2,j}$ on the reference frame,

$$f^{m}_{i,j} = f(w^{m}_{i,j}) \quad , \quad \tilde{f}^{m+1}_{i,j} = f(\tilde{w}^{m+1}_{i,j})$$

and the same for g. The vector $\tilde{w}$ is a predictor also computed by the formula (5) with

$$\mathcal{F}^{m+1/2}_{i+1/2,j} = f^{m}_{i+1,j} \, \Delta y_{i+1/2,j} - g^{m}_{i+1,j} \, \Delta x_{i+1/2,j} \; .$$

In the calculations with (5), we have used the maximum local time step $\Delta t = \Delta t_{i,j}$ allowed by the Courant-Friedrichs-Lewy condition for each cell $\Omega_{i,j}$. This procedure allows a substantial reduction of the computing time necessary to reach the steady-state.

3.2 Computation of boundary fluxes

(a) Extrapolation technique

On a rigid wall, the slip condition requires that the normal velocity be zero. If we denote by $\Omega_{i,1}$ a cell having its side $\Gamma_{i,1/2}$ on the wall, the exact average flux across the wall is given by (3) with $j = 0$ and

$$f(w)\, dy - g(w)\, dx = \pi \begin{bmatrix} 0 \\ dy \\ -dx \end{bmatrix}$$

In the numerical method, this flux is approximated by

$$\mathcal{G}^{m+1/2}_{i,\,1/2} = \pi_{i,\,1/2} \begin{bmatrix} 0 \\ \Delta y_{i,1/2} \\ -\Delta x_{i,1/2} \end{bmatrix} \tag{6}$$

where the pressure $\pi_{i,1/2}$ on the wall is obtained by a parabolic extrapolation of the pressures $\pi^{m}_{i,1}$, $\pi^{m}_{i,2}$ and $\pi^{m}_{i,3}$ when (6) is used in the calculation of a predictor or of the pressures $\tilde{\pi}^{m+1}_{i,1}$, $\tilde{\pi}^{m+1}_{i,2}$ and $\tilde{\pi}^{m+1}_{i,3}$ when (6) is used in the calculation of definitive values. The first-order accuracy in time of this simple procedure is not a shortcoming since the method is pseudo-unsteady.

On a subsonic inflow boundary, the flux is computed from the entropy and velocity direction data and an extrapolation for the density. On a subsonic outflow boundary the flux is computed from the pressure data and an extrapolation for the momentum components.

(b) Compatibility relations technique

The boundaries have also been treated by using the compatibility relations technique devised by Viviand and Veuillot /6/ in the finite-difference approach.

On inflow or outflow boundaries, the choice of the conditions to prescribe and of the compatibility relations to satisfy is the same as in /6/. On the wall, the best results have been obtained in the finite-volume

144

approach by replacing the compatibility relation relative to the tangential
momentum (see /6/ p. 34) by an extrapolation of the tangential velocity.

3.3 Numerical results

As a first application, we consider the internal flow through a
parallel channel having a 4.2% thick circular arc "bump" on the lower wall.
The ratio of static downstream pressure to total upstream pressure corresponds
to a Mach number of 0.85 in isentropic flow and the distance between the
walls is 2.07 times the chord length of the bump. The computational mesh
consists of 71 × 20 cells and is very close to the standard mesh of the
Workshop. The extrapolation technique has been used to compute the boundary
fluxes. The isomach lines are shown on Fig. 2. The entropy wake downstream
of the shock appears clearly in this numerical result. The entropy error on
the bump has been found to be lower than 0.5%. The correct choice of upstream
and downstream boundary conditions is very important. Thus, we have noticed
that if all the freestream conditions are prescribed upstream, the solution
exhibits very large oscillations upstream of the bump.

A second application of the method is the calculation of the flow
over the NACA 0012 airfoil at zero incidence for a freestream Mach number
M_∞ = 0.85. The computational mesh (shown partly on Fig. 3) has been
constructed by Chattot, Coulombeix and da Silva Tomé /7/ from the transformation of Jameson /8/. The number of mesh cells is 188 × 24. The outer boundary is located between 4 and 7 chord lengths from the airfoil. Calculations
have been performed with the two treatments described for the boundary conditions. We show on Fig. 4 the isomach lines computed in each case and on Fig. 5
and 6 the pressure and entropy distributions on the airfoil. The computed
pressures do not depend very much on the technique used to prescribe the
boundary conditions. However concerning the entropy error, the technique of
compatibility relations leads to a real improvement.

Finally, we have calculated the flow over the NACA 0012 airfoil with
an angle of attack α = 1°25 and M_∞ = 0.8. The numerical results (shown in Fig. 7
and 8) have been obtained in the previous mesh with the extrapolation technique at boundaries. Two shock waves are present in the numerical solution. We
note that the Rankine-Hugoniot relations are satisfied better across the upper shock than across the lower shock which is the weakest one. Upstream of
the shocks, the entropy error is about 2.5% on the upper surface of the
airfoil and 1.5% on the lower surface but this error decreases rapidly as
one goes away from the airfoil.

4 - FINITE-VOLUME METHOD BASED ON A SEVEN-POINT SCHEME $\mathcal{S}^\alpha_{\beta_1,\beta_2}$.

4.1 **Computation of interior fluxes.** For hyperbolic systems of conservation laws in one space variable, Lerat and Peyret /9/, /10/ have constructed and studied the class containing all the three-point schemes of
second-order accuracy in predictor-corrector form. This class of schemes
depends on two parameters α and β (schemes $\mathcal{S}^\alpha_\beta$). In particular $\mathcal{S}^1_0$ is the
Mac Cormack scheme, $\mathcal{S}^{1/2}_{1/2}$ is the Richtmyer scheme and the scheme cor-
responding to α = 1 + $\sqrt{5}/2$ and β = 1/2 has been found to be "optimal" with
regard to the problem of spurious oscillations in the numerical solutions.
Recently the schemes $\mathcal{S}^\alpha_\beta$ have been extended in two space variables (to be
published). The class of nine-point schemes of second-order accuracy in
predictor-corrector form has been constructed but we shall consider here
only the subclass of seven-point schemes. In general this subclass depends
on 6 parameters. For a linear hyperbolic system, it still depends on 3
parameters (contrary to the one-dimensional case where all the $\mathcal{S}^\alpha_\beta$ schemes
would be identical). One can choose these "linear parameters" in order to
get schemes which reduce to the Mac Cormack scheme for a linear system.
However in preference we have chosen the linear parameters to achieve better
linear stability properties (see /11/). So, we have obtained various schemes
depending on three "nonlinear parameters" α, β_1 and β_2 (schemes $\mathcal{S}^\alpha_{\beta_1,\beta_2}$).

The predictor of such a scheme is a first-order approximation of the solution at time $t^n + \alpha \Delta t$ and at point $x_i + \beta_1 \Delta x$, $y_j + \beta_2 \Delta y$. In the present calculations we have used $\alpha = 1+\sqrt{5}/2$ and $\beta_1 = \beta_2 = 1/2$.

The finite-volume method based on the scheme $\mathcal{J}^\alpha_{1/2,1/2}$ can be written in the form (5) with

$$\tilde{\mathcal{F}}^{m+1/2}_{i+1/2,j} = \left[\tfrac{1}{4}\left(\tilde{f}^m_{i,j} + \tilde{f}^m_{i+1,j-1}\right) + \tfrac{\alpha-1}{4\alpha}\left(\tilde{f}^m_{i+1,j} + \tilde{f}^m_{i,j+1}\right) + \tfrac{1}{2\alpha} \tilde{f}^{m+\alpha}_{i+1/2,j+1/2} \right] \Delta y_{i+1/2,j}$$
$$- \left[\tfrac{1}{4}\left(\tilde{g}^m_{i,j} + \tilde{g}^m_{i+1,j-1}\right) + \tfrac{\alpha-1}{4\alpha}\left(\tilde{g}^m_{i+1,j} + \tilde{g}^m_{i,j+1}\right) + \tfrac{1}{2\alpha} \tilde{g}^{m+\alpha}_{i+1/2,j+1/2} \right] \Delta x_{i+1/2,j} \qquad (7)$$

and an analogous expression for $\mathcal{G}^{m+1/2}_{i,j+1/2}$.

The predictor is computed as follows :

$$\tfrac{1}{2}\left(A_{i+1,j} + A_{i,j+1}\right) \tilde{w}^{m+\alpha}_{i+1/2,j+1/2} = \tfrac{1}{2}\left(A_{i+1,j}\, w^m_{i+1,j} + A_{i,j+1}\, w^m_{i,j+1}\right)$$
$$- \alpha \Delta t \left(\tilde{F}^m_{i+1,j+1/2} - \tilde{F}^m_{i,j+1/2} + \tilde{G}^m_{i+1/2,j+1} - \tilde{G}^m_{i+1/2,j}\right)$$

with for instance

$$\tilde{F}^m_{i+1,j+1/2} = f^m_{i+1,j}\left(y_{i+1,j+1} - y_{i+1,j}\right) - g^m_{i+1,j}\left(x_{i+1,j+1} - x_{i+1,j}\right)$$

where $(x,y)_{i,j}$ are the coordinates of the centre of the cell $\Omega_{i,j}$.

4.2 <u>Computation of boundary fluxes</u>

With the same notations as in §3.2, the flux on a rigid wall is approximated by (6) in which $\uparrow_{i,1/2}$ is obtained by a parabolic extrapolation at time t^n .
Furthermore the numerical flux (7) must be modified for $j = 1$ since the vector $w^m_{i+1,j-1}$ is not available for this value of j. So, the calculation of $\mathcal{F}^{m+1/2}_{i+1/2,1}$ has been made by replacing the term

$$\tfrac{1}{4}\left(f^m_{i,1} + f^m_{i+1,0}\right) \quad \text{by} \quad \tfrac{1}{4}\left(f^m_{i,1/2} + f^m_{i+1,1/2}\right)$$

and likewise the term

$$\tfrac{1}{4}\left(g^m_{i,1} + g^m_{i+1,0}\right) \quad \text{by} \quad \tfrac{1}{4}\left(g^m_{i,1/2} + g^m_{i+1,1/2}\right)$$

where the vectors $w^m_{i,1/2}$ (on the wall) are obtained from parabolic extrapolations. However, one observes that the necessary predictors can be computed like any interior cells.

On inflow or outflow boundaries, the procedure is similar to the one in §3.2,a.

4.3 <u>Numerical results</u>

We consider the application of the method based on the scheme $\mathcal{J}^{1+\sqrt{5}/2}_{1/2,1/2}$ to the calculation of the flow past the NACA 0012 airfoil at zero incidence with $M_\infty = 0.85$. The computational mesh is the same as previously. The isomach lines in the flowfield and the pressure and entropy distributions on the airfoil are shown in Fig. 9, 10, 11. It is interesting to compare these

146

results with those given by the first method based on the Mac Cormack scheme
with an analogous treatment of boundary conditions (extrapolation technique).
In both methods, the artificial viscosity term described in /5/ has been
added in the momentum equations. This correction maintains the conservation
form and also the order of accuracy of the method. The dimensionless coefficient
χ of artificial viscosity has been taken equal to 0.8 in both cases (*). Under
these conditions, it can be seen that the second method reduces the spurious
oscillations on the airfoil and in the flowfield. The second method gives also
an appreciable reduction in the entropy error : the maximum value of
$(p/p_\infty)/(\rho/\rho_\infty)^\gamma - 1$ upstream of the shock decreases from 1.4% to 0.4%.

REFERENCES

/1/ Mac Cormack, R.W., and Paullay, A.J. - Computational Efficiency
Achieved by Time-Splitting of Finite-Difference Operators. AIAA Paper
72-154 (1972).

/2/ Rizzi, A.W. and Inouye, M. - A Time-Split Finite-Volume Technique for
Three-Dimensional Blunt-Body Flow. AIAA Journal 11, p. 1478-1485 (1973).

/3/ Mac Cormack, R.W., Rizzi, A.W., and Inouye, M. - Steady Supersonic
Flowfields with Embedded Subsonic Regions. In "Computational Methods
and Problems in Aeronautical Fluid Dynamics" edited by Hewitt, B.L.
et al., Academic Press (1976).

/4/ Lax, P.D., and Wendroff, B. - Systems of Conservation Laws. Comm. Pure
Appl. Math. 13, p. 217-237 (1960).

/5/ Lerat, A., and Sidès, J. - Numerical Simulation of Unsteady Transonic
Flows Using the Euler Equations in Integral Form. 21st Israel Annual
Conference on Aviation and Astronautics. Feb. 1979. To appear. Pro-
visional edition : TP ONERA n° 1979-10.

/6/ Viviand, H., and Veuillot, J.P. - Méthodes pseudo-instationnaires pour
le calcul d'écoulements transsoniques. Publication ONERA n° 1978-4
(1978).

/7/ Chattot, J.C., Coulombeix, C., and da Silva Tomé, C. - Calculs d'écou-.
lements transsoniques autour d'ailes. La Recherche Aérospatiale n° 1978-4
p. 143-159 (1978).

/8/ Jameson, A. - Three Dimensional Flows Around Airfoils with Shocks.
Lecture Notes in Computer Science 11, p. 185-212 (1974).

/9/ Lerat, A., and Peyret, R. - The Problem of Spurious Oscillations in the
Numerical Solution of the Equations of Gas Dynamics. Lecture Notes in
Physics 35, p. 251-256 (1975).

/10/ Lerat, A., and Peyret, R. - Propriétés dispersives et dissipatives d'une
classe de schémas aux différences pour les systèmes hyperboliques non
linéaires. La Recherche Aérospatiale n° 1975-2, p. 61-79 (1975).

/11/ Livne, A. - Seven-Point Difference Schemes for Hyperbolic Equations.
Math. Comput. 29, p. 425-433 (1975).

/12/ Lerat, A., and Sidès, J. - Calcul numérique d'écoulements transsoniques
instationnaires. AGARD Conf. Proceed. 226, p.15.1-15.10 (1977). English
translation available as TP ONERA 1977-19E.

(*) Another form of artificial viscosity /12/ has been added to the first method
when using the compatibility relations technique for the treatment of boundary
condition.

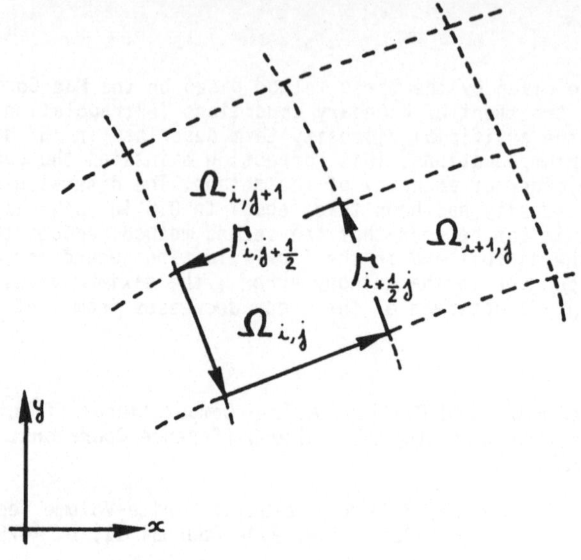

Fig. 1 - Typical mesh cell $\Omega_{i,j}$

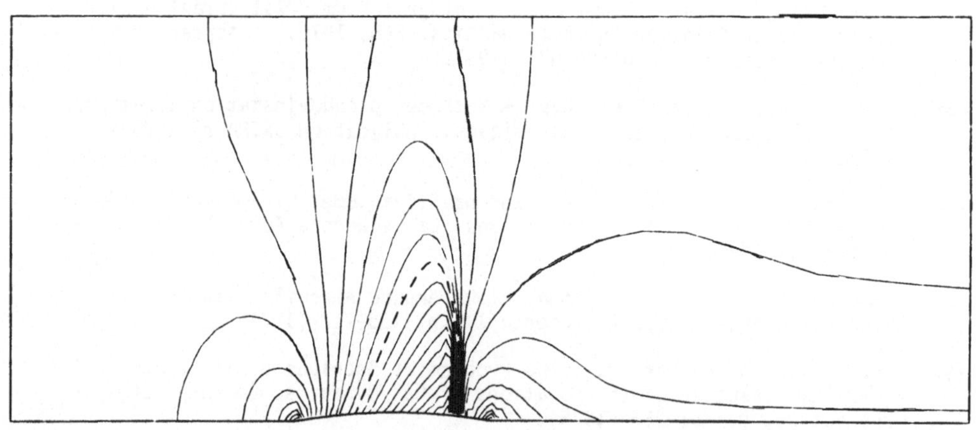

Fig. 2 - Isomach lines for the internal flow computed with the first method.

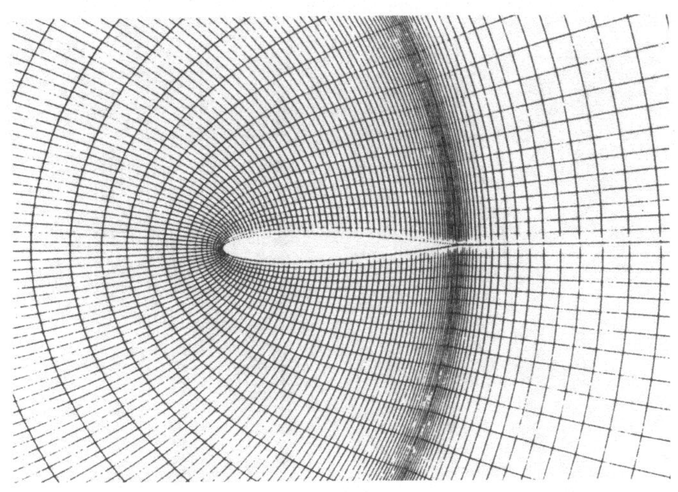

Fig. 3 - Computational mesh around the NACA 0012
airfoil (partial view)

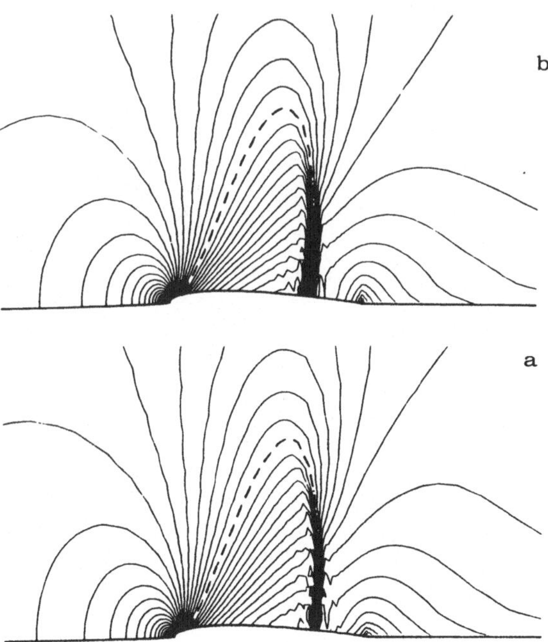

Fig. 4 - Isomach lines around the NACA 0012
airfoil at $M_\infty = 0.85$ and $\alpha = 0°$ computed
with the first method and two treatments of
boundary conditions :

a) extrapolation technique

b) compatibility relations technique

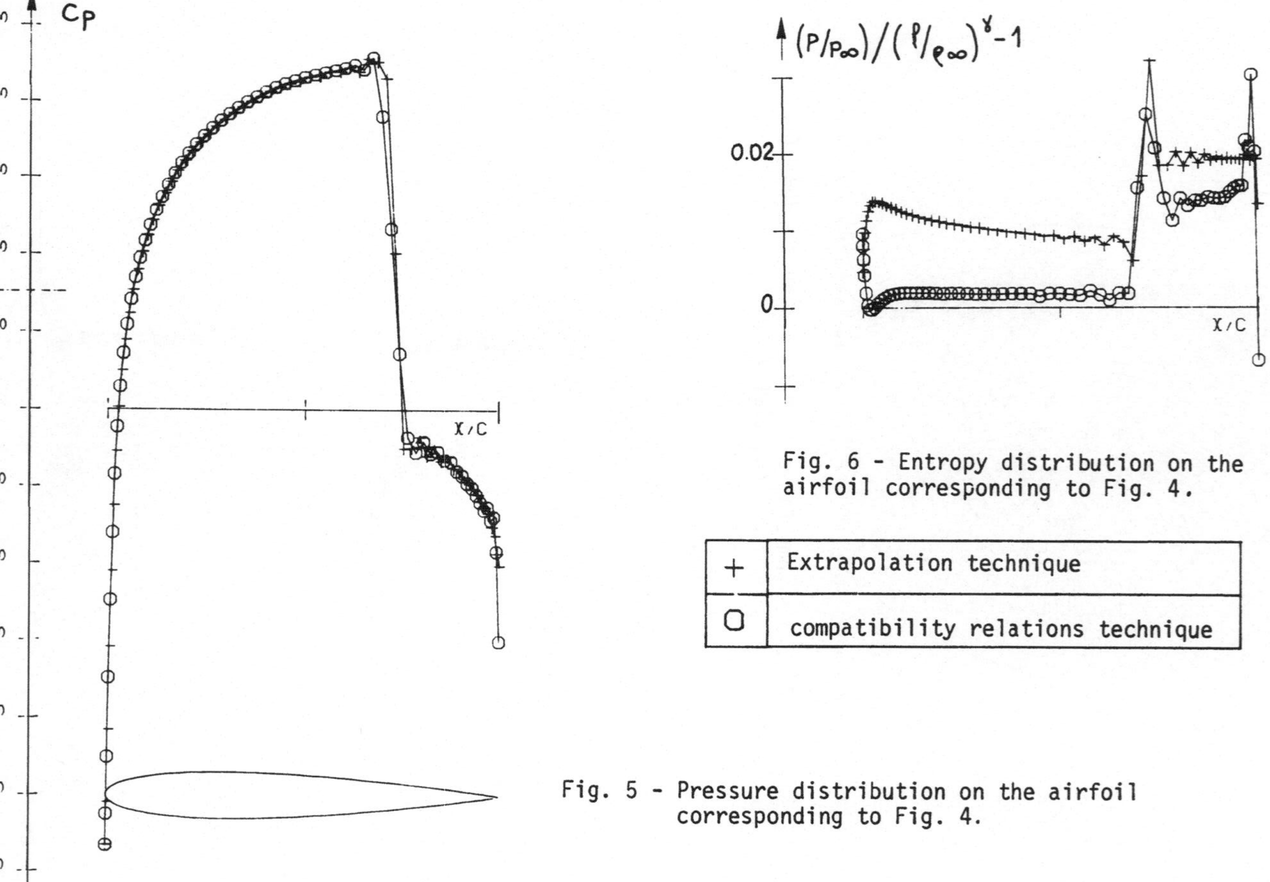

Fig. 6 - Entropy distribution on the airfoil corresponding to Fig. 4.

+	Extrapolation technique
O	compatibility relations technique

Fig. 5 - Pressure distribution on the airfoil corresponding to Fig. 4.

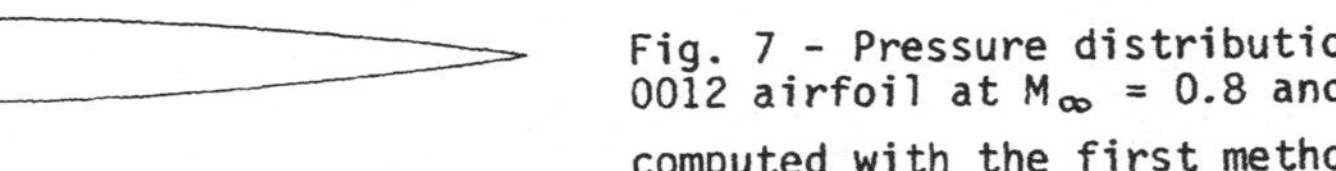

Fig. 7 - Pressure distribution on the NACA 0012 airfoil at M_∞ = 0.8 and $\propto$ = 1°25 computed with the first method.

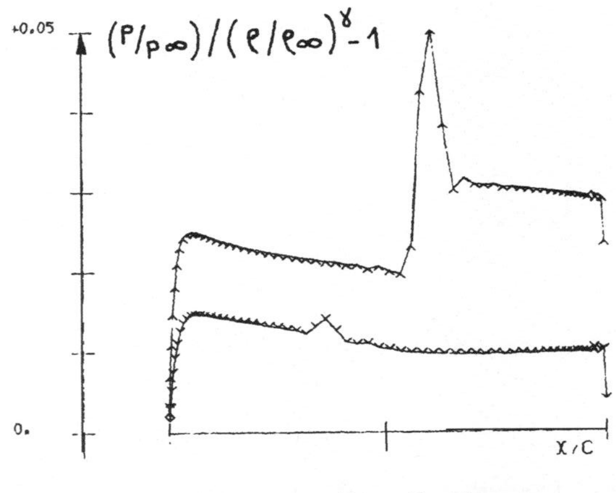

Fig. 8 - Entropy distribution corresponding to Fig. 7.

Upper side	Lower side
⌃	⌄

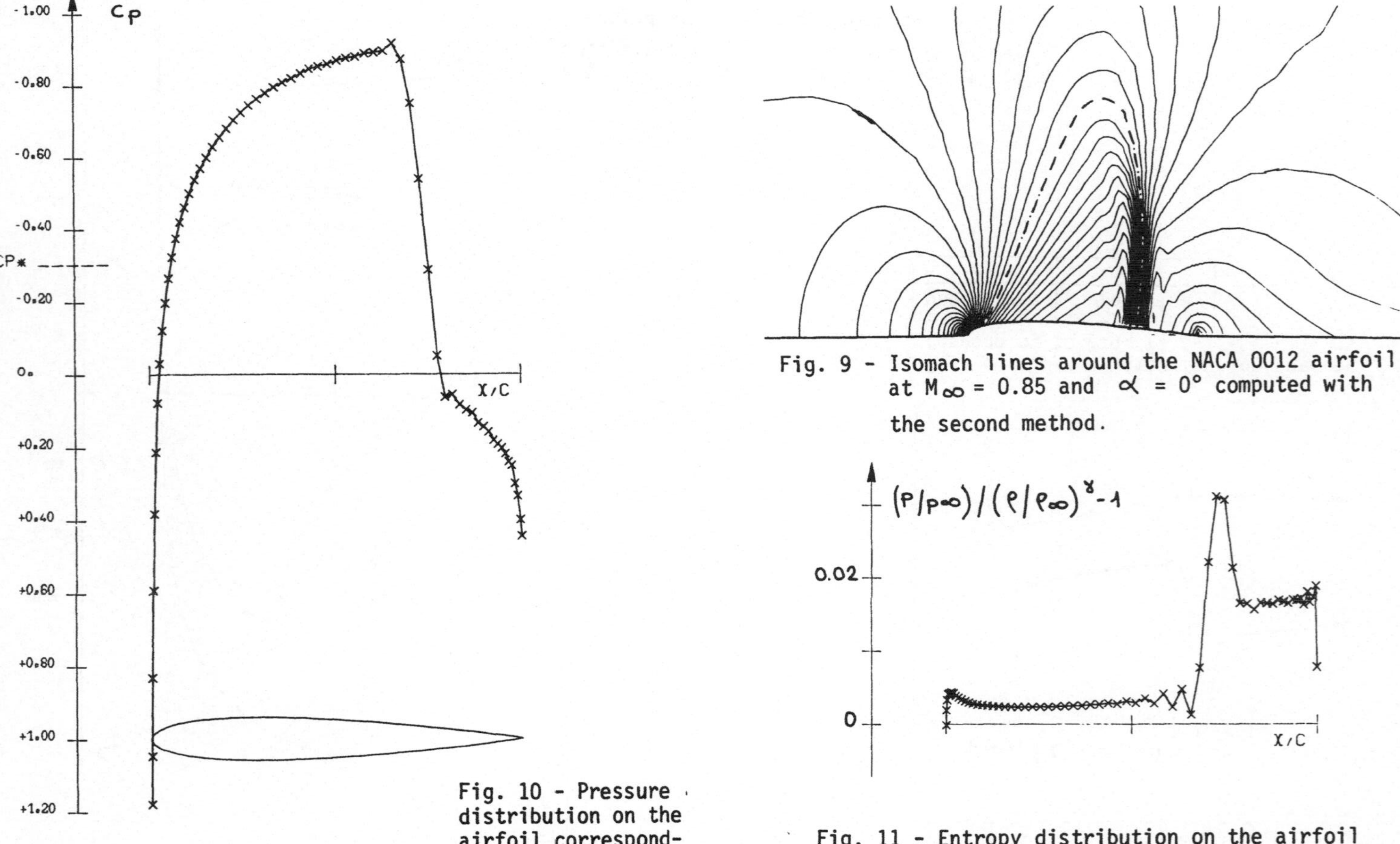

Fig. 9 - Isomach lines around the NACA 0012 airfoil at M_∞ = 0.85 and α = 0° computed with the second method.

Fig. 10 - Pressure distribution on the airfoil corresponding to Fig. 9.

Fig. 11 - Entropy distribution on the airfoil corresponding to Fig. 9.

COMPUTATION OF ROTATIONAL TRANSONIC FLOW

Arthur Rizzi

FFA The Aeronautical Research Institute of Sweden

S-161 11 Bromma, Sweden

SUMMARY

A pseudo-time dependent, split, explicit finite-volume procedure that captures discontinuities in the solution is applied
to Workshop Problem A, B, C and E. The boundary conditions
used in the method are described in detail, and special features
of the algorithm for dealing with discontinuities are discussed.
The slip line following from the trailing edge of the body is
considered a discontinuity and captured. No special treatment
is employed at the trailing edge. An attempt to condition the
underlying matrix problem in order to influence favorably the
convergence process produced unsatisfactory results even though
the calculation was stable. A solution to Problem B, for which
the shock wave was fitted, compares reasonably well with the one
in which the shock was captured. Some computational particulars
are given for the calculations.

COMPUTATIONAL METHOD

The method used to compute my solutions to the Workshop's Test
Problem A, B, C, E and F is a so-called finite-volume or flux
procedure which has been developed over several years[1-3]. In
its current state it is a pseudo-time dependent procedure in that
the steady state is attained asymptotically whereas the steady
energy equation has been integrated to give a relation between
pressure p, density ρ and velocity V, and is thus removed from
the system.

In strong conservation form the equations (Euler) governing inviscid rotational flow are

$$\frac{\partial}{\partial t} \int U \, dvol + \oint_S \vec{H} \cdot \hat{n} \, ds = 0 \tag{1}$$

where $U = \begin{bmatrix} \rho \\ \rho u \\ \rho v \end{bmatrix}$ has the elements of mass and Cartesian components of momentum and $\vec{H}(U) = \begin{bmatrix} U \vec{V} + p(\hat{i}x \atop \hat{i}y) \end{bmatrix}$ is the vector flux of U which is transported across cell surfaces S whose normal is $\hat{n}$. No ambiguity exists in this representation of the continuous problem. However, the same does not hold for the discrete approximation of Eg. (1).

The computational domain is first discretized using the standard mesh recommended for the Workshop which defines small computational cells upon which Eq. (1) is balanced. One plausible interpretation of Eg. (1) then is that the discrete dependent variables U_{ij} are volumetric averages located at the center of the cells (solid dots in Fig. 1) and that the corresponding flux H_{ij} is evaluated at the cell surfaces S_i and S_j. In this way, if the cells are lined up along boundaries, boundary conditions apply only to H and not U. The functional dependence of H on U defines the specific integration scheme. In my case the discrete solution is carried out using the time-split Mac Cormack[1-3] scheme where $H_{ij} = H(U_{ij})$ for the forward predictor and $H_{ij} = H(U_{i-i,j})$ for the backward corrector in the i-direction.

BOUNDARY CONDITIONS

Solving Eq. (1) in the interior of the domain is the least difficult part of the computation. There the difference scheme can be applied straight-forwardly, and the predictions of Fourier stability analysis seem to bear up even for this nonlinear problem. What happens at or near boundaries is another matter.

Due to the distinction made in the finite-volume approach between the location of the flow properties U and that of the flux H, the boundary conditions are somewhat unusual. For example, U is considered an interior property, and no conditions are given for it. Boundary conditions are applied only on H and enter either the predictor or corrector stage depending upon their backward or forward character. The conditions used are summarized in Figure 2.

At the upstream inflow boundary Ω_1 all three flux components in the forward predictor are set to their freestream values $H(U_\infty)$ even though one characteristic of the governing equation points out of the domain. Although this can cause undesirable reflections of waves, the corrector step does provide some adjustment, and in the computed examples reasonable results are achieved. Downstream at Ω_2 where it exits, the flow in the wake has been disturbed by the body and freestream conditions do not apply. Instead I set $p = p_\infty$ to account for the characteristic pointing into the domain. The other flow properties require extra conditions which I take to be

$$\hat{v} \cdot \text{grad} \left(\tfrac{\rho}{V}\right) = 0 \quad \text{and} \quad \hat{v} \times \vec{V} = 0$$

where $\hat{v}$ is the direction of the freestream.

On the body the physical conditions $\vec{V} \cdot \hat{n} = 0$ means that there H depends only on p for which we must obtain a value by differencing some auxiliary equation. The one I choose to difference is the normal component of the quasilinear momentum equation[4]. With it the pressure on the body is deduced from the interior values U in a formally first-order procedure which according to Gustafsson[5] should not reduce the accuracy of the overall solution. The other properties, namely density and velocity are not needed for the computation, but are desirable information since the practical goal of the calculation is to find the effect of the flow on the body. I determine the velocity components u and v by extrapolating the magnitude of the interior velocity vectors to the body and apportion u and v so that the tangency condition is met. The density then is obtained from the relation expressing the constancy of total

enthalpy. Since the full system of equations is not being
solved on the body, this extrapolation procedure no doubt gives
rise to the deviations in entropy which my computations display.

The wake of course is a viscous phenomenon which in inviscid
theory can only be represented as an infinitesimally thin contact discontinuity, i.e. a slip stream. Since my approach is
designed to capture shock waves, I treat the contact discontinuity in the same manner, namely to difference all flow properties across both shock and slip stream in the usual way without special treatment and to designate those regions of steep
but continuous gradients as approximations to the expected discontinuities.

SPECIAL FEATURES

The presence of shock waves and the wake in transonic flow present perhaps the most demanding challenge to any computational
method. One reason for this is that the method must obtain the
weak solution to the governing equations which may depend very
much on the form of the equations[6] as well as the method itself.
Techniques specifically designed for the calculation of weak
solutions and the capturing of discontinuities in computational
flow fields are numerous[7]. For controlling the nonlinear instabilities and spurious oscillations induced by flow discontinuities I use a type of artificial dissipation which is added
to the difference scheme in two ways depending on the local
gradients of pressure and velocity.

Flux-corrected expansion

Mac Cormack[8] has observed that in regions where the flow is ex-
panding and the velocity changes sign the nonlinearity of the
convective term $U \vec{V} \cdot \hat{n}$ in H causes the loss of information
about the sign of this term which can lead to a nonphysical increase of negative momentum offset by an equal increase of posi-

tive momentum. One possible correction to this feature is to
replace in the i-direction operator the evaluation of the convective part of the predicted flux $U_{ij}(\vec{V} \cdot \hat{n})_{ij}$
by $\frac{1}{2} U_{ij} (\vec{V}_{ij} + V_{i-ij}) \cdot \hat{n}$ and the corrected

flux $U_{i-1j}(\vec{V} \cdot \hat{n})_{i-1j}$ by $\frac{1}{2} U_{i-1j}(\vec{V}_{ij} + \vec{V}_{i-1j}) \cdot \hat{n}$

if $(\vec{V}_{ij} - \vec{V}_{i-1j}) \cdot \hat{n} > 0$ together with a similar replacement
in the j-direction operator. Its effect is diffusive but helps
to restore some of the lost information.

Switched Shuman filter

Across a shock wave gradients of U are large and the computed
solution usually is nonmonotonic which can excite nonlinear un-
stabilities in the calculation. In order to control these Mac
Cormack[9] follows the concept although not the exact procedure
of Harten and Zwas[10] and introduces artificial dissipation into
the difference scheme in the form of a switched filter, the idea
being to add it selectively to the regions where the flow is not
smooth. In my case the original flux terms in the corrector
step of the i-direction operator is replaced by

$$\tilde{H}_{ij} \leftarrow \tilde{H}_{ij} + \beta_{ij}\,\theta_{ij}\,(U_{ij} - U_{i-1j}) \tag{2}$$

where $\theta_{ij} = \left| P_{i+1j} - 2P_{ij} + P_{i-1j} \right| / (P_{i+1j} + 2P_{ij} + P_{i-1j})$

is a normalized switch that is $0(1)$ in regions of large pressure gradients and $0(\Delta x^2)$ elsewhere. The quantity $\beta_{ij} =$
$= 1/4 \, (\left| \vec{V} \cdot \hat{n} \right| + a)_{ij} / \Delta t$ is a dimensional coefficient necess-
are for the consistency of units. The overall effect is an eddy-
viscosity-like term added to the difference equations which remain second-order accurate where the flow is smooth but are reduced to first order elsewhere.

The round symbols in Fig. 3 are the values of C_p for the
NACA 0012 airfoil calculated with the method incorporating these
special features. Ahead of the shock the variation is smooth

and monotonic but downstream of it rather severe oscillations
develop. This I attribute to the fact that the artificial dissipation has been added so far only to points differenced in
the interior of the domain. The pressure on the body, which is
used to calculate the C_p we see here, derives from a one-
sided difference of the momentum equation in quasi conservative
form, a process that could cause the oscillations. My remedy,
consistent with the computation in the interior, is to apply
the Shuman filter operator (2) to the pressure obtained on the
body, but with $\theta = 1$ so that it is one order higher than the
body pressure calculation. The triangular symbols are the result of this calculation; virtually unchanged ahead of the shock
except at the stagnation point, but smoother behind it. It is
interesting to note that the small re-expansion just behind the
shock, bearing some resemblance to the Oswatitsch-Zierep singularity, is in any case a low frequency phenomenon since it passes
through the filter.

Matrix conditioning

To determine when the steady state is reached in a time-dependent calculation is not a trivial matter. Being only slowly
dissipated by the difference scheme, transient waves travel
across the domain, and at boundaries reflect back into the flow
field until dissipation ultimately removes them. Where the
local flow speed is just sonic, their speed of propagation is
extremely slow. Perhaps the best indication of the absence of
transients is a check on the residues of U, that is the mag-
nitude of $\partial / \partial t \int U \, dvol$ which gives an indication of the balance of the steady part of Eg. (1).

Because of the slow decay of transients, particularly those of
long wavelength, it is desirable to advance the computation as
far forward in time, and with as large a time step, as possible.
The concept of matrix conditioning can be an economical way of
doing this. Consider a discrete representation of the steady
problem $\Lambda U = B$ whose solution U is to be obtained iterat-

ively by some pseudo-time integration and B are the boundary
conditions. Matrix conditioning consists of pre-multiplication
of this equation by some matrix M which should approximate
the inverse of Λ. The idea is that since pre-multiplication
by an exact inverse would provide the desired solution without
any iteration, multiplication by an approximate inverse should
produce a new system requiring less iteration. In the simplest
case if M is taken as the identity matrix, the original time
integration is recovered. Another simple choice for M is a
diagonal matrix whose entries are the spatially varying local
time steps. The effect of this pre-multiplication defines a
new time integration but leaves the character of the equations
unchanged. It scales the eigen-values of Λ so that each computational cell is advanced with a CFL number of unity. The
integration is not consistent of course, and the stability and
accuracy of the results are difficult to predict. And the situation is complicated further by the splitting process. Figure 4
presents one example in which the solution in Fig. 3 (triangles)
has been iterated an additional 2000 cycles using the local time
step. Although stable, the wavy character of the body pressure,
which I could not eliminate, is not very accurate. The technique requires further development.

Shock fitting

Because of the simple character of the mesh suggested for Problem B, the implementation of a shock-fitting procedure is not
too difficult a matter. The computer program of Ref. 1, which
in addition to fitting the shock wave also integrates the time
dependent energy equation, has been adapted to Problem B. In
Fig. 5 the results are compared with those obtained by the present shock-capturing procedure. That the captured shock is
virtually coincident with the fitted one is reassuring, but the
reason for the difference in pressure levels ahead of the shock
is not clearly understood. Nor do I have a good explanation
for the slight discepancies upstream and downstream of the bump.

PARTICULARS OF THE COMPUTATIONS

For each of the test problems the freestream flow was used as
the initial conditions for the calculations which began on a
mesh that was twice as coarse in each coordinate direction as
the Standard Workshop Mesh. The coarse solution was then interpolated to the (fine) Standard Mesh and the computation pro-
ceeded to the final results. The advancement in time of the
computation, the number of iterations, the mean and maximum
residues, and the expended Cyber 175 computer time are listed
in Table 1.

REFERENCES

1.   Rizzi, A.W.: Transonic Solutions of the Euler Equations
      by the Finite Volume Method. In Symposium Transsonicum II,
      eds. Oswatitsch, K. and Rues, D., Springer-Verlag, Berlin,
      1976, pp. 567-574.

2.   Rizzi, A.W.; and Schmidt, W.: Finite Volume Method for
      Rotational Transonic Flow Problems. Proceedings 2nd Gamm-
      Conference on Numerical Methods in Fluid Mechanics, eds.
      E.H. Hirschel and W. Geller, GAMM, Köln, 1977, pp. 152-161.

3.   Rizzi, A.W.; and Schmidt, W.: Study of Pitot-Type Super-
      sonic Inlet-Flowfields Using the Finite-Volume Approach.
      AIAA Paper No. 78-1115, 1978.

4.   Rizzi, A.W.: Numerical Implementation of Solid-Body Bound-
      ary Conditions for the Euler Equations. ZAMM, Vol. 58, 1978,
      pp. T301-T304.

5.   Gustafsson, B.: The Convergence Rate for Difference Approxi-
      mations to Mixed Initial Boundary Value Problems. Math Comp.
      v. 29, April 1975, pp. 396-406.

6.   Courant, R.; and Hilbert, D.: Methods of Mathematical Physics
      Vol II, Interscience, 1962, p. 490.

7.   Sod, G.: A Survey of Several Finite Difference Methods for
      Systems of Nonlinear Hyperbolic Conservation Laws, J. Comp.
      Phys., v. 27, 1978, pp. 1-31.

8.   Mac Cormack, R.W.; and Panllay, A.J.:   The Influence of the
      Computational Mesh on Accuracy for Initial Value Problems
      with Discontinuous or Non-unique Solutions, Computers &
      Fluids, v. 2, 1974, pp. 339-361.

9.   Mac Cormack, R.W.; and Baldwin, B.S.:   A Numerical Method
      for Solving the Navier - Stokes Equations with Application
      to Stock - Boundary Layer Interactions, AIAA Paper No. 75-1,
      Pasadera, 1975.

10.   Harten, A.; and Zwas, G.:   Switched Numerical Shuman Filters
      for Shock Calculations, J. Eng. Math. , v. 6, April 1972,
      pp. 207-216.

TEST PROBLEM		TIME (chords/V_∞)	ITERATIONS		RESIDUES		CYBER 175 cpu TIME (Sec.)
			Coarse mesh	Fine mesh	rms	max.	
A	I	i	58.5	3000	5100	.052	.46	858
		ii	76.3	2000	7800	.010	.22	2466
	II	i	62.7	2000	5900	.020	.29	938
		ii	68.6	2000	6900	.007	.12	1083
		iii	64.7	–	5100	.021	.34	874
		iv	55.8	2000	4900	.006	.088	1610
		v	56.7	2000	5000	.004	.028	1639
B			41.1	–	4000	.003	.026	476
C	I		65.3	2000	6900	.027	.39	2207
	II		87.6	4000	7500	.033	.56	2530

Table 1. Computational particulars.

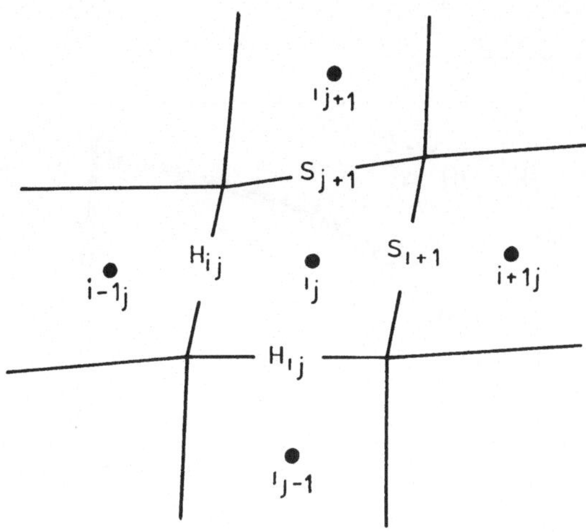

Fig. 1. Flow properties U located in
center of computational cell.

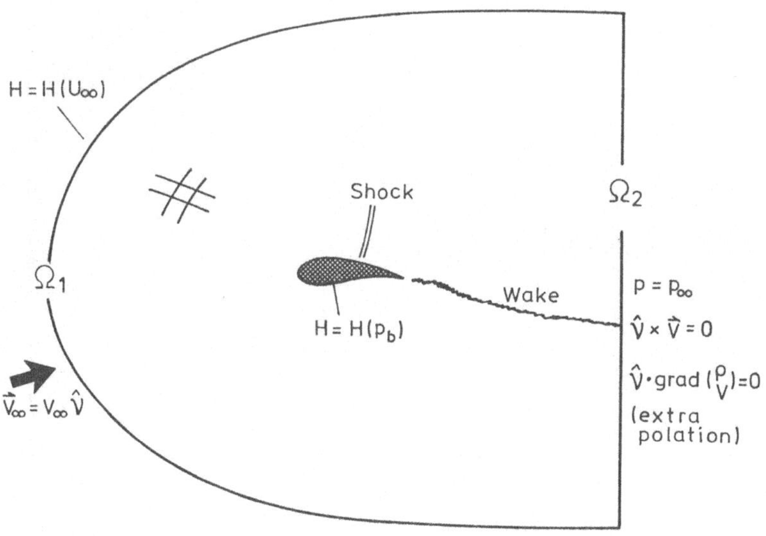

Fig. 2. Boundary conditions on flux H.

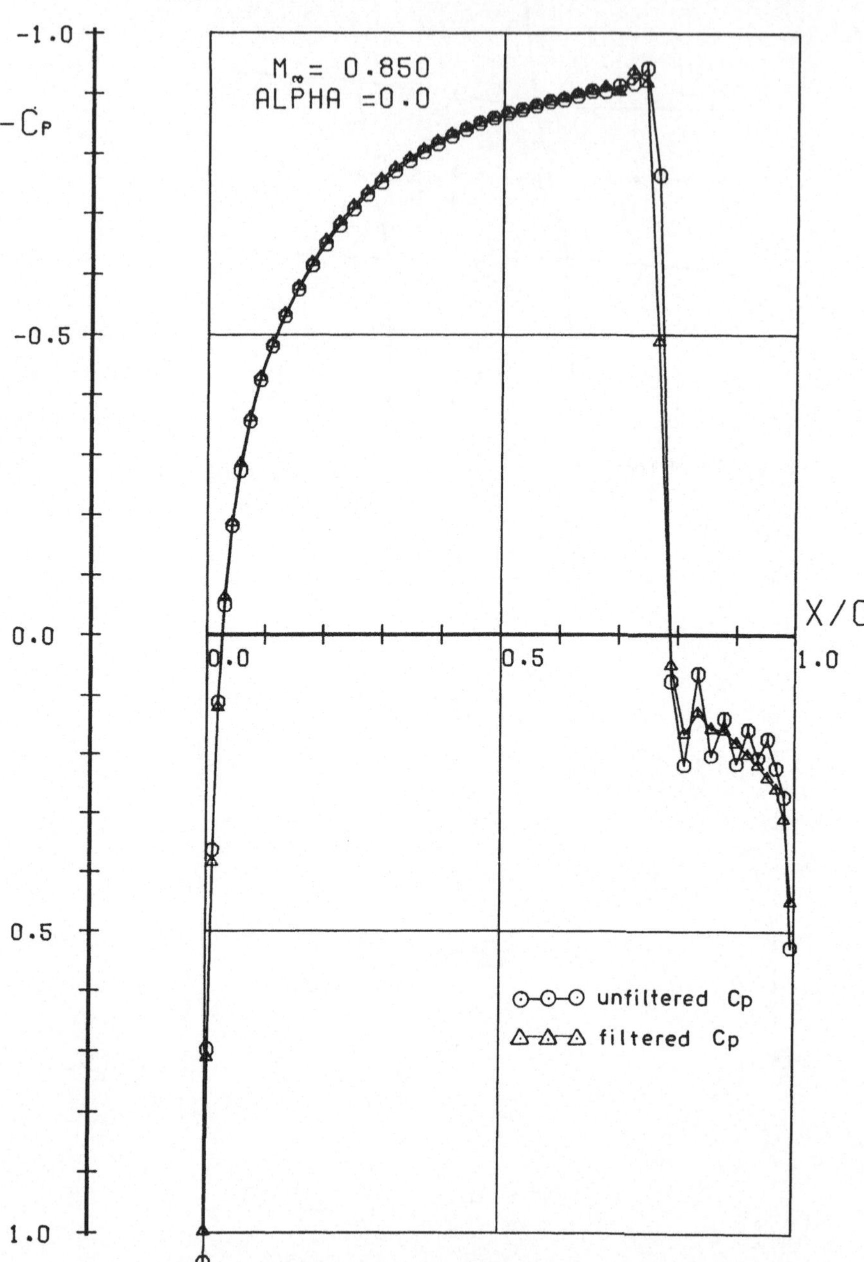

Fig. 3. The effect of applying the Shuman filter to C_p on the airfoil, i.e. the boundary value. $M_\infty = 0.85$ and $\alpha = 0$.

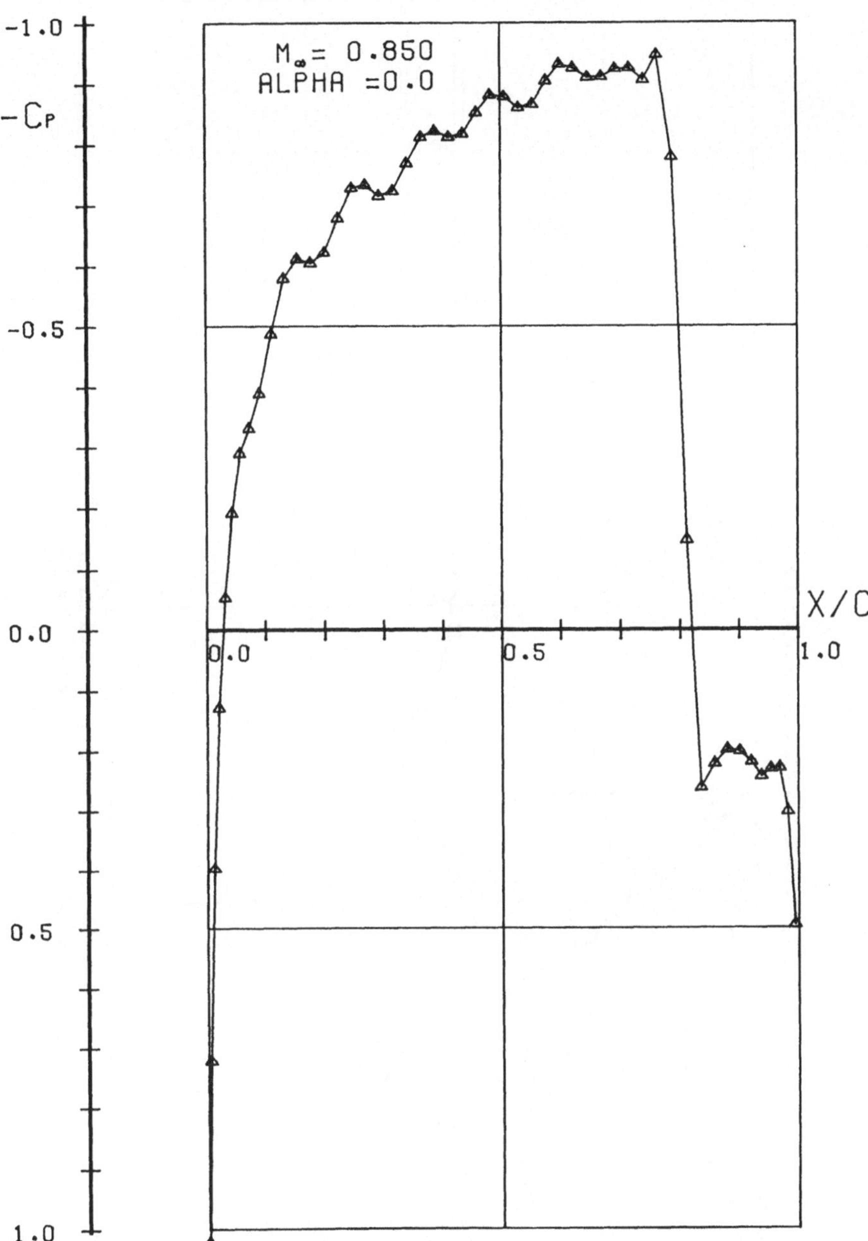

Fig. 4. The solution after 2000 iterations using a local-time-step form of matrix conditioning. $M_\infty = 0.85$ and $\alpha = 0$.

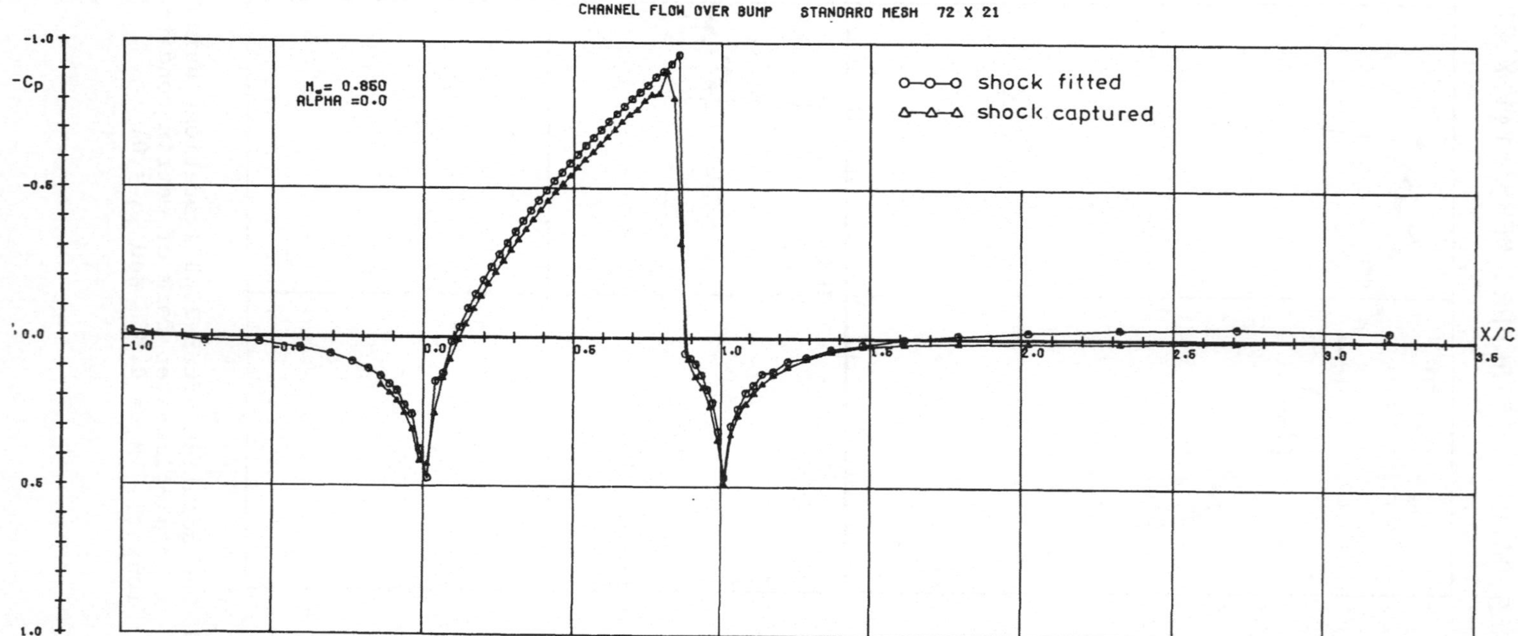

Fig. 5. Comparison between a shock-fitted solution
and a shock-captured solution for flow at
$M_\infty = 0.85$ over a bump in a channel (Prob. B).

COLLECTIVE COMPARISON OF THE SOLUTIONS TO THE WORKSHOP PROBLEMS

Arthur Rizzi
FFA   The Aeronautical Research
      Institute of Sweden
S-161 11 BROMMA, Sweden

&

Henri Viviand
ONERA
F-92320 CHATILLON, France

INTRODUCTION

It is not possible here to discuss in complete detail the
various computational methods used to solve the Workshop Problems
and all their results, so we must restrict ourselves to more
general comments and observations. We begin by classifying the
various major methods (to which we give the author's name) into
three basic types according to the underlying mathematical model:

a) fully-conservative potential methods
b) non-conservative potential methods
c) fully-conservative Euler methods

plus two others which must be distinguished as subtypes:

d) one fully-conservative transonic small perturbation
   (TSP) method
e) one isentropic Euler method.

The detailed grouping is given in Table 1. Because of the large
number of solutions received we cannot display and discuss them
all here. They are presented without discussion, however, in an
FFA Technical Note which can be obtained upon request to the FFA
library. Here we restrict ourselves to the results from Test
Problems A and C since they adequately illustrate all the conclusions that we can draw.

We have not made any comparisons of entire flowfield solutions
because of their unwieldiness, instead we focussed our attention
on the pressure distributions computed on the airfoil, the most
interesting aerodynamic quantity and probably the most sensitive
one from the computational viewpoint since it is obtained at a
boundary where special procedures usually are used and where
flow gradients are largest. Thus we condense an entire flowfield

AUTHOR/METHOD	CLASSIFICATION
1. JAMESON 2. HOLST 3. CHATTOT-COULOMBEIX 4. VEUILLOT-VIVIAND	Fully conservative potential solutions (FCPOT)
5. SCHMIDT	TSP solution
6. HIRSCH 7. EBERLE 8. BAKER 9. CARLSON 10. LOCK	Non-conservative potential solutions (NCPOT)
11. ZANNETTI	Isentropic solution
12. SELLS 13. SIDES-LERAT 14. VEUILLOT-VIVIAND 15. RIZZI	Fully conservative solutions to Euler equations (EULER)

Table 1. Classification of the major methods that
were applied to the Test Problems.

to a single plot of coefficient of pressure C_p versus X.
It is helpful to go even one step further and look at integrals
of these plots, the familiar lift and drag coefficients C_L
and C_D. We then have two numbers to represent an entire solution, which can give us a general overview of how the various
solutions compare. Figures 1 and 2 show, respectively, the
general trend of the disparity between C_D and C_L computed
for Problem A by the various methods, a disparity much larger
than generally expected by the participants. It can be attribut-
ed to two possible origins:

i) differences in the various mathematical models used
ii) discretization errors in a broad sense, that is, the
difference between the numerical solution and the
theoretical solution of the mathematical model used,
assuming that the latter is unique.

The latter accounts for all the particular details of a given
method like the finite differences used, its treatment of boundary conditions and trailing-edge condition, and the extent of
the mesh and its spacing. We illustrate item (i) in Figs. 1
and 2 by a different gray shading to distinguish between the

solutions given by the three different mathematical models.
Differences between solutions obtained for the same mathematical
model, i.e., within each gray band, presumably are due to the
discretization error.

For subcritical flows (cases M_∞ = .72 and .63), the theoretical
solution is identical for all the mathematical models (except
subtype d), and indeed the shaded bands merge together indicat-
ing that the differences between the corresponding groups of
numerical solutions as well as the differences between solutions
in the same group are relatively small. And, although small,
these differences can be taken as a general measure of the overall discretization error associated with the least accurate of
all the solutions presented. We adopt this as our rule-of-thumb
for the highest degree of accuracy that we can expect from the
methods as they were presented at the Workshop.

For supercritical flows, on the contrary, we observe large
differences between the three groups, and these differences become increasingly marked as the flow conditions M_∞ and α
increase. The difference between solutions within each group
(the length of each band) also grows larger with increasing flow
conditions, and, it is interesting to observe, this growth is
largest for the conservative potential methods and smallest for
the Euler methods. The one mildly transonic flow case,
M_∞ = .80 and $\alpha = 0^\circ$, displays reasonable agreement for all
solutions and marks a transition from the subcritical to the
strongly supercritical cases. For these latter cases the assumption of potential flow is clearly violated, and the tendency for
the nonconservative potential solutions to follow the Euler solutions as closely as they do is unexpected.

THEORETICAL SOURCES OF THE DISAGREEMENT

We have just seen how the appearance of a shock wave in the solutions marks the onset of divergence between the three mathematical models and initiates as well an increase in discretization
error for all methods. The reader should bear in mind the

following theoretical considerations that may underlie these
effects.

Mathematical model

Discontinuous solutions (or "weak" solutions) cannot be defined
theoretically for the non-conservative equations, that is to say
without explicitly specifying the jump conditions independently
of the equations, whereas for a conservative system the jump conditions result directly from the conservative form of the equations. The discontinuities in the Euler solutions are of two
kinds: either shock waves which satisfy the familiar Rankine-
Hugoniot relations, or slip surfaces which admit a discontinuous
tangential velocity and density but not pressure. The discontinuities in the potential solutions (based on the continuity
equation in conservative form) are isentropic shocks which
satisfy the conservation of mass (isentropic jump conditions).
The relationship between a "Rankine-Hugoniot" shock and an isentropic shock is shown in Fig. 3 in terms of the Mach number
ahead of the shock M_1 and p_2 the pressure just downstream of
it. (Figures 7 d and 11 d show the relationship in terms of C_p
for the two cases $M_\infty = .80$ and $M_\infty = .85$.) Although small,
differences begin even for $M_1 = 1.1$. This of course is a well-
recognized difference between these two models.

Another difference, less appreciated and harder to assess, occurs
at a trailing edge and is illustrated in Fig. 4. Inviscid theory
requires a subsonic flow to stagnate at concave corner points
where the slope of the streamline is discontinuous. Furthermore
it demands that the flow properties on the streamline leaving the
upper surface and the one leaving the lower surface satisfy the
tangential discontinuity condition, the only admittable one in
this theory. Without a shock wave the potential and Euler solutions are identical. As indicated in Fig. 4 a the upper and
lower surface streamlines both pass through a stagnation point
at the trailing edge, adjoin, and leave the point along the bi-
sector of the trailing-edge angle, the flow properties remaining

continuous across this streamline downstream. When a shock is
present, the situation remains unchanged for the potential solution, but the Euler solution now registers a loss in the total
pressure p_t downstream of the shock. Either of two possibil-
ities, shown in Fig. 4 b , can take place, depending upon whether
the loss is greater on the upper or lower side. In either case
consideration of the loss in total pressure and the conditions
for a tangential discontinuity require only one streamline to
stagnate and the other to leave the trailing edge smoothly, thus
becoming a slip line downstream, a situation different from that
for potential flow. Because of the strong effect that flow near
the trailing edge has on circulation and lift, this difference
may be significant.

Discretization error

We know, of course, that for a given method the discretization
error is directly related to the mesh upon which the differencing is carried out. However, the influence of the mesh on the
solution is difficult to assess in any precise way because of
the many factors involved: mesh topology, extent of the computation domain, total number of mesh points, variation of mesh
size (in both directions) in the flow field in relation to flow
gradients... We believe that these aspects of any given mesh
play a crucial role in the accuracy of the results that can be
obtained upon it. As we shall see in the next section, the few
comparisons which can be made with results obtained using the
same method but various different meshes indicate that the solution is very sensitive to these factors.

COMPARISON OF PRESSURE DISTRIBUTIONS. NACA 0012 AIRFOIL

So far we have observed general trends in the results based on
the integrated values C_D and C_L. In this section we compare
the actual pressure distributions on the NACA 0012 airfoil
turned in by the participants. Two quantities that characterize

the nature of these distributions and which aid us in carrying
out the comparisons are the minimum value of the pressure coefficient C_{p_m} and the coordinate χ_s of the shock position,
measured from the leading edge and defined as the middle point
of the jump in the C_p curve. When oscillations in the C_p
curve obscure the selection of the minimum value, we take for
C_{p_m} an average value extrapolated from the nearest smooth section of the curve in the upstream direction. For the lifting
cases the additional subscript U or L denotes the value on
either the upper or lower surface of the airfoil.

A number of participants submitted the entire sequence of solutions that they obtained while successively refining or altering
their mesh, and we took their "best" solution when comparing the
entire C_p curves with others. But in the comparisons of C_{p_m}
and χ_s we use the small letters: a, b, c and a', b', c'...
in front of the symbols in a given column in order to refer to
results obtained by the same author/method with various meshes.
The order a, b, c corresponds to an increasing total number of
points in a mesh of the same type. Two different meshes with
the same total number of mesh points are denoted a, a' or
b, b'... . Results obtained with the workshop standard mesh are
characterized by a small oblique dash attached to the symbol: $\mathbf{\acute{c}}$.
Appendix B provides some drawings of the meshes used and information about the number of points they contain.

Subcritical flow

The "best" C_p distributions given by 15 participants for the
case $M_\infty = 0.72$ and $\alpha = 0$ are plotted on top of one another
in Fig. 5a. The greatest discrepancy occurs at the minimum
value. This disparity is better enumerated in Fig. 5b where
the minimum value C_{p_m} and drag coefficient C_D (which theoretically is zero) are plotted for each of the participants. It
also contains the additional information on the effect of various
meshes when used with the same method. A surprising feature
here, and even more striking in the cases that follow, is the
disparity in results obtained by a given method when two equally

172

dense but topologically different meshes are used. For example,
the difference between the two values of C_{p_m} given by
Veuillot's method (14) in conjunction with mesh a and a'
is practically as large as the differences between the results
from any of the other EULER methods. Similarly, the disagreement between Jameson's results for the two different meshes c
and c' is larger than expected for such extremely fine dis-
cretizations. (And the corresponding disparity in Fig. 8 b is
even more surprising.) It is worth noting, however, that when
a given mesh is successively refined, as Holst did, the value
that C_{p_m} approaches is - .66 which probably is an accurate
value.

When lift is added to the flow as for the case in Fig. 6 we
find the largest disagreement again at C_{p_m} on the upper surface, but the differences in C_{p_m} on the lower surface are no
larger than elsewhere on the airfoil. If we compare the three
basic groups, we can say that no one group displays a larger
disparity among its own members than any other group. And the
total overall disparity, the envelopes of the curves in Figs.
5 a and 6 a are an indication of the range of the discretization
errors for these 15 methods and meshes. Furthermore it is
worthwhile observing in Figs. 5 and 6 that, if we disregard
Zannetti's solution in Fig. 6 a because his mesh is unusually
coarse, there appears to be no significant difference between
the disparity of the lifting and nonlifting solutions. Evidently for all methods the errors associated with the treatment of
the trailing edge are of the same order as the errors in the
rest of the flowfield.

Supercritical flow

The mildly supercritical case (M_∞ = 0.80 and α = 0°) begins
to display in Figs. 7 and 8 increasing differences between solutions in the same group as well as between groups. When lift
is added (M_∞ = 0.80 and α = 1.25°) the differences become
striking in Figs. 9 and 10. We see here that the discrepancies

between the solutions within the NCPOT group and within the
EULER group are not too great, nor is the disparity between
these two groups very large. The values of C_L given by both
are in the range 0.29 to 0.37.   Whereas the FCPOT solutions
do not agree among themselves nor with the other two groups.
The values of C_L ranged from 0.55 to 1.1.   On the upper
surface, the shock was in some solutions near the trailing edge
and in others at the trailing edge (Holst & Jameson).

If we return to a non-lifting case, but at the higher Mach number
$M_\infty = 0.85$ Figs. 11 and 12 indicate that the degree of scatter
in the results is reduced, although the general trend of smallest
scatter among the EULER solutions, and largest among the FCPOT,
with the NCPOT results being intermediate, still remains. And
this trend holds when lift is added ($M_\infty = 0.85$ and $\alpha = 1$)
which can be seen in Figs. 13 and 14. Now for this case all
methods produce a shock on the lower surface, and on the upper
surface all of the FCPOT solutions, except Chattot's, locate the
shock at the trailing edge.

The last case for this airfoil is with the flow conditions
$M_\infty = 0.95$ and zero incidence and the comparison differs somewhat from the others. In general, all the solutions (Figs. 15
and 16) agree on the airfoil surface with an oblique shock standing at the trailing edge. Thus, for all the methods (except the
TSP method of Schmidt and the finite-element method of Eberle),
the theoretical solutions give the same flow properties on the
airfoil, so that the differences observed in the numerical solutions for the other supercritical cases are in fact not observed
for this case. But this case is also quite interesting to study
the influence of the mesh, not so much concerning the flow on
the profile as concerning the flow downstream of the trailing
edge. Indeed a fish-tail shock configuration (oblique shock at
the trailing edge and a normal shock standing some distance downstream[*]) exists. But some numerical solutions show that the
position of the normal shock depends strongly on the extent of

---

[*] It is not quite clear whether such a configuration exists in
all the potential solutions.

the computation domain[*], which is not too surprising in view of
the high value of the free-stream Mach number and of the resulting supersonic flow region which is extremely large. The exact
position of this normal shock really cannot be inferred from the
solutions presented.

COMPARISON OF PRESSURE DISTRIBUTIONS.   RAE 2822 Airfoil

In this section we compare the C_p curves given by the various
methods in order to bring out what effect the change of geometry
has on the comparison of the methods. And we find that the
effect is rather small. All of the general trends discerned
from the solutions for the classical NACA 0012 airfoil are
equally apparent in the solutions for this modern airfoil.

All the C_p curves for the subcritical case $M_\infty = 0.676$ and
$\alpha = 1^\circ$ drawn in Figs. 17 theoretically should agree, and in
fact they do to about the same degree of accuracy as the lifting
subcritical flow past the NACA 0012 (see Fig. 6) which is about
10 %. The maximum error here also occurs around the point of
C_{p_m}, but now within each group we see very close agreement
between the members of NCPOT with somewhat larger but roughly
an equal degree of disparity between the solutions within each
of the other two groups. However, when we turn to the strongly
supercritical case $(M_\infty = 0.75$ and $\alpha = 3^\circ)$ in Figs. 18, we
find a striking growth in the disparity, largest for the FCPOT
group with three results showing the shock at the trailing edge
and one (Chattot) at about 65 % of the chord. The differences
between the groups NCPOT and EULER are amazingly small for such
strong shocks and also the disparity between solutions within
each of these groups is smaller than that for FCPOT, the degree
of scatter within NCPOT being intermediate between that of FCPOT
and EULER.

---

[*] See the individual papers by Holst, Carlson, and Veuillot and
Viviand.

CONCLUSIONS

The participants in this Workshop are to be thanked for contributing such a large number of computed solutions to the test problems
we proposed. It is probably the first time in the field of computational transonic flow that so many solutions to a number of
test problems have been computed by quite different methods under
reasonably controlled conditions, and that they have been collect-
ed and made available for comparison. In fact the number of results and the two-day duration of the Workshop prohibited us from
examining in detail the entire computed flowfields, but the comparison of pressure distributions on the airfoil sufficed to
bring out the important general conclusions which we summarize
here. The reader, however, should bear in mind that many of our
observations relate to supercritical flows with not so weak, and
in some cases rather strong shock waves, for which many of the
methods presented here were not specifically designed.

The solutions group themselves according to the mathematical
models used (FCPOT, NCPOT, or EULER). Nevertheless the disagreement between solutions of the same mathematical model, as well
as the general differences between solutions of the different
models, are larger than any of us expected. We observe that, in
general, the FCPOT methods give numerical solutions which contain
larger supersonic zones and stronger shocks than the NCPOT methods
and the EULER methods, which may account for the FCPOT solutions
being more sensitive to the topology of the mesh than the other
methods. But perhaps the most striking fact is the degree of
similarity between the NCPOT and EULER solutions even for cases
with strong shock waves, whereas the difference between these
two groups of solutions and the FCPOT solutions is very substantial, especially for the lifting cases. Also the scatter
of results given by the conservative potential methods is appreciably larger than the scatter of results given by the other two
groups of methods. There appears to be a paradox in the fact
that the non-conservative potential methods, for which shock relations are not theoretically known or defined, lead to less scatter
of results than the conservative potential methods for which the
shock relations are precisely defined. Theoretical investigations

of the numerical methods are required to resolve this paradox.

The NACA 0012 airfoil at conditions $M = 0.8$, $\alpha = 1.25^\circ$ appears to be a very difficult and a very interesting case. In the FCPOT solutions, the position of the shock wave on the upper surface varies from 0.75 to 1, and the lift coefficient varies from 0.55 to 1.1, whereas in the NCPOT and the EULER solutions these quantities vary respectively from 0.5 to 0.7, and from 0.28 to 0.42. It is worth pointing out that all the potential solutions, except one, give no shock on the lower surface, whereas all the EULER solutions give one.

Uniqueness of the theoretical solution, which certainly does not exist for the NCPOT formulation, has also been questioned by Jameson for the FCPOT formulation, when the shock stands at the trailing edge. And subsequent to the Workshop he has found an example where his method can produce two distinct but bona fide numerical solutions to an identical problem (see Appendix C).

Our understanding of the influence that the choice of mesh bears on the accuracy of the resulting solution is, unfortunately, limited, but it is a topic that deserves further study. We have detected some indications of the influence of mesh topology on the accuracy of the solution, but further tests under better controls must be carried out before any precise conclusions can be reached. In fact a new workshop might well be devoted to a thorough investigation of how the specific details of the mesh affect the accuracy of the computed solution. It seems, in any case, safe to say that extremely fine meshes with computational domains extending very far away from the airfoil would be required in order to get a high degree of accuracy (say relative errors in C_{p_m} and in C_L of 1 %, and absolute error in shock position of 5 % of the chord), and such accuracy evidently cannot be obtained with the methods used here in conjunction with the meshes like the standard mesh of 141 x 21 nodal points, since even for the two subcritical cases, the scatter of the results on C_{p_m} and C_L is of the order of 10 % in relative error.

In any vital field of research today's perplexing question becomes tomorrow's obvious answer, and we trust that after further work and development the conclusions which we can only grope at here will become apparent in the near future.

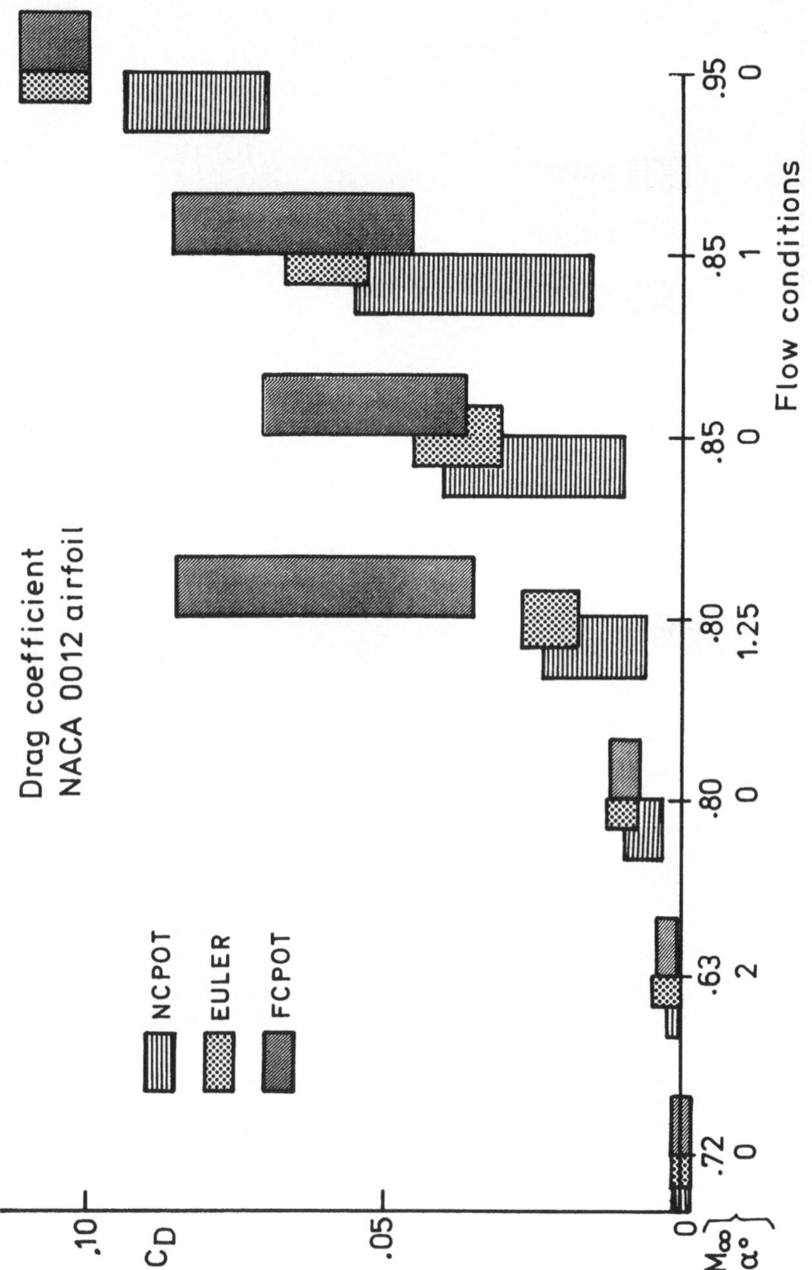

Figure 1. Overview of how the drag coefficient C_D computed by the various methods compare. Problem A.

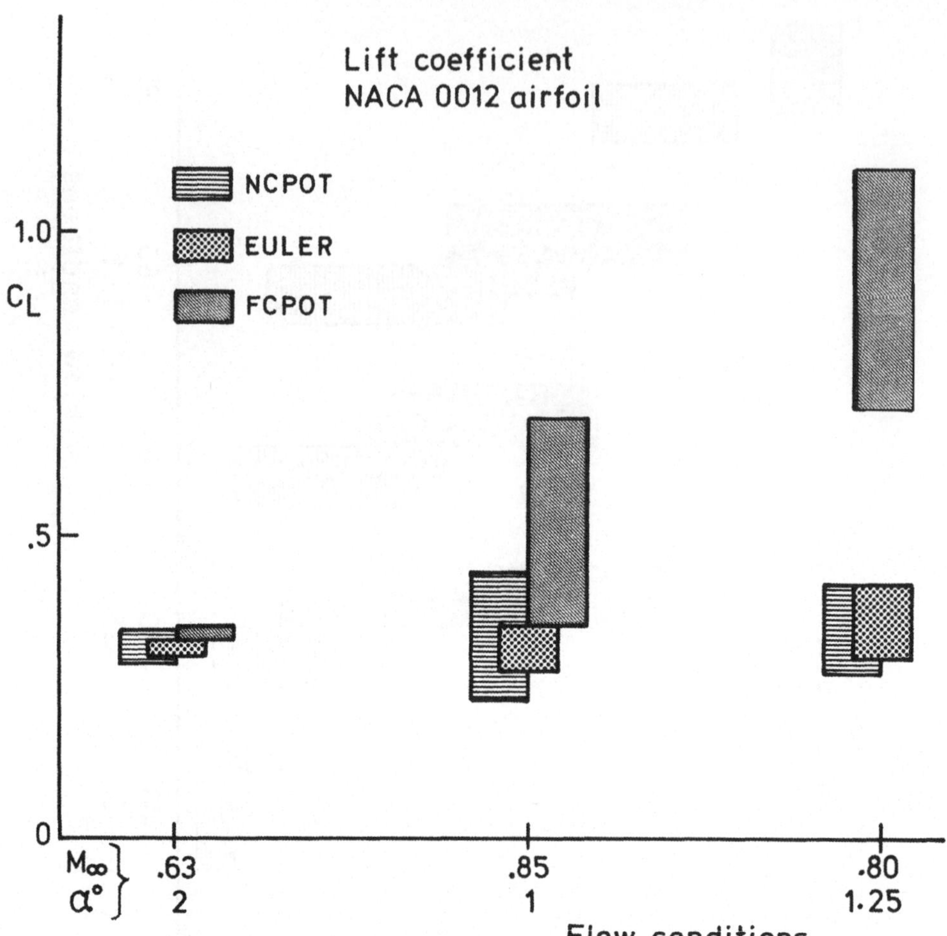

Figure 2. Overview of how the lift coefficient C_L computed by the various methods compare. Problem A.

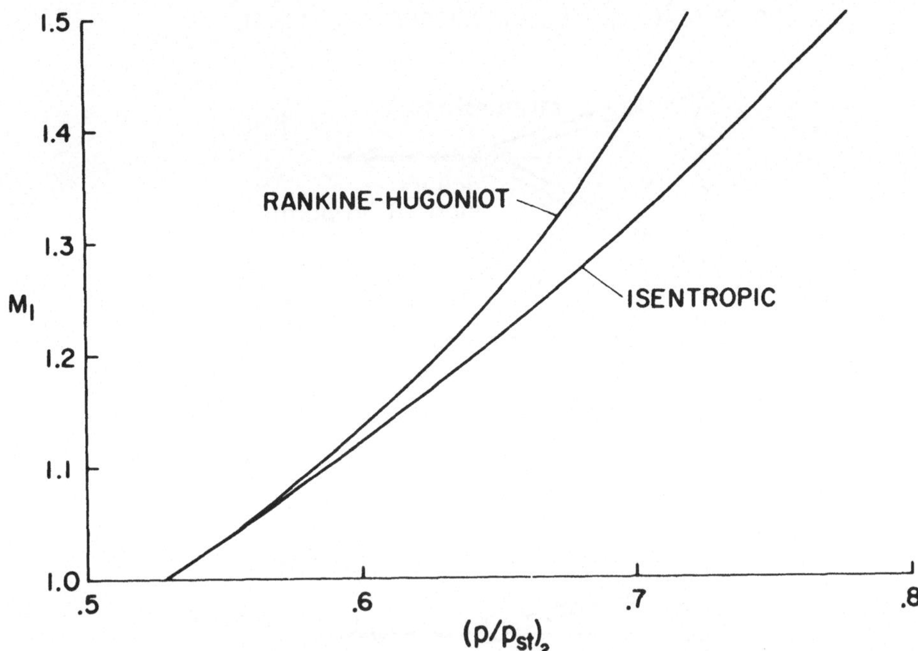

Figure 3.   Relationship between the isentropic and the Rankine-Hugoniot normal shock condition for a perfect gas.

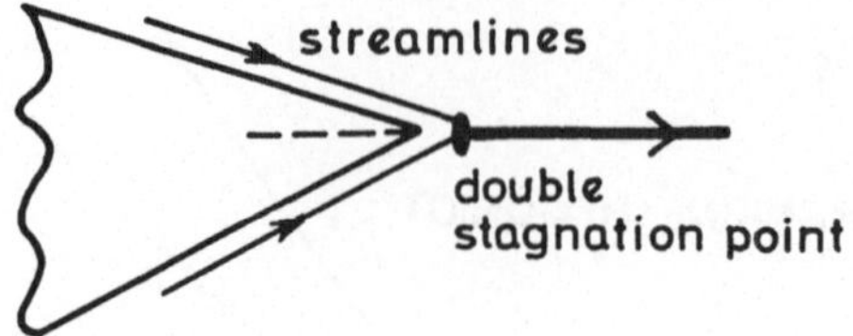

(A)   FLOW WITH NO SHOCK WAVES

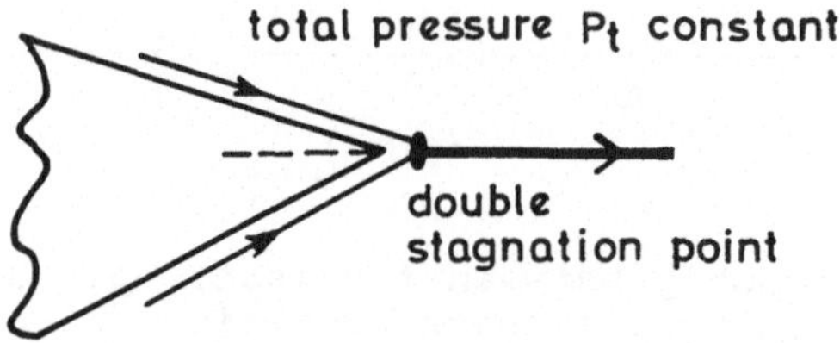

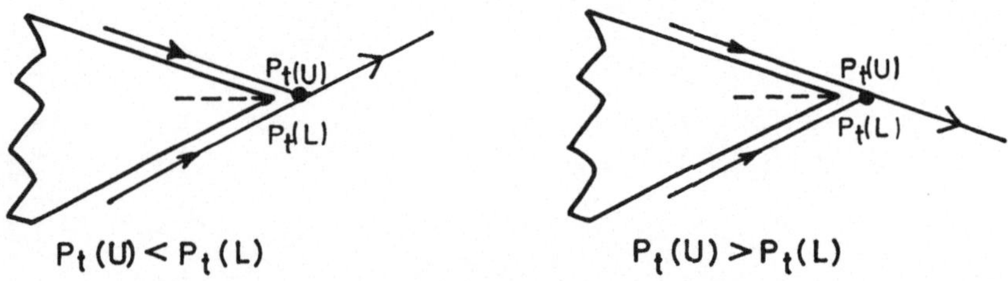

(B)   FLOW WITH SHOCK WAVES

Figure 4.   Similarity and difference between the potential model and the Euler equations locally at the trailing edge of an airfoil.

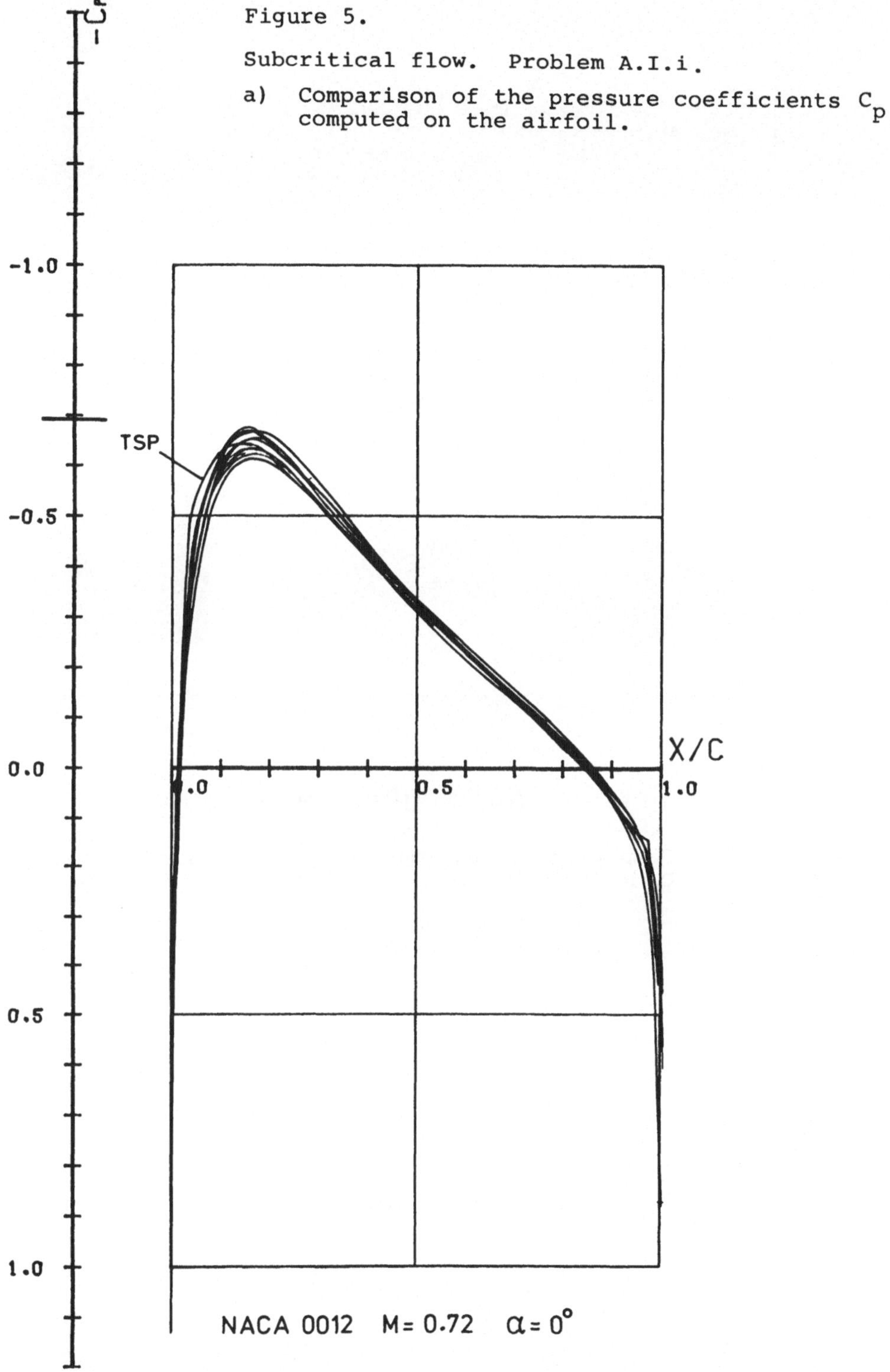Figure 5.
Subcritical flow.   Problem A.I.i.
a)   Comparison of the pressure coefficients C_p computed on the airfoil.
C_p
-1.0
-0.5
0.0
0.5
1.0
TSP
X/C
.0
0.5
1.0
NACA 0012   M= 0.72 $\alpha = 0°$

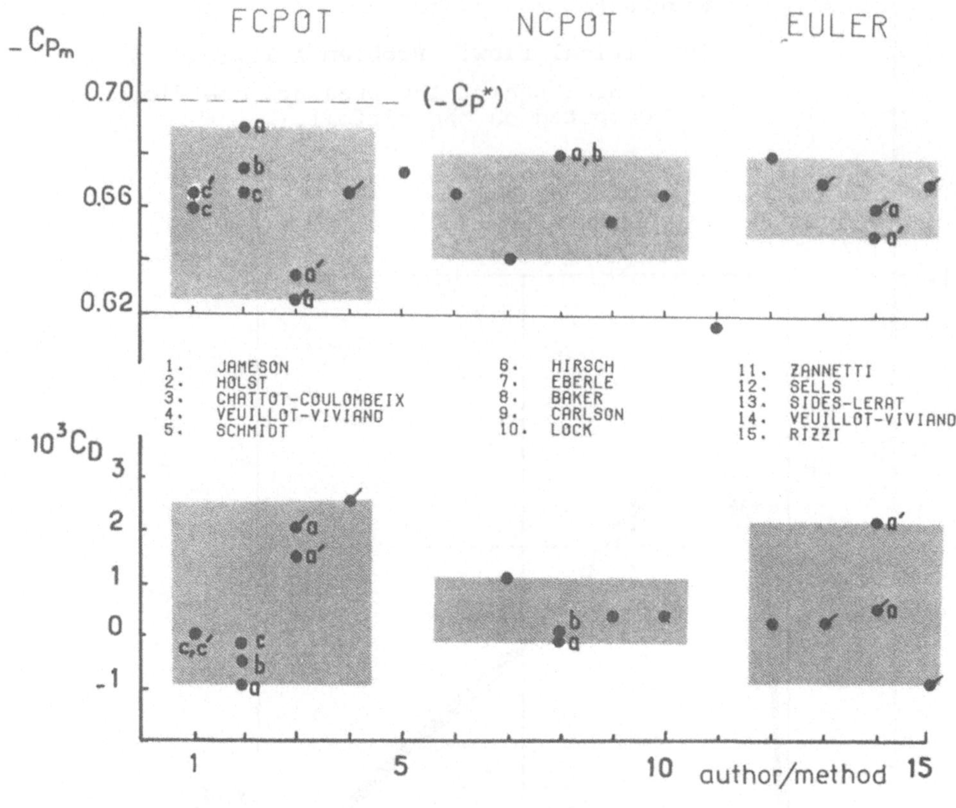

NACA 0012 M = 0.72 $\alpha = 0°$

Figure 5.   b)   Comparison of minimum C_p and drag coefficient C_D.

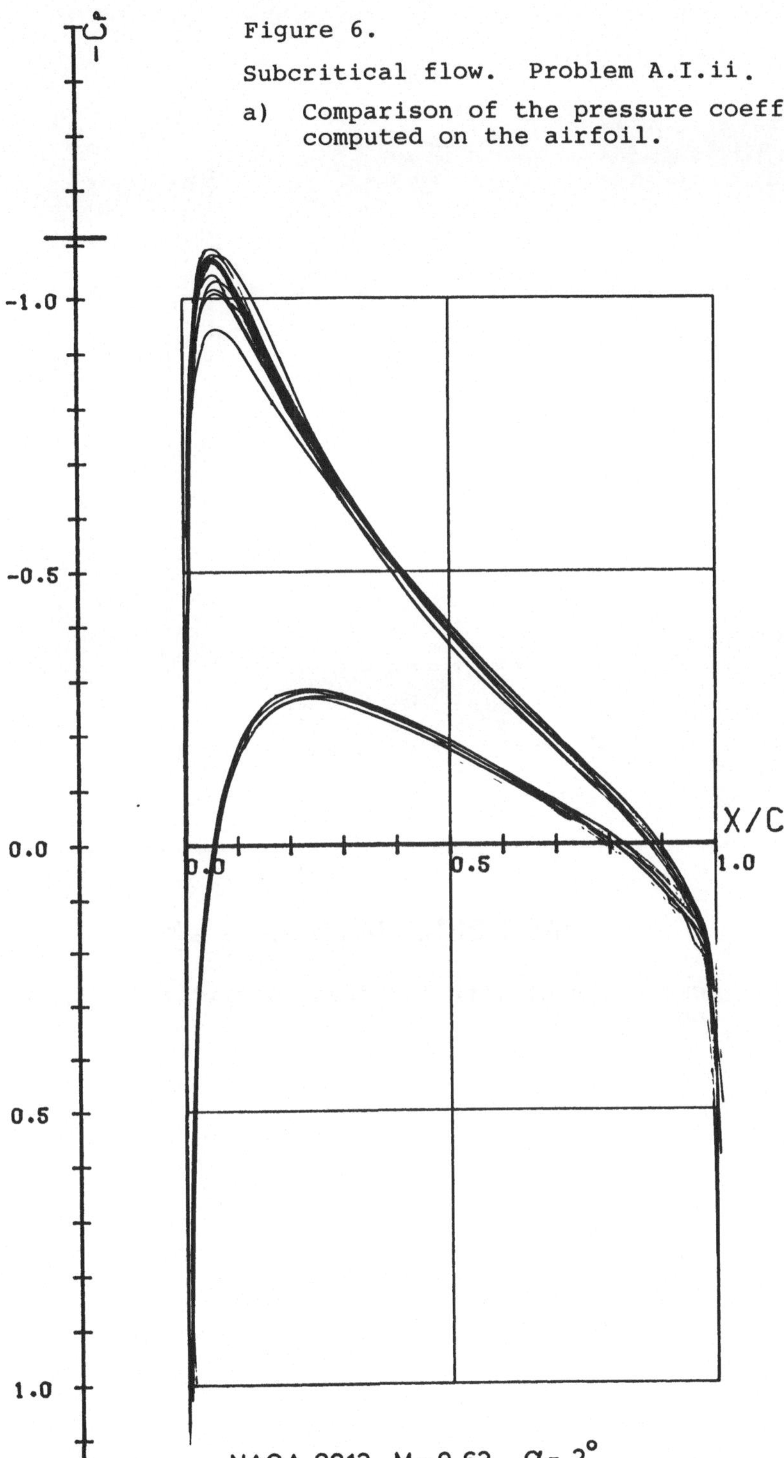

Figure 6.

Subcritical flow. Problem A.I.ii.

a) Comparison of the pressure coefficients C_p
   computed on the airfoil.

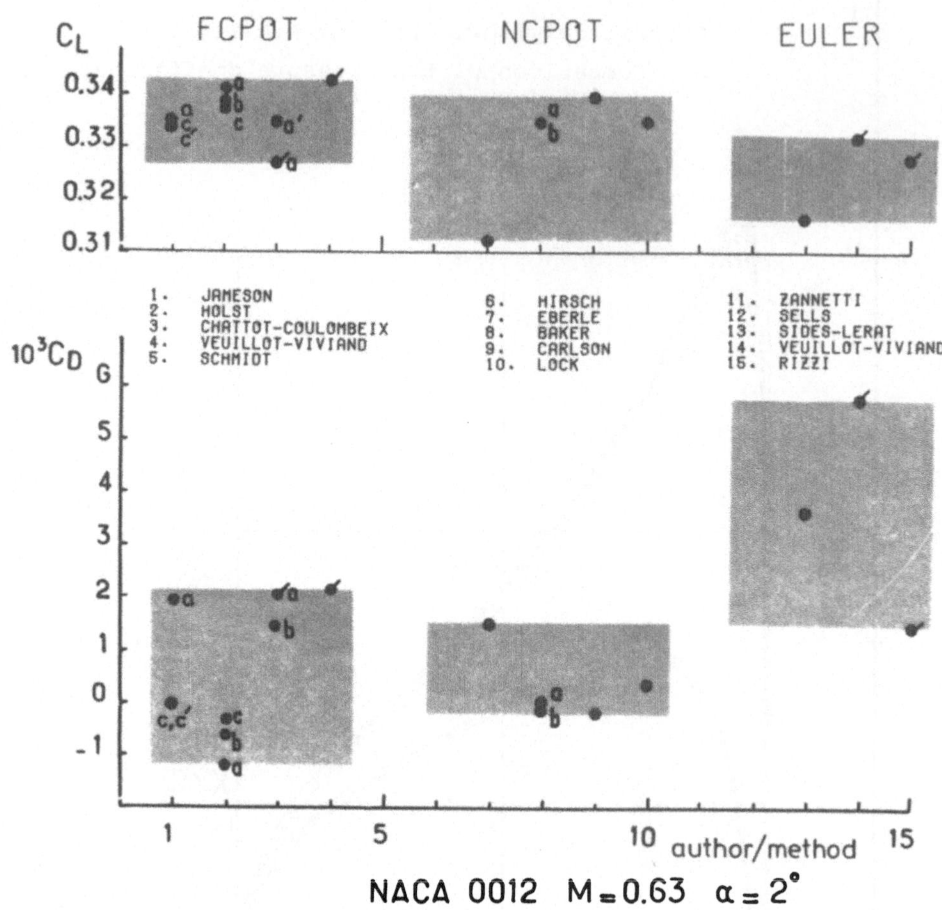

Figure 6. b) Comparison of lift and drag coefficients C_L and C_D.

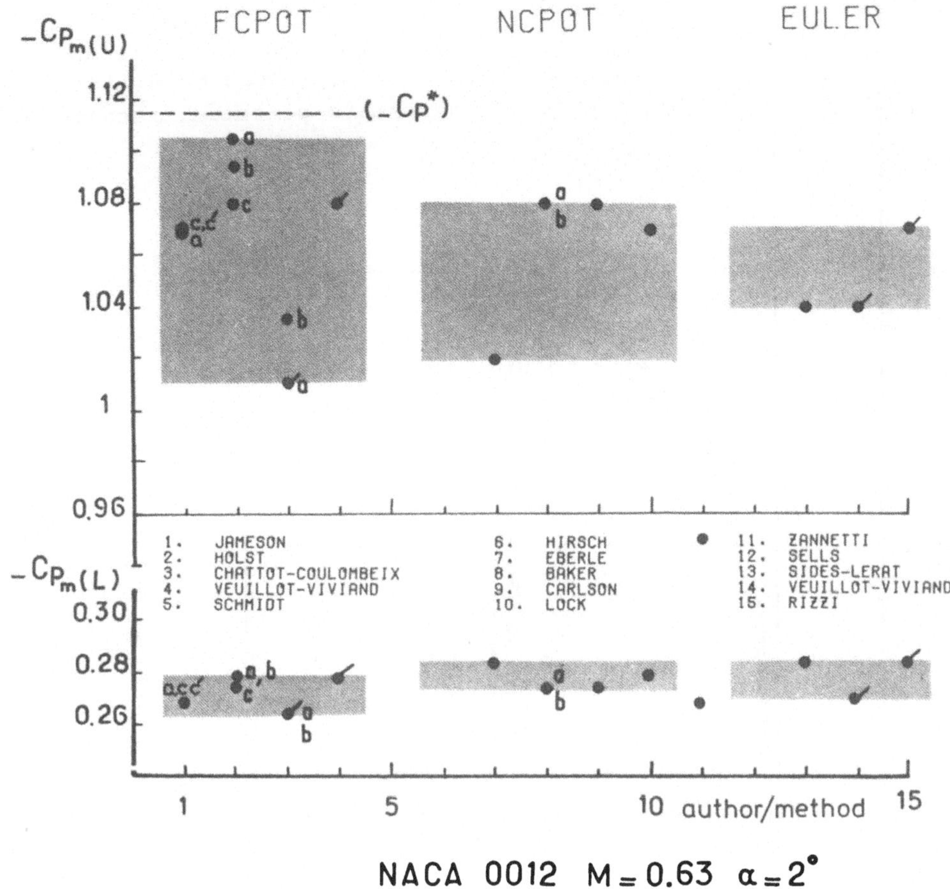

Figure 6.   c)   Comparison of minimum C_p on the upper and lower surface.

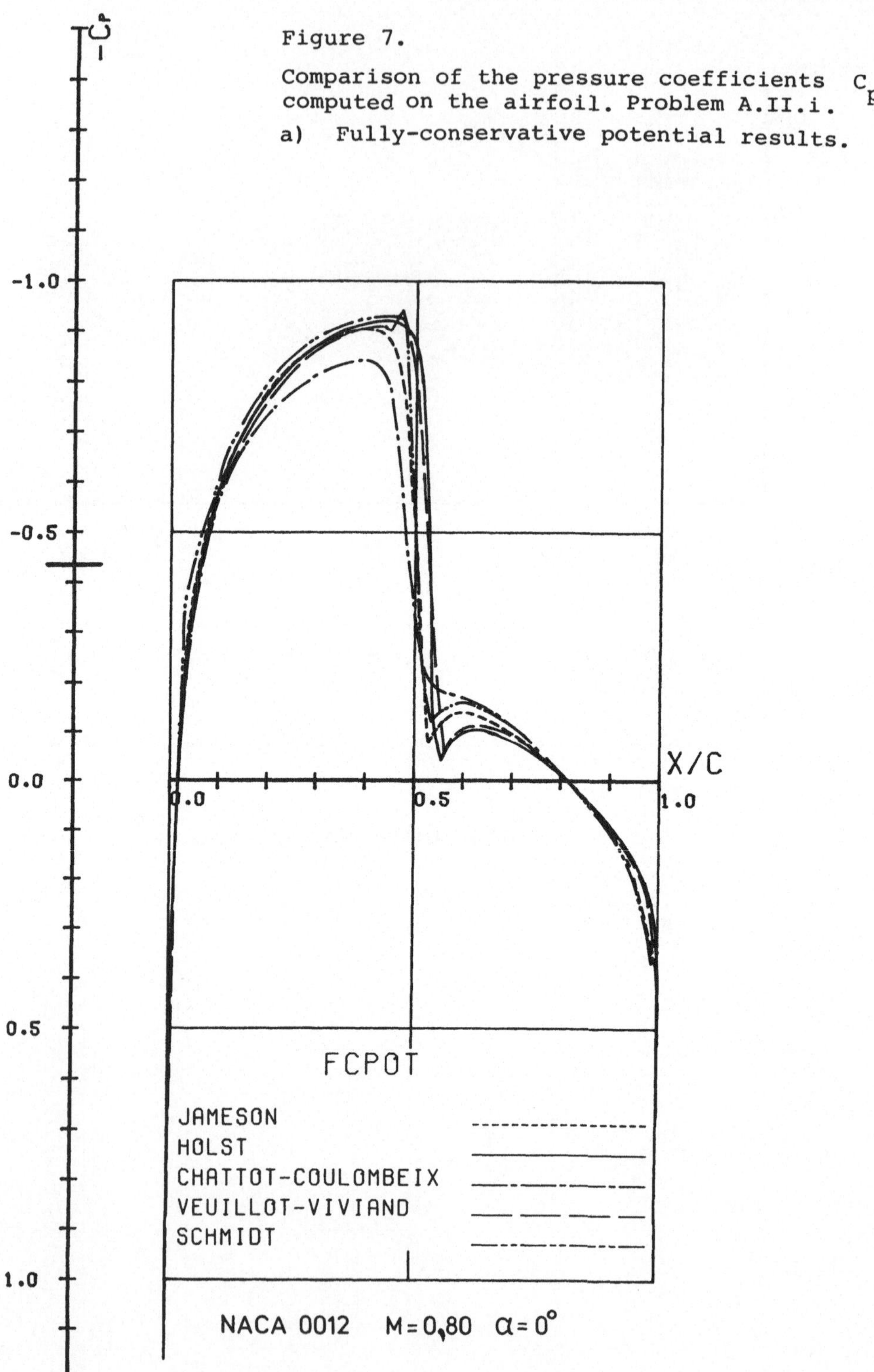

Figure 7.

Comparison of the pressure coefficients C_p computed on the airfoil. Problem A.II.i.

a) Fully-conservative potential results.

Figure 7.

b)   Non-conservative potential results.

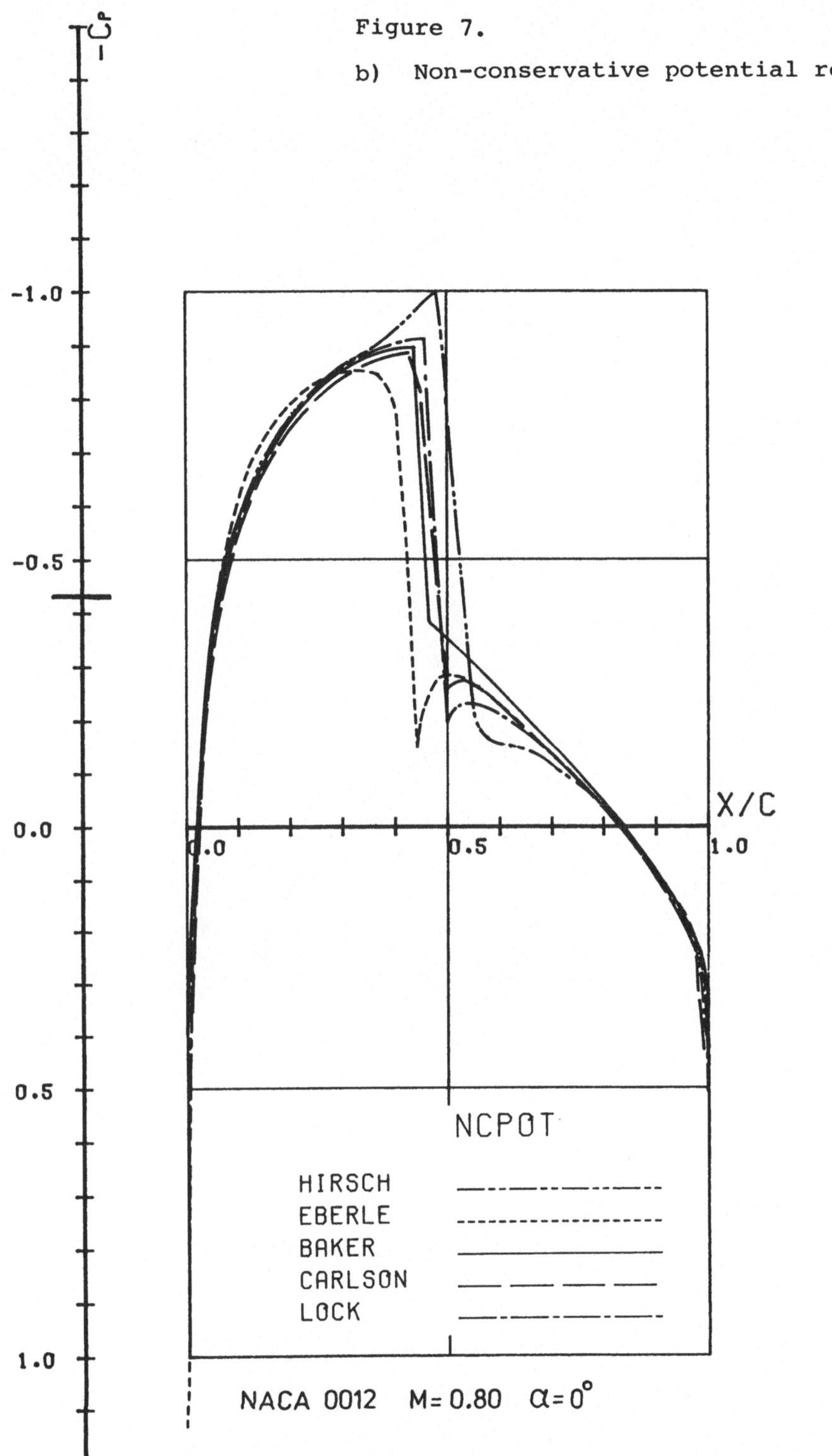

-Cp
X/C
NCPOT
HIRSCH
EBERLE
BAKER
CARLSON
LOCK
NACA 0012   M= 0.80   α= 0°

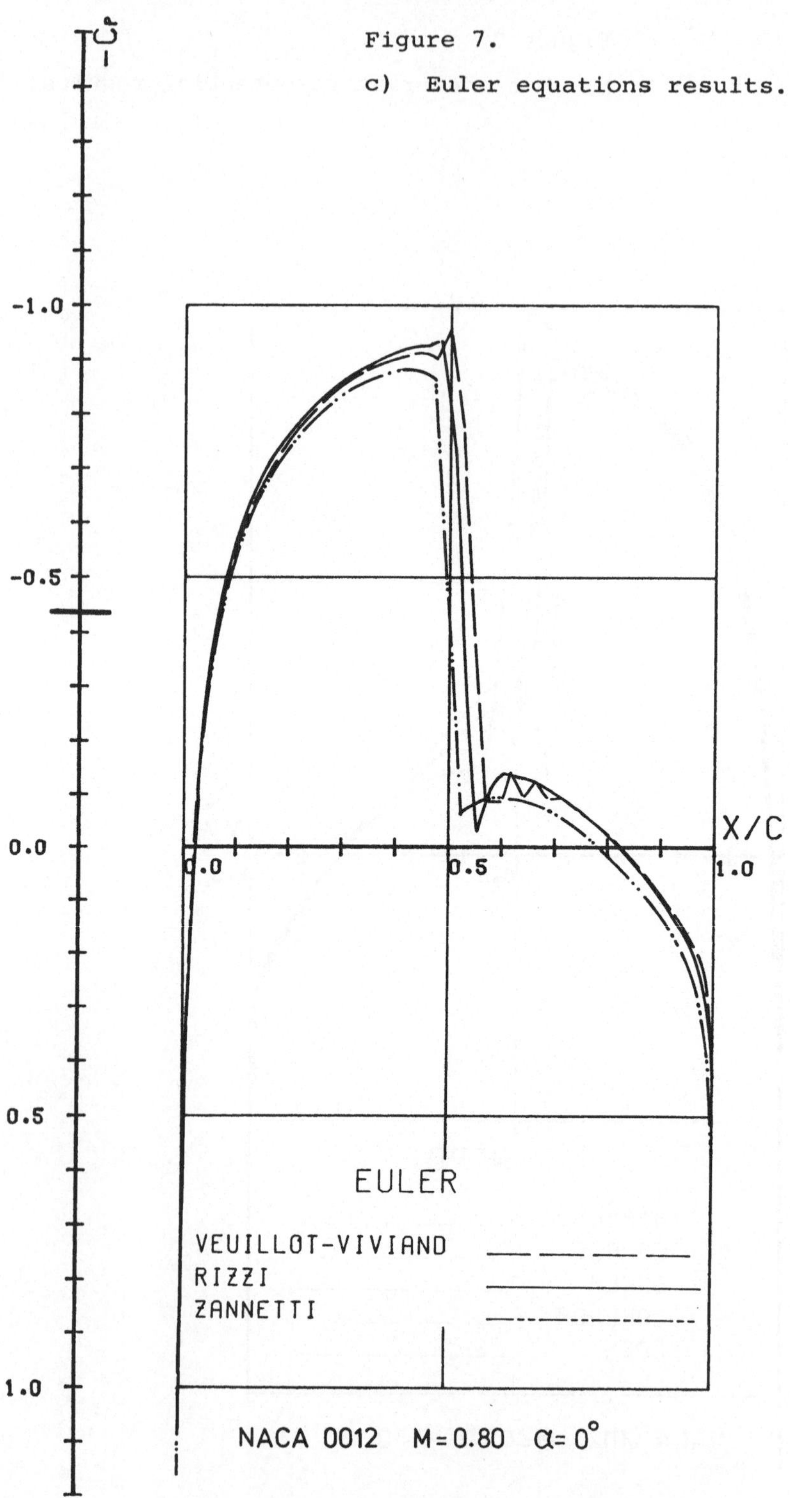

Figure 7.
c)   Euler equations results.
-C$_p$
-1.0
-0.5
0.0
0.5
1.0
0.0
0.5
1.0
X/C
EULER
VEUILLOT-VIVIAND
RIZZI
ZANNETTI
NACA 0012   M=0.80   α= 0°

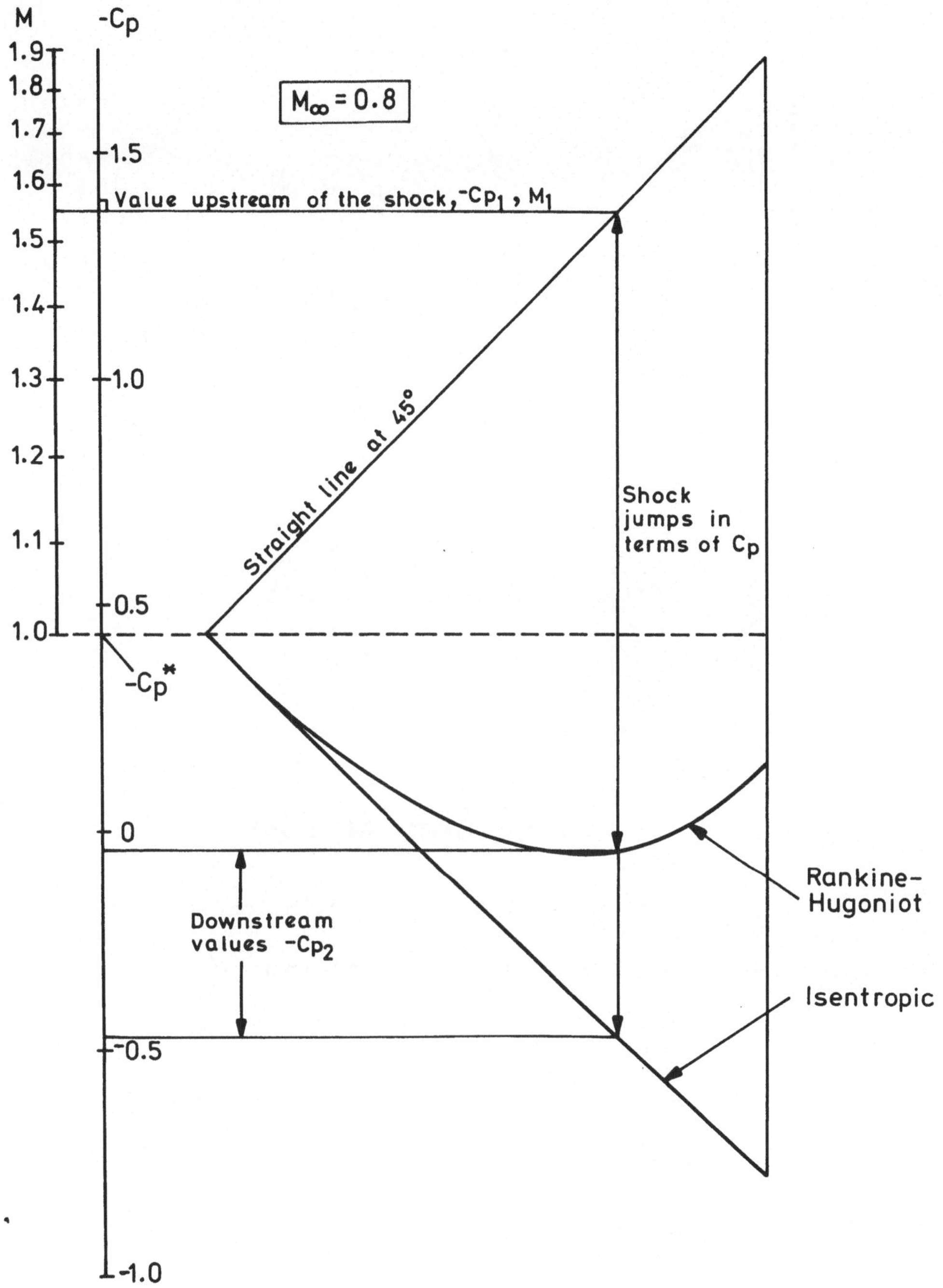

Figure 7.

d) Jump in C_p across a normal shock for both the Rankine-Hugoniot and isentropic shock conditions at $M_\infty = .80$.

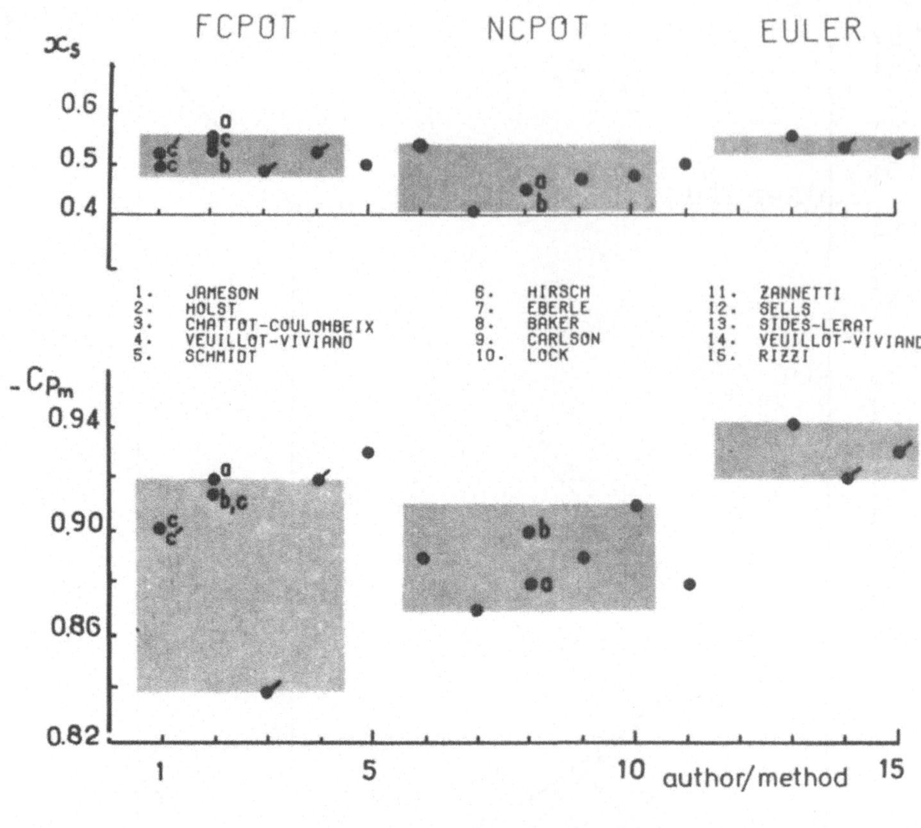

Figure 8.   Indicative quantities of the solutions to Problem A.II.i.

a)   Shock position χ_s and minimum C_p.

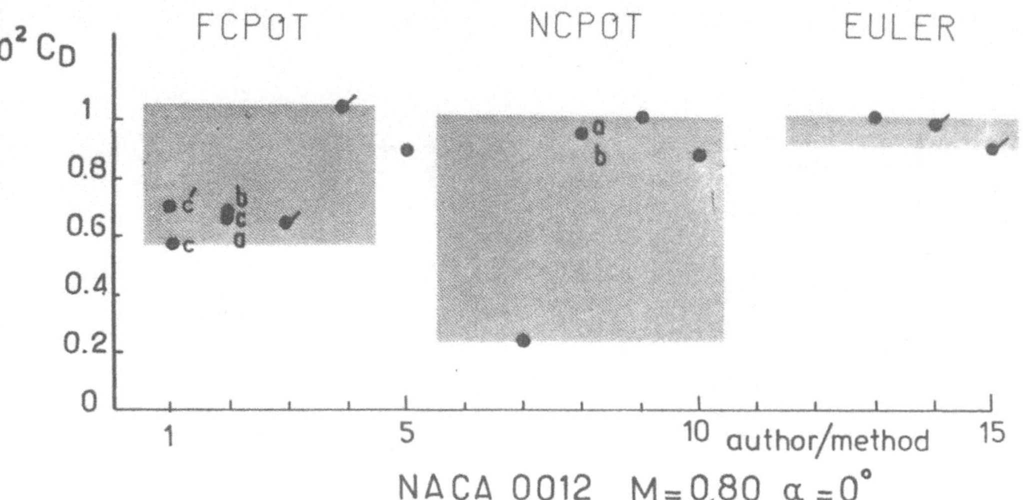

Figure 8.   b)   Drag coefficient C_D .

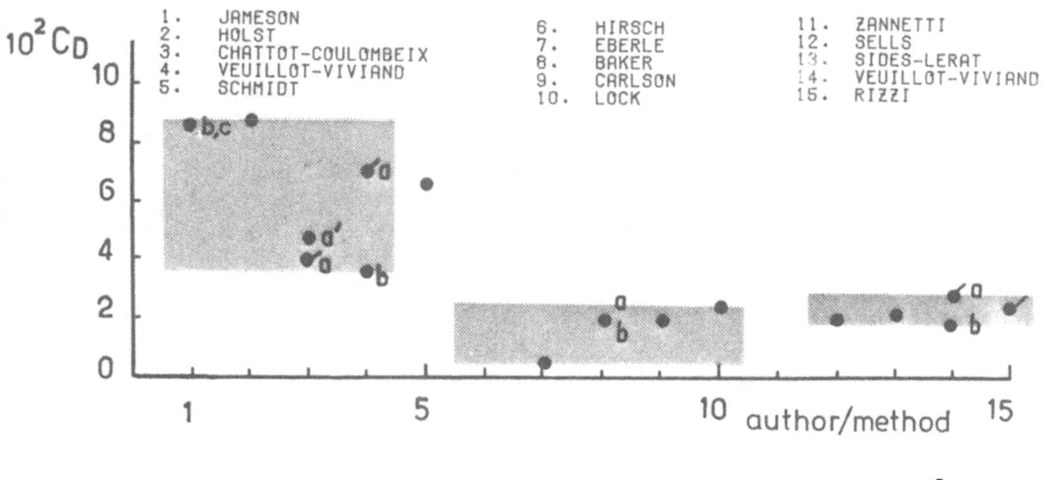

Figure 9.   Indicative quantities of the solutions to Problem
A.II.iv.

a)   Drag coefficient C_D .

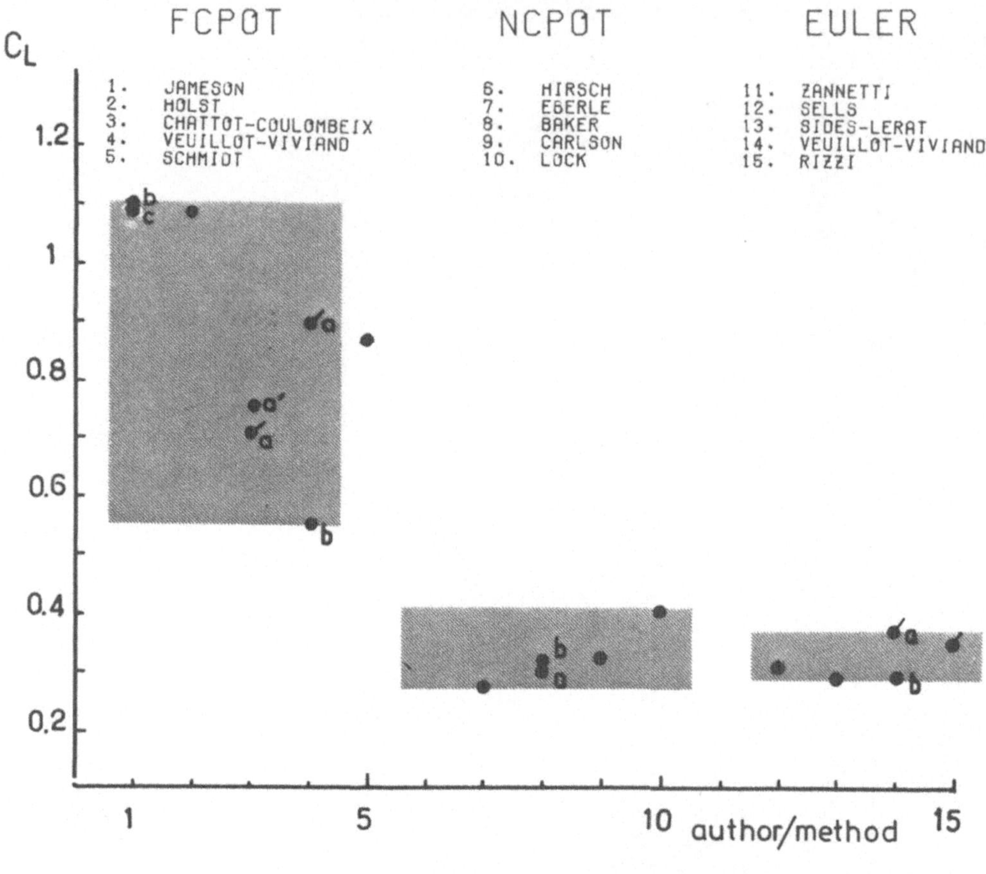

Figure 9.   b)   Lift coefficient C_L .

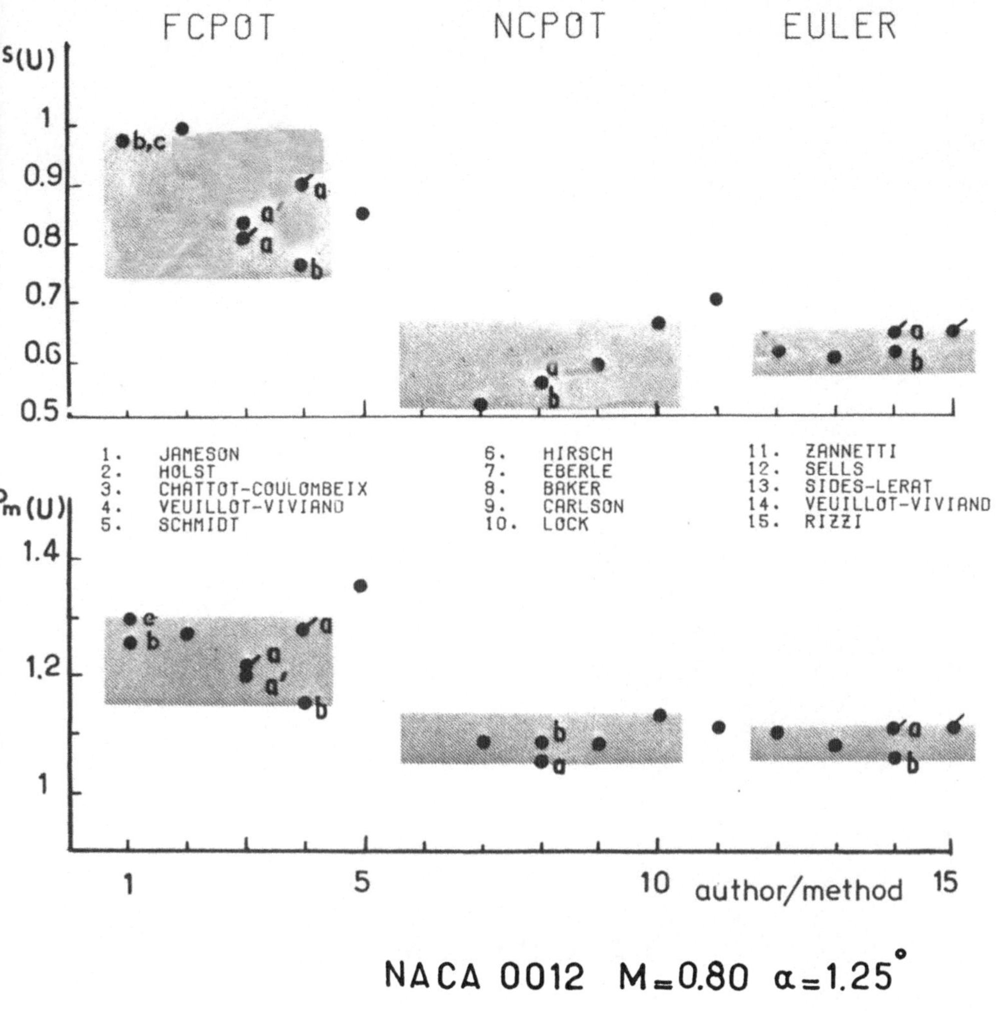

Figure 9. c) Upper surface shock position X_s and minimum C_p.

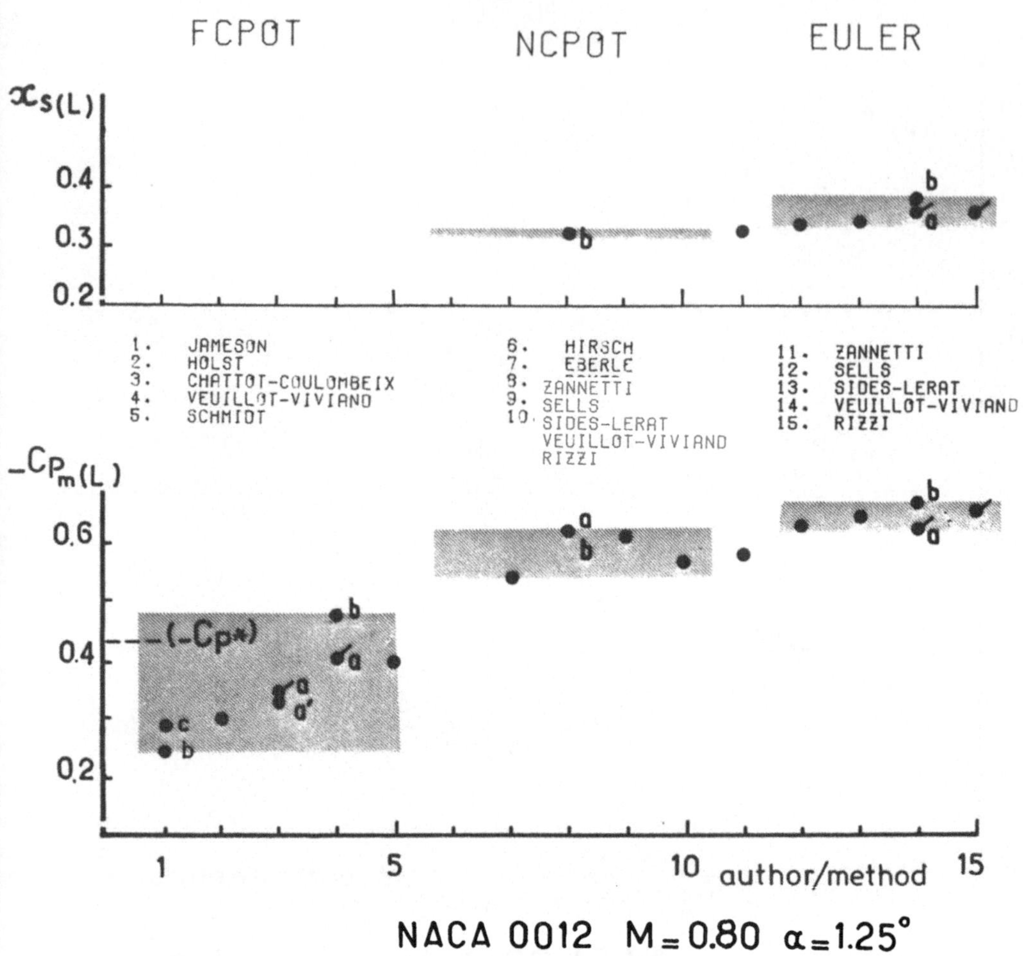

NACA 0012 M $=0.80$ $\alpha = 1.25°$

Figure 9.   d)   Lower surface shock position χ_s and minimum C_p.

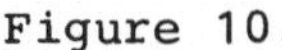

Figure 10.

Comparison of the pressure coefficients C_p computed on the airfoil. Problem A.II.iv.

a) Fully-conservative potential results.

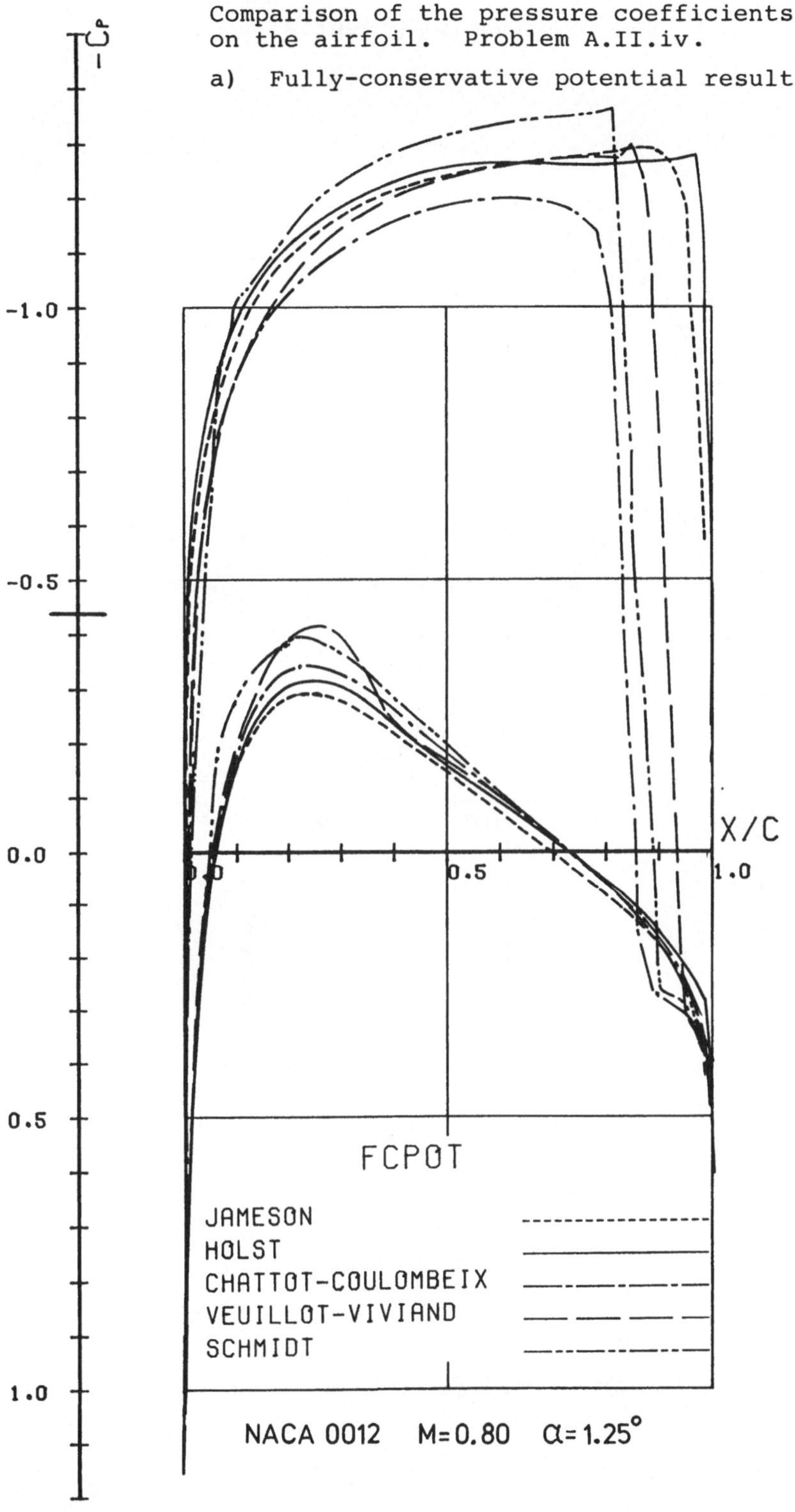
-Cp
-1.0
-0.5
0.0
0.5
1.0
X/C
0.0
0.5
1.0
FCPOT
JAMESON
HOLST
CHATTOT-COULOMBEIX
VEUILLOT-VIVIAND
SCHMIDT
NACA 0012 M=0.80 α=1.25°

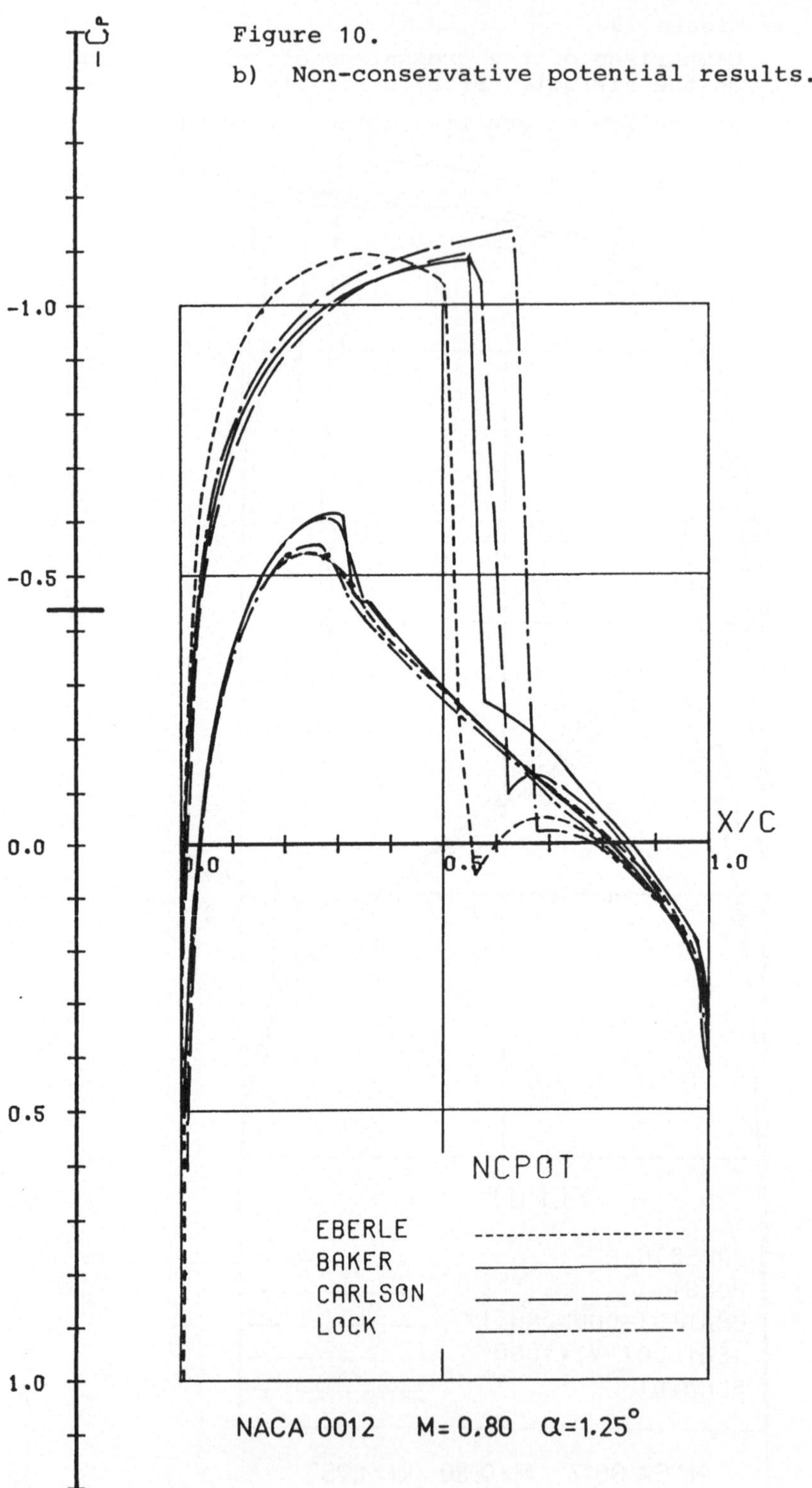

-Cp
Figure 10.
b)   Non-conservative potential results.
-1.0
-0.5
0.0
0.5
1.0
0.0
0.5
1.0
X/C
NCPOT
EBERLE
BAKER
CARLSON
LOCK
NACA 0012   M= 0.80   α=1.25°

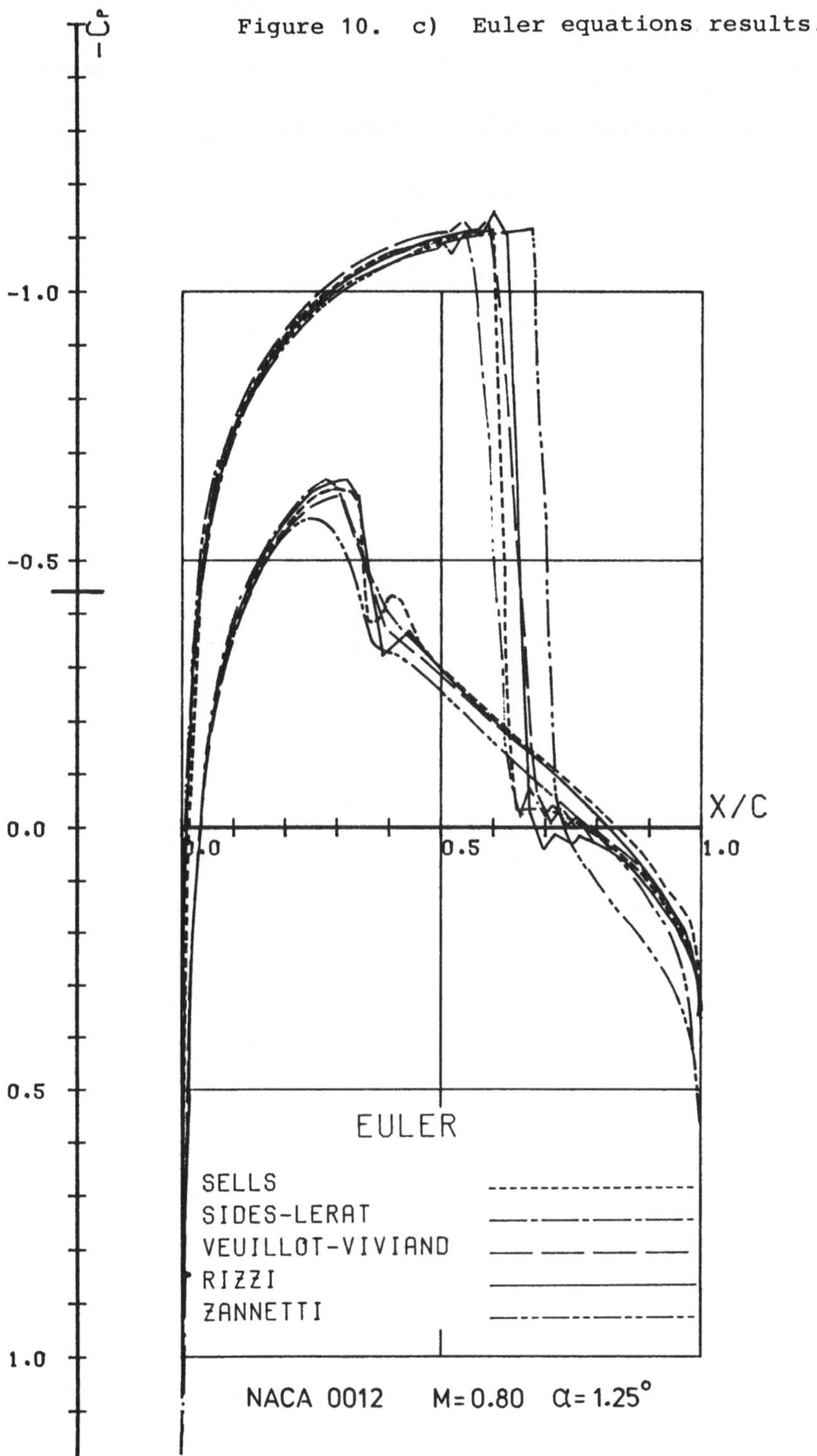

$-C_p$
Figure 10. c) Euler equations results.
-1.0
-0.5
0.0
0.5
1.0
X/C
0.0
0.5
1.0
EULER
SELLS
SIDES-LERAT
VEUILLOT-VIVIAND
RIZZI
ZANNETTI
NACA 0012   M=0.80   α=1.25°

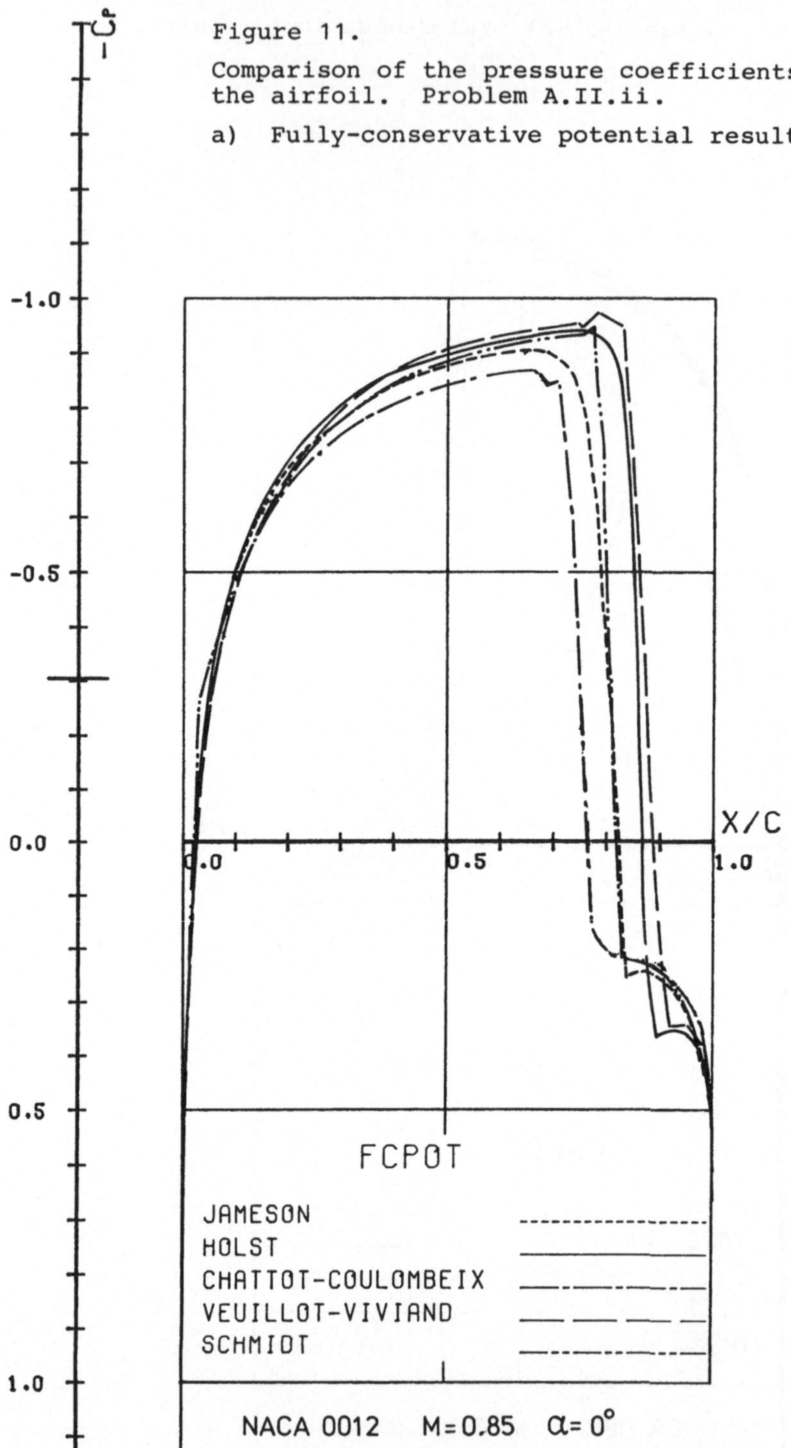

Figure 11.
Comparison of the pressure coefficients computed on the airfoil. Problem A.II.ii.
a) Fully-conservative potential results.
X/C
FCPOT
JAMESON
HOLST
CHATTOT-COULOMBEIX
VEUILLOT-VIVIAND
SCHMIDT
NACA 0012   M=0.85   α= 0°

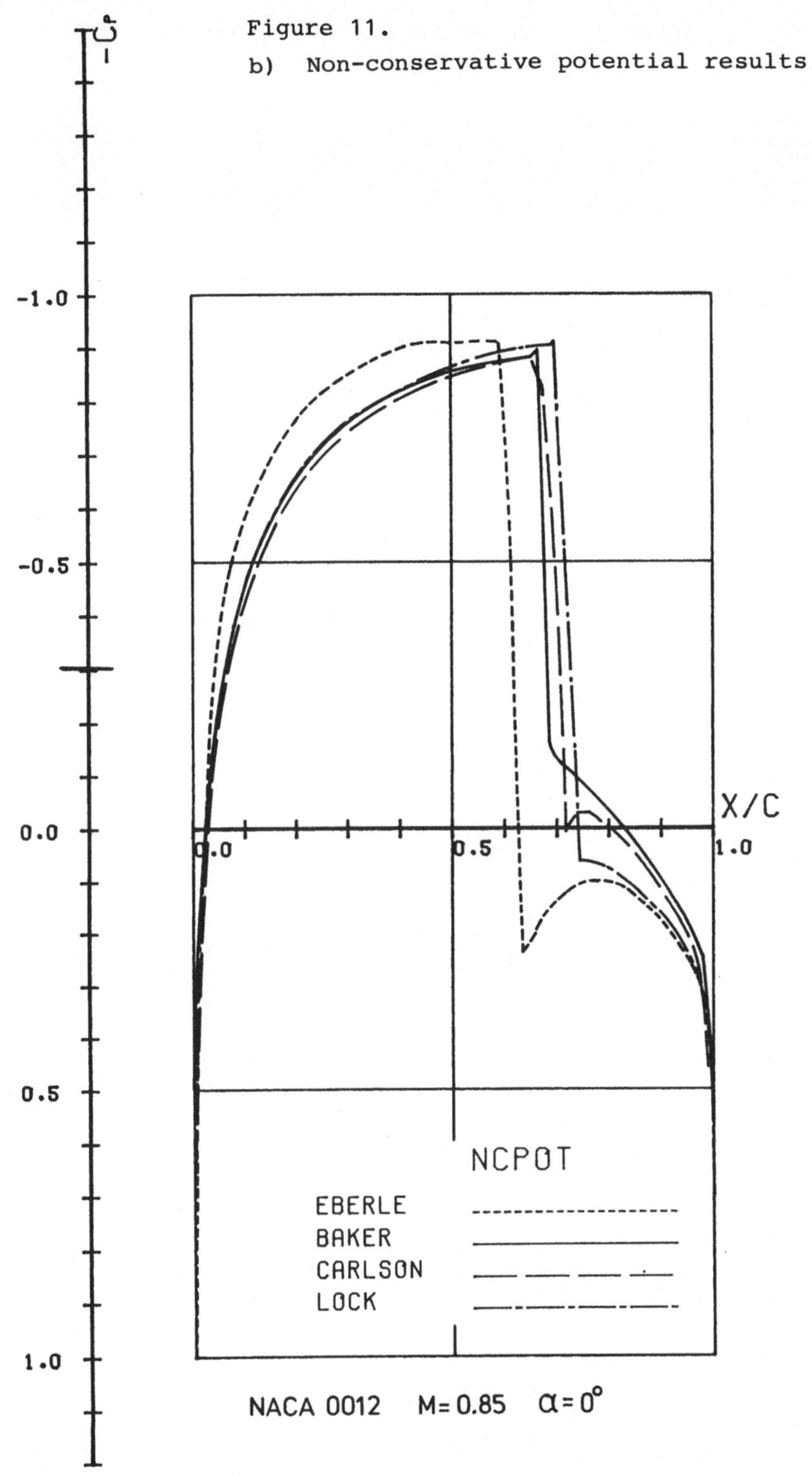

Figure 11.

b) Non-conservative potential results.

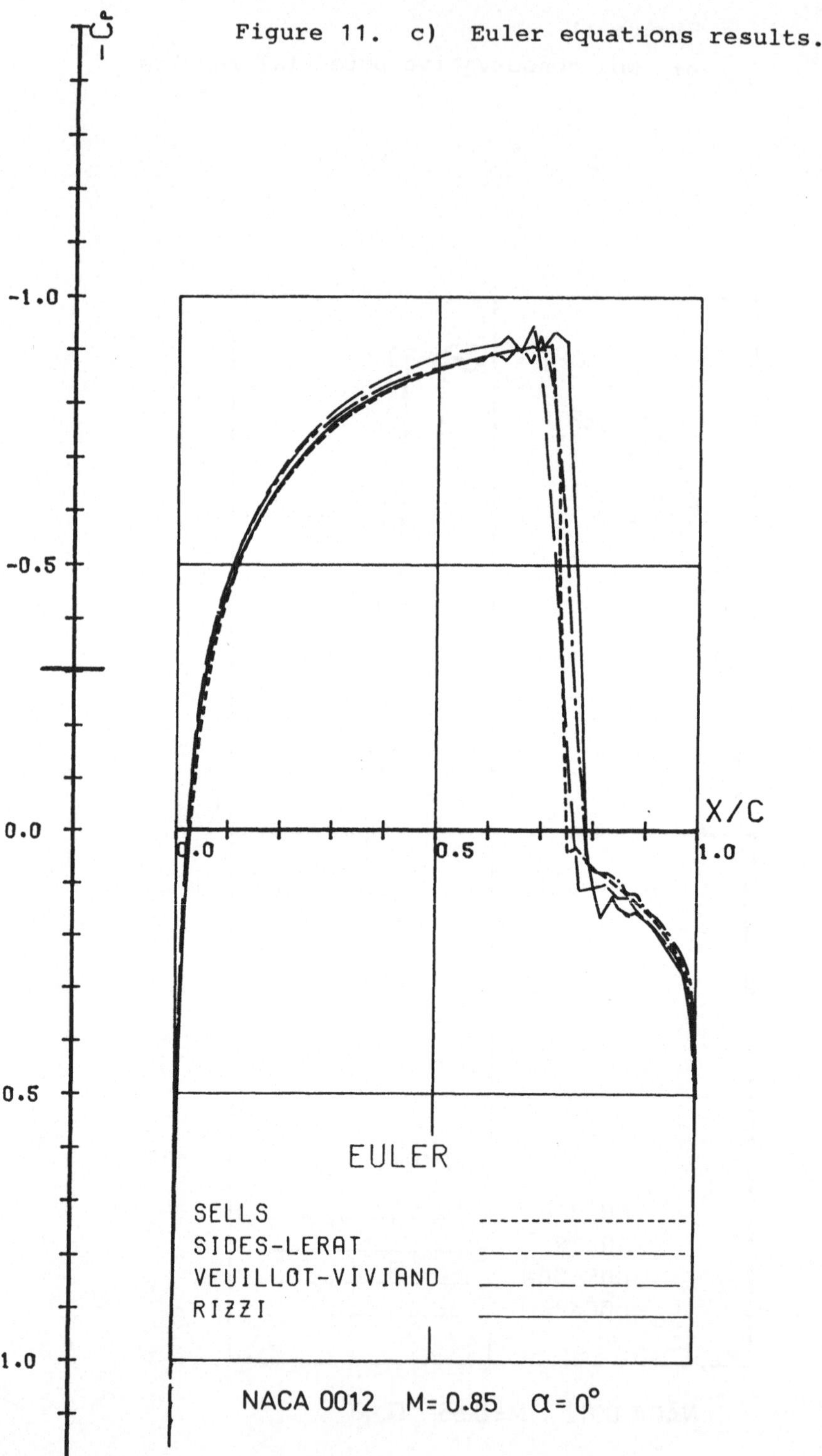

Figure 11. c) Euler equations results.

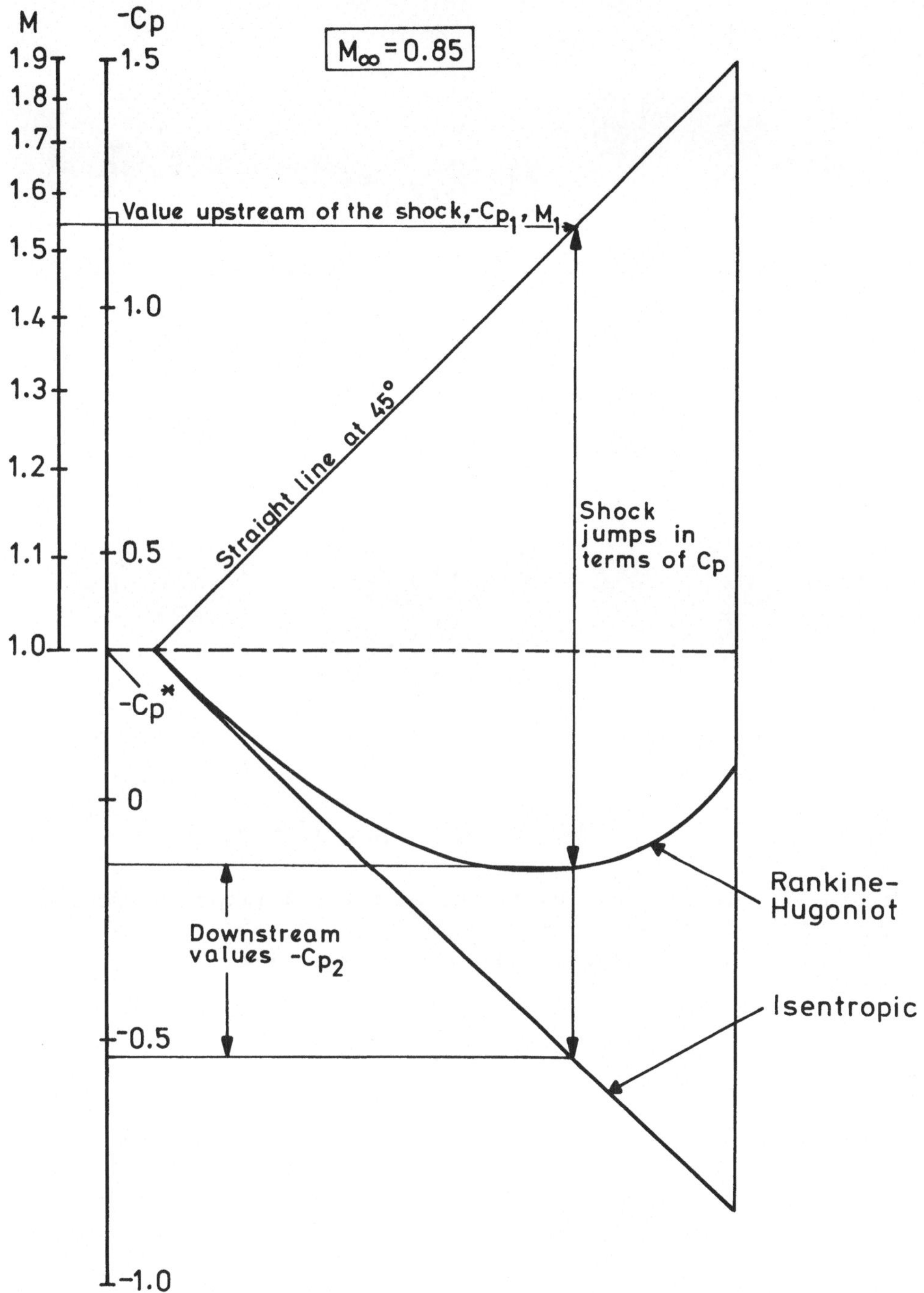

Figure 11. d) Jump in C_p across a normal shock for both the Rankine-Hugoniot and isentropic shock conditions at $M_\infty = .85$.

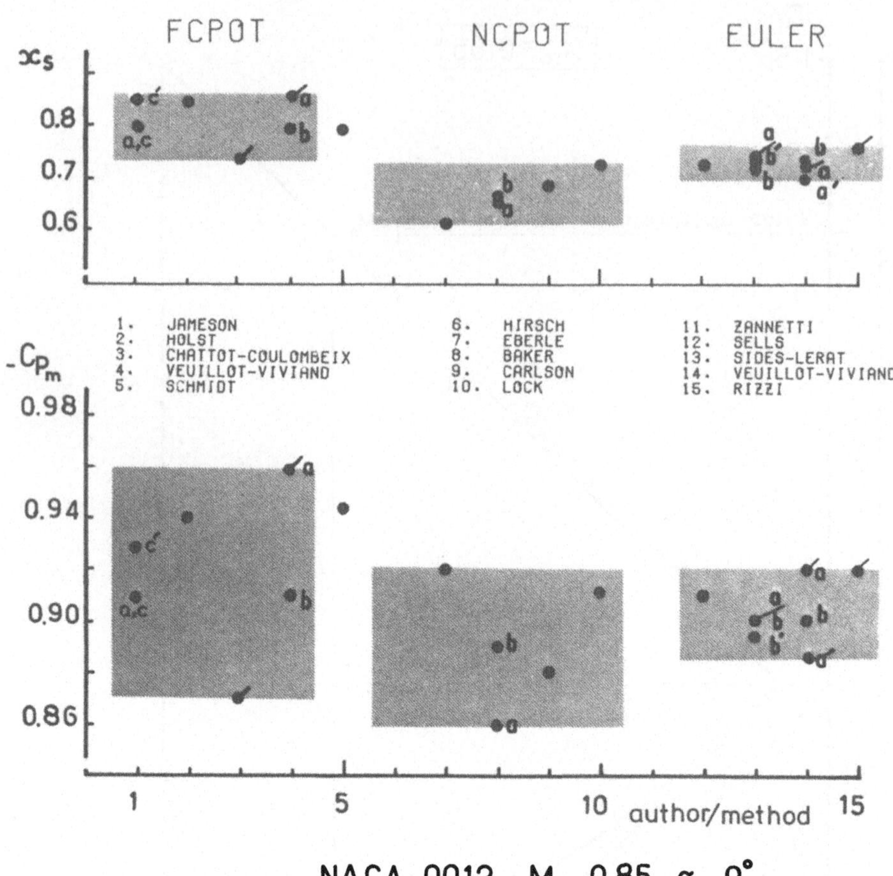

Figure 12.   Indicative quantities of the solutions to Problem A.II.ii.

a)   Shock position x_s and minimum C_p.

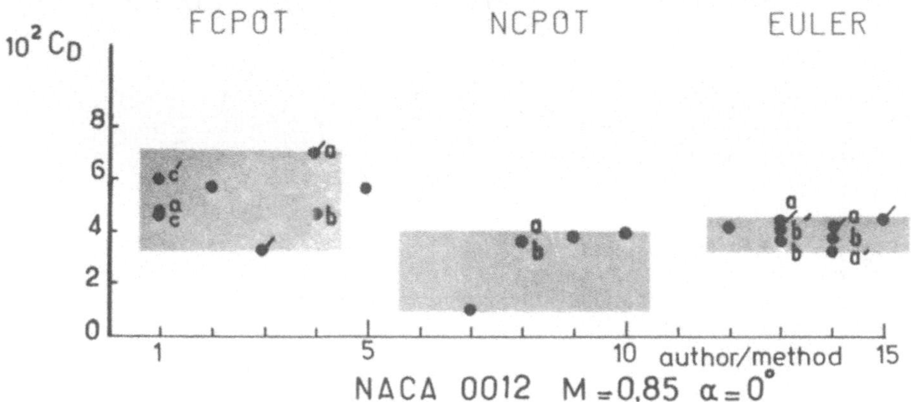

Figure 12. b) Drag coefficient C_D.

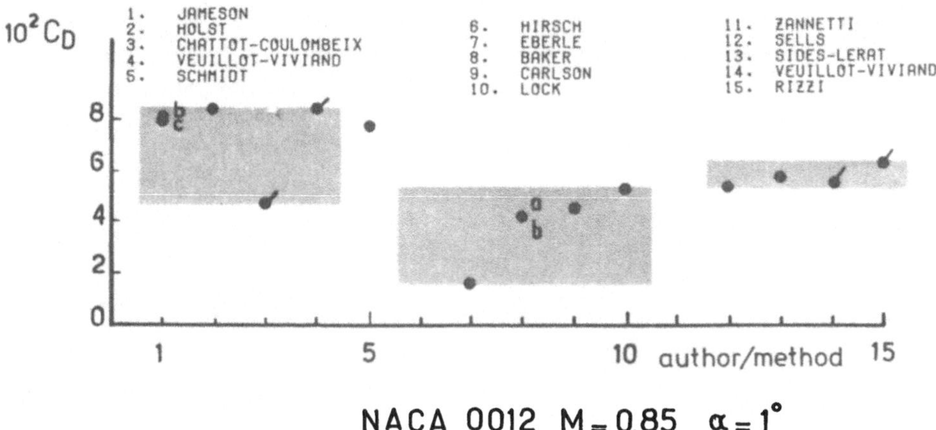

Figure 13. Indicative quantities of the solutions to Problem
A.II.v.

a) Drag coefficient C_D.

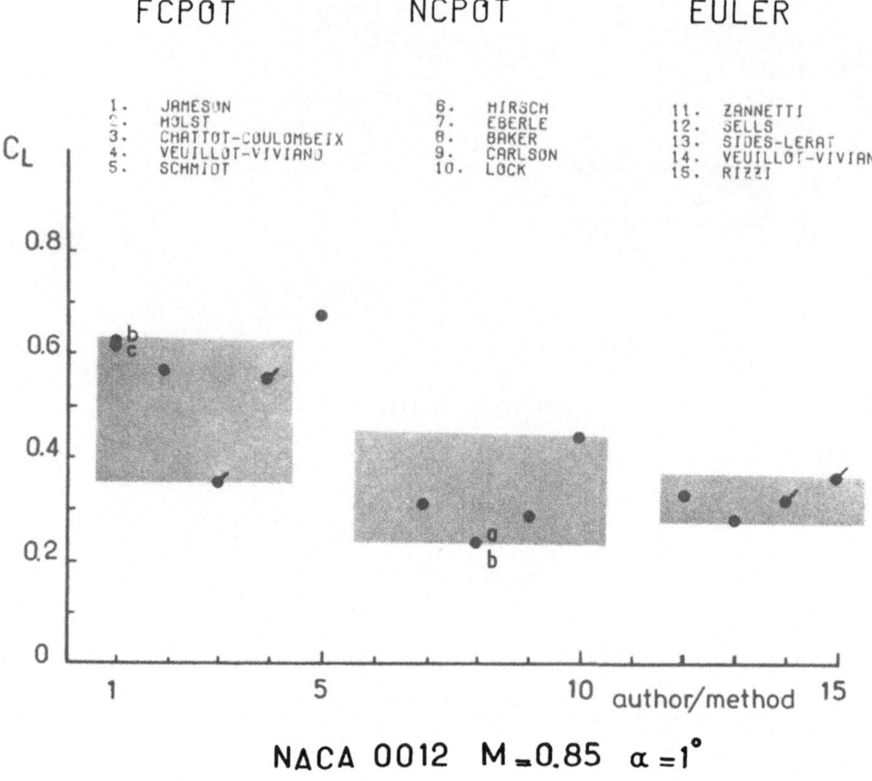

Figure 13.   b)   Lift coefficient C_L.

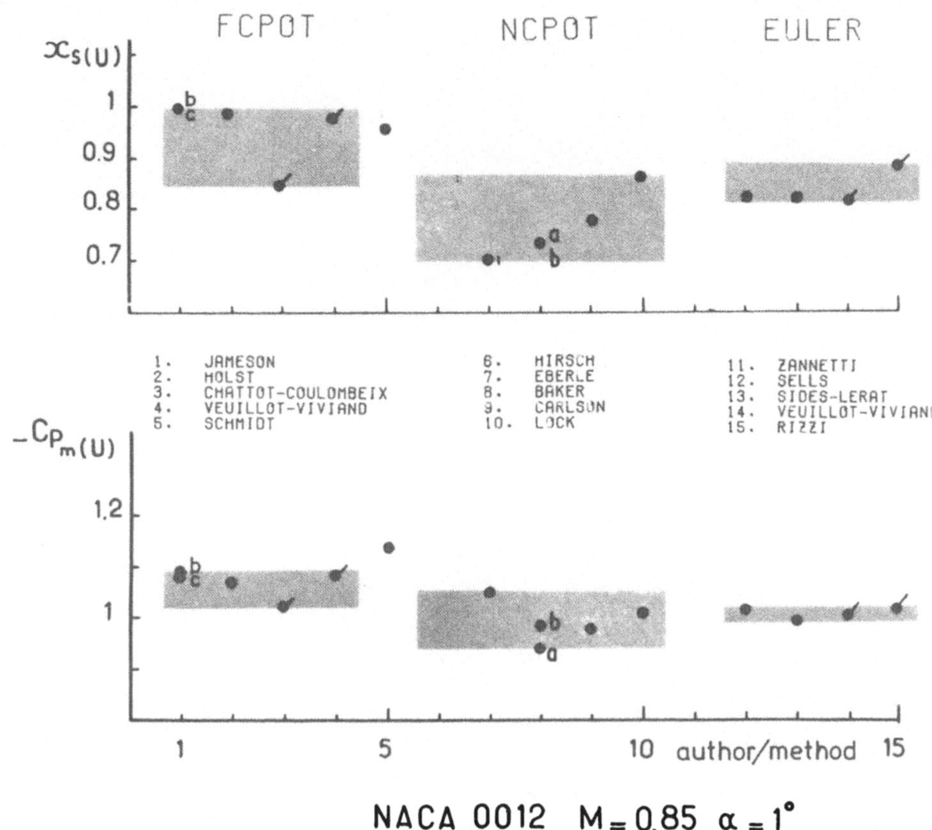

NACA 0012 M = 0.85 α = 1°

Figure 13. c) Upper surface shock position χ_s and minimum C_p.

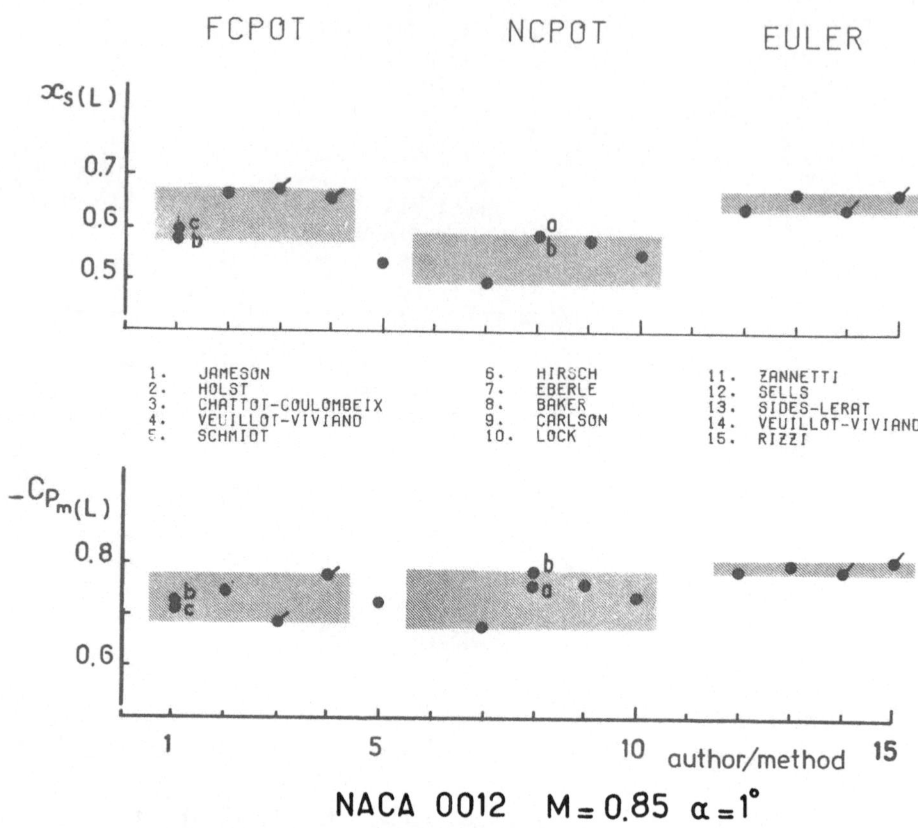

Figure 13.   d)   Lower surface shock position χ_s and minimum C_p.

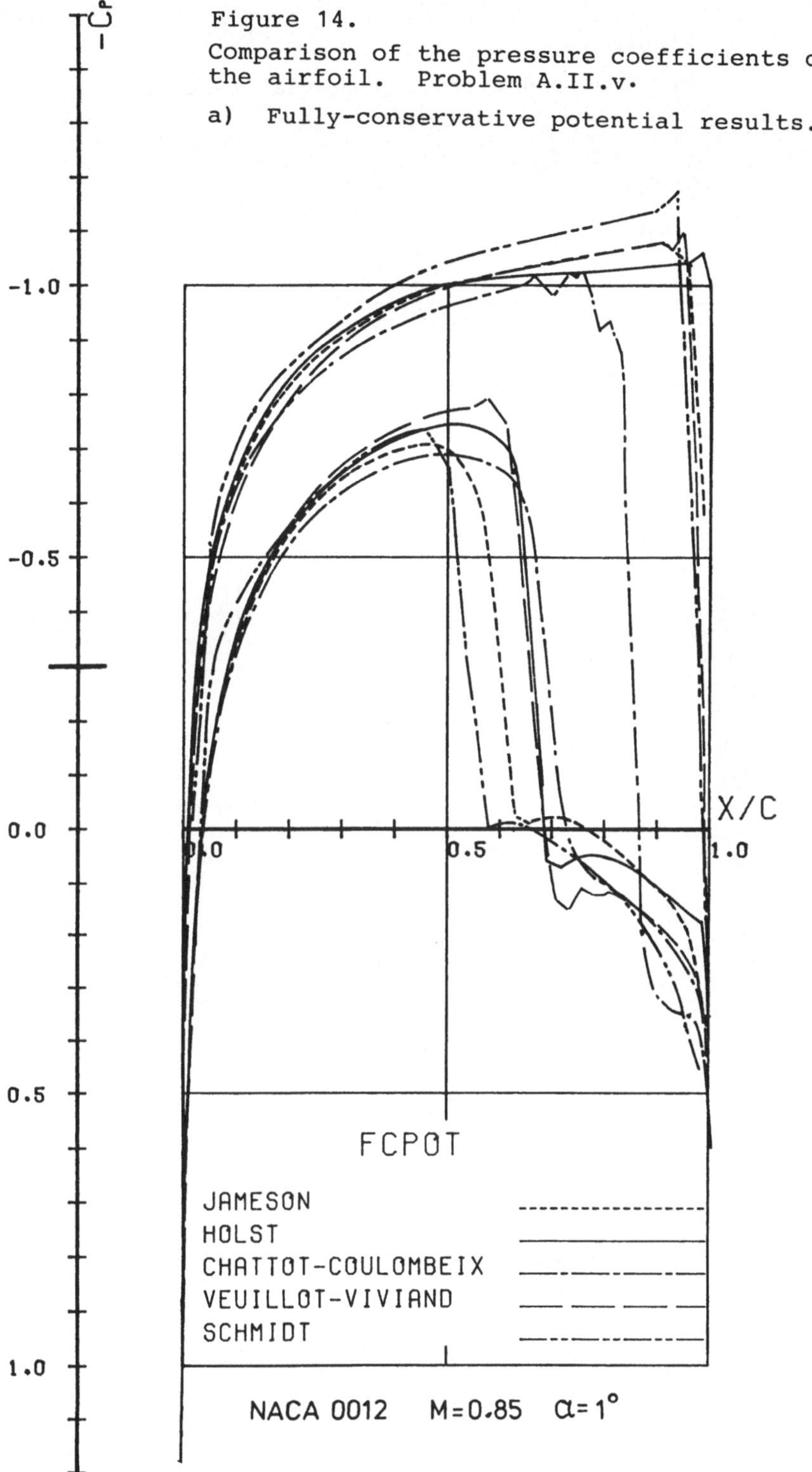

Figure 14.
Comparison of the pressure coefficients computed on the airfoil. Problem A.II.v.

a) Fully-conservative potential results.

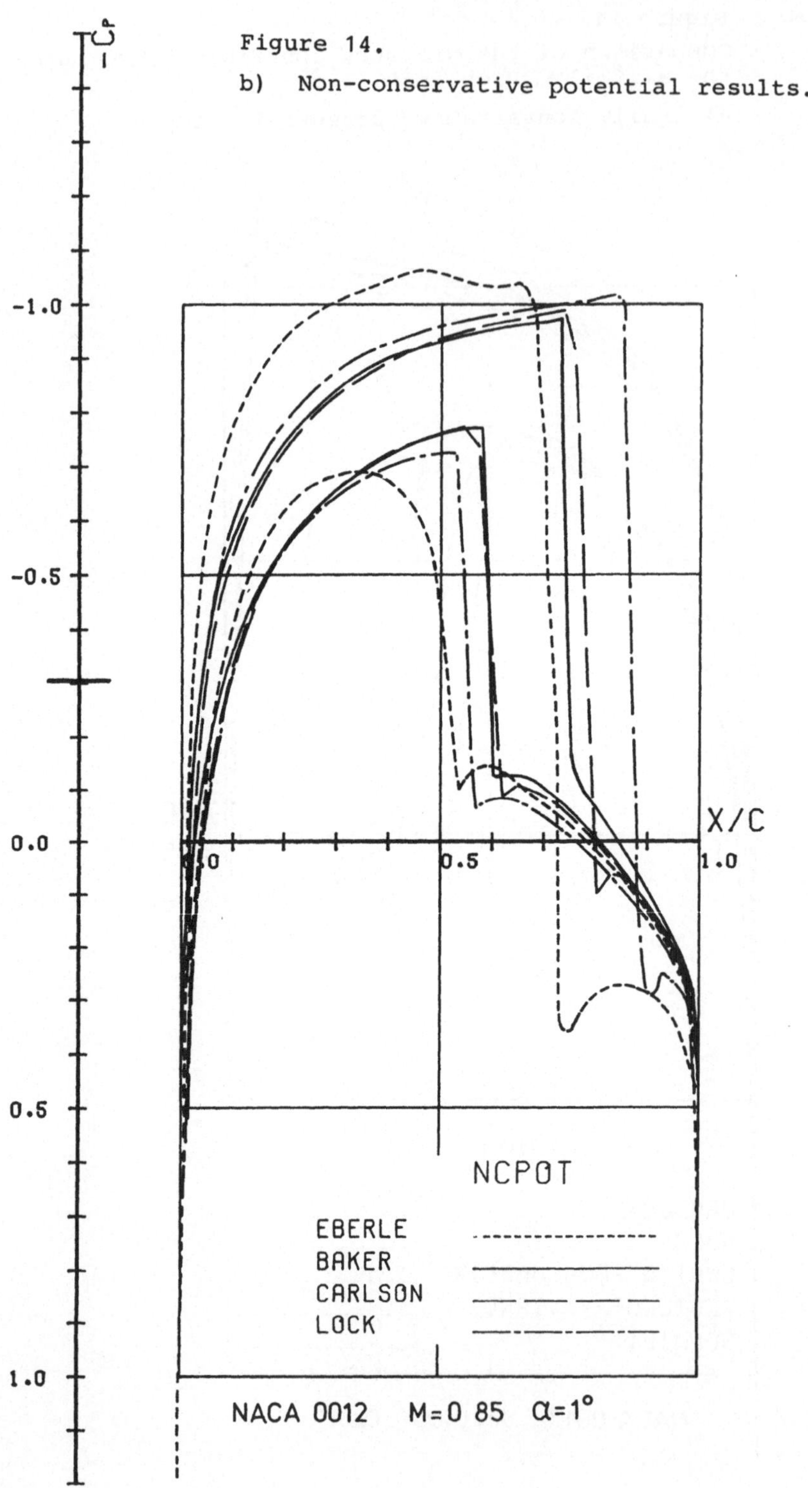

Figure 14.

b)   Non-conservative potential results.

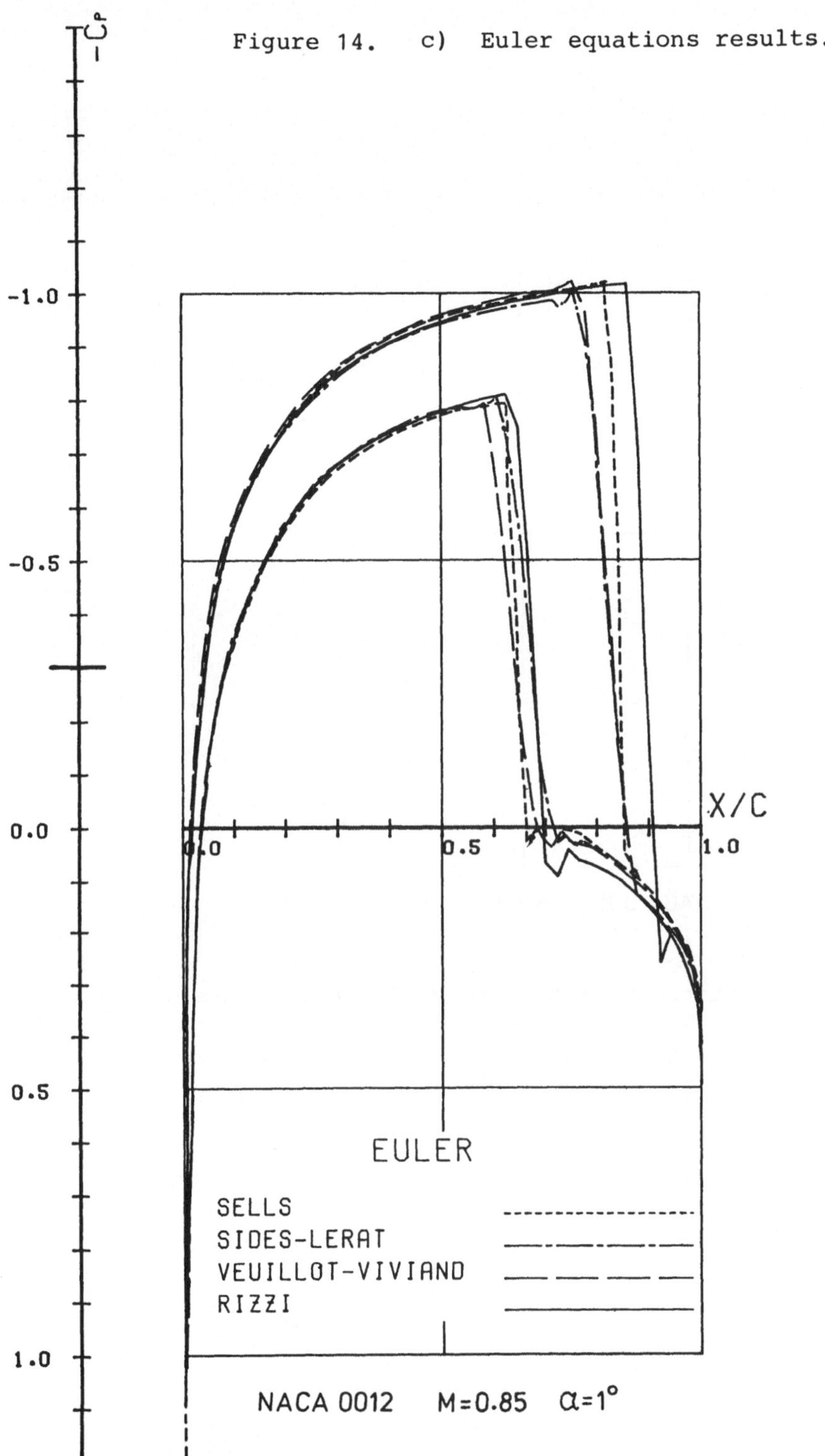

Figure 14.   c)   Euler equations results.

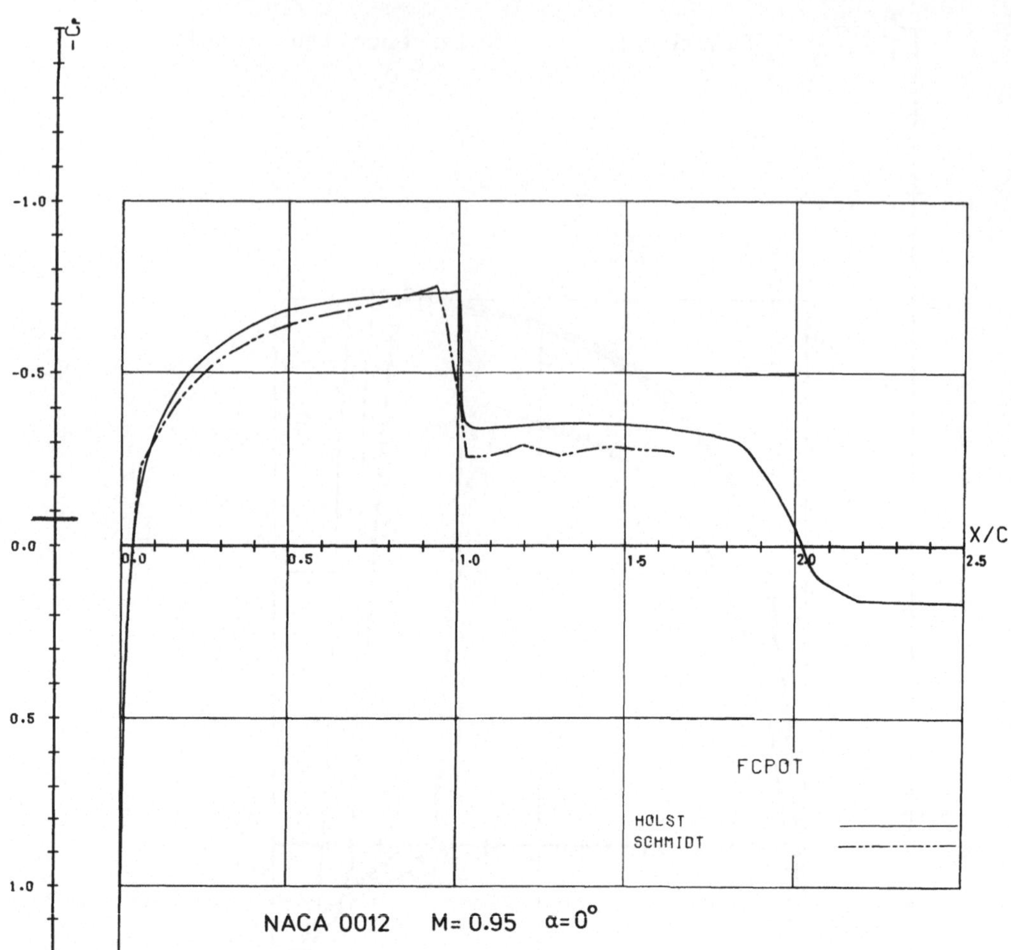

Figure 15.   Comparison of the pressure coefficients computed on the airfoil.   Problem A.II.iii.

a)   Fully-conservative potential results.

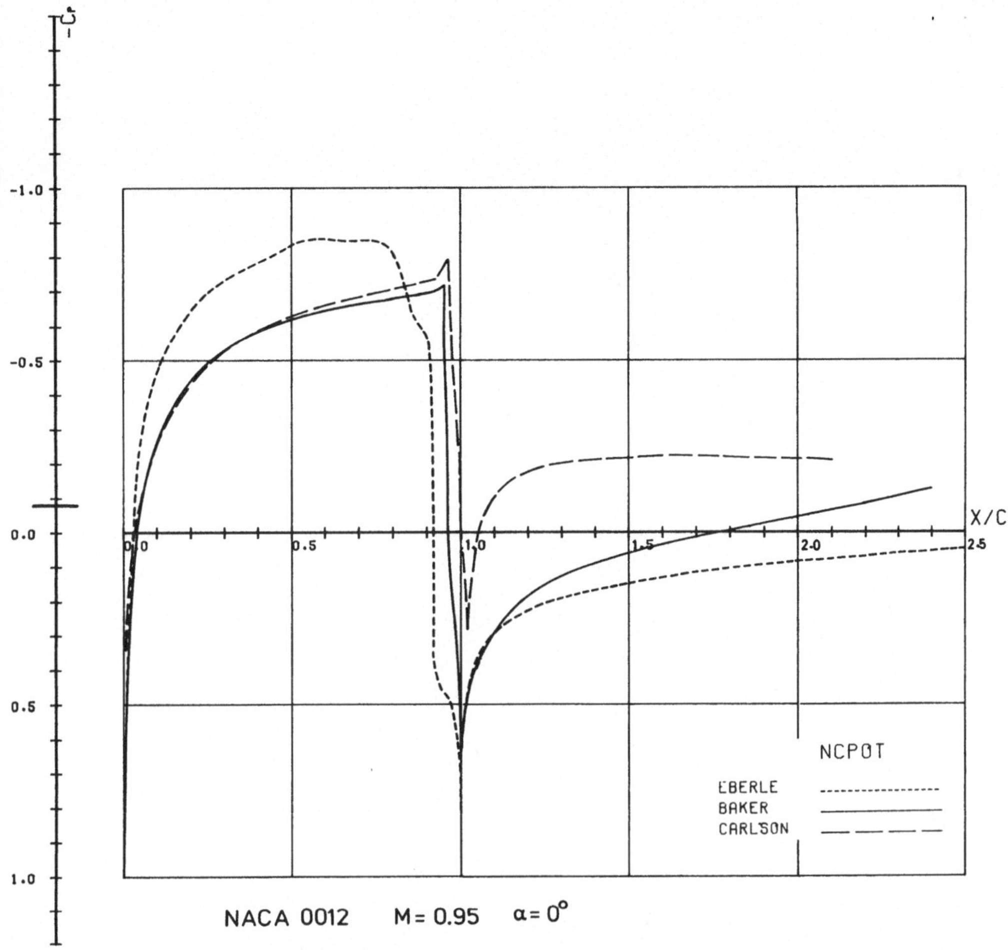

Figure 15.   b)   Non-conservative potential results.

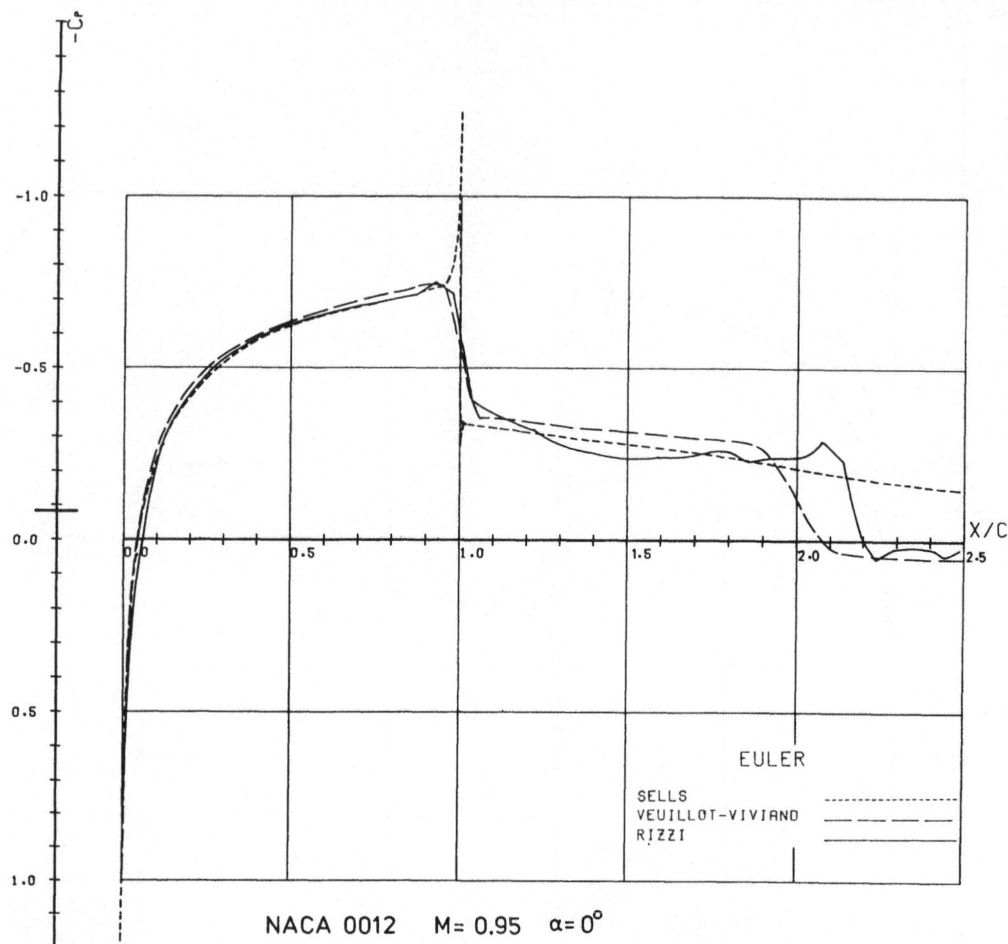

Figure 15.   c)   Euler equations results.

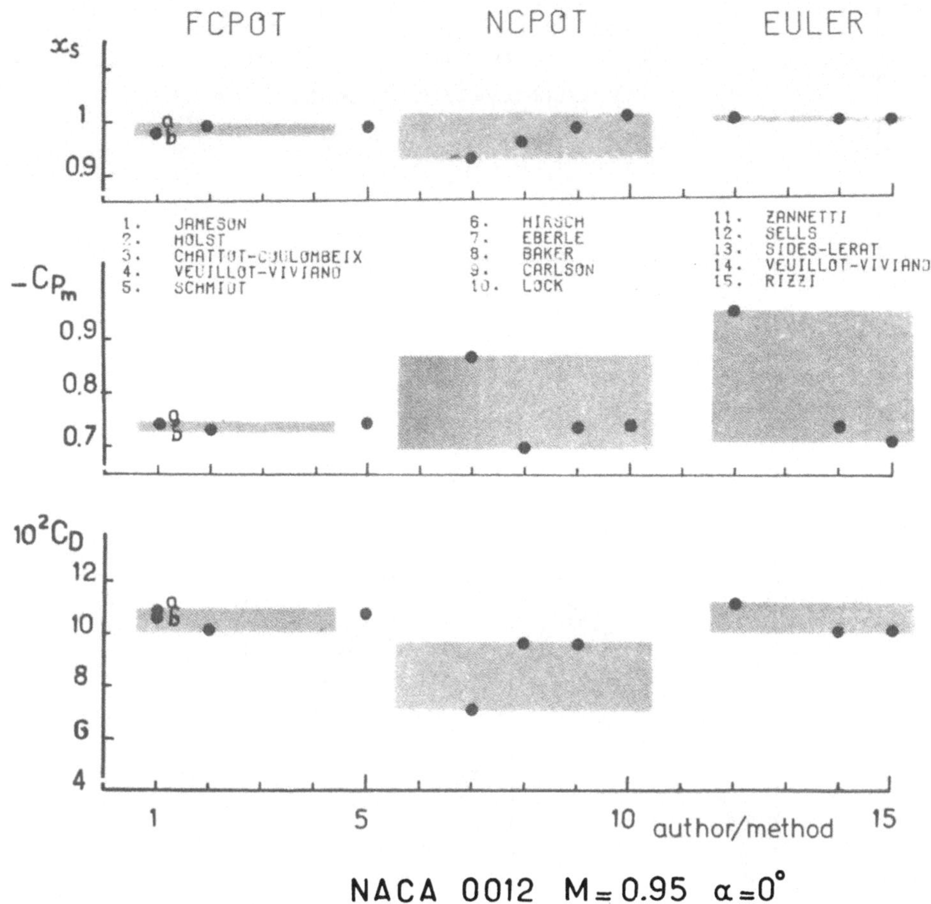

Figure 16.   Indicative quantities of the solutions to Problem A.II.iii.   Shock position χ_s, and drag coefficient C_D.

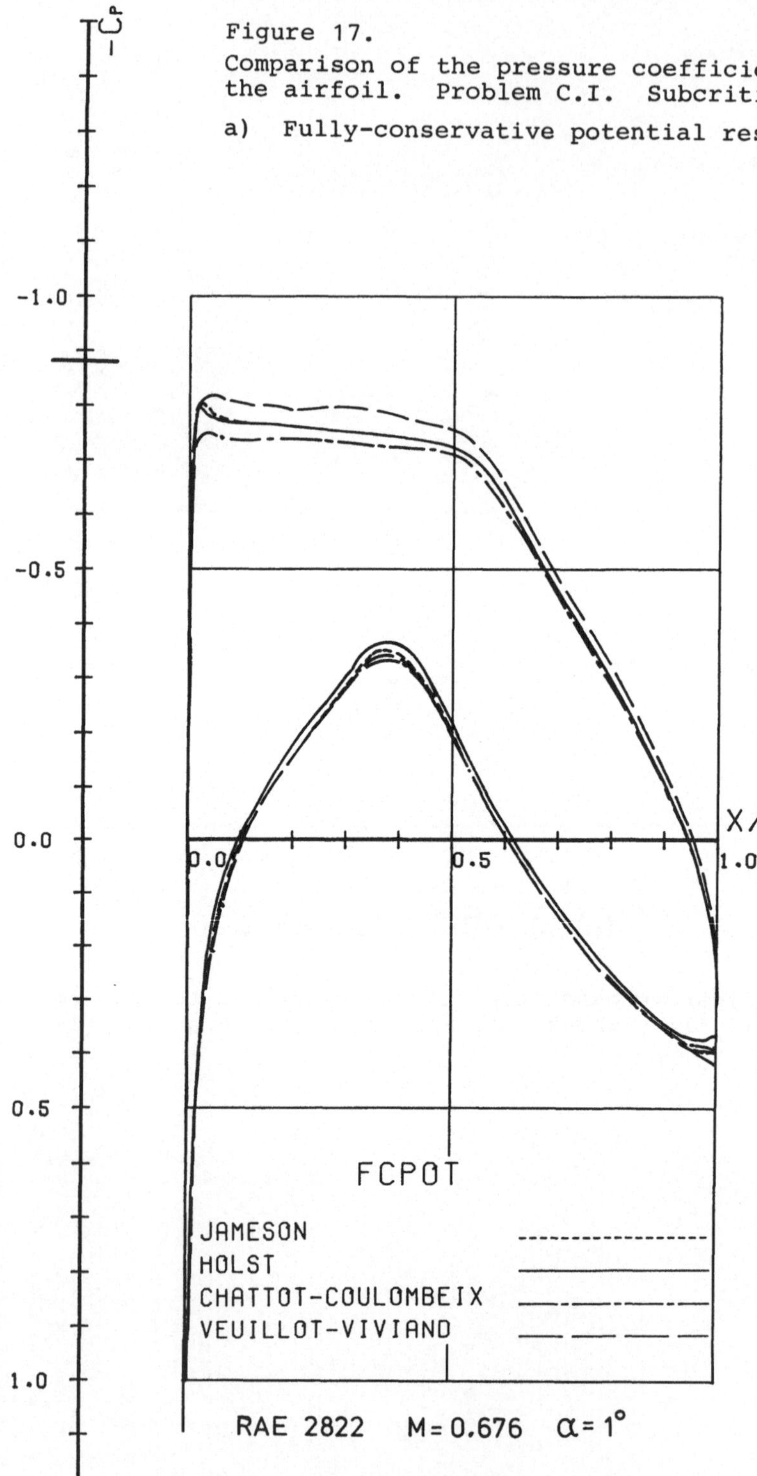

Figure 17.
Comparison of the pressure coefficients computed on the airfoil. Problem C.I. Subcritical flow.

a) Fully-conservative potential results.

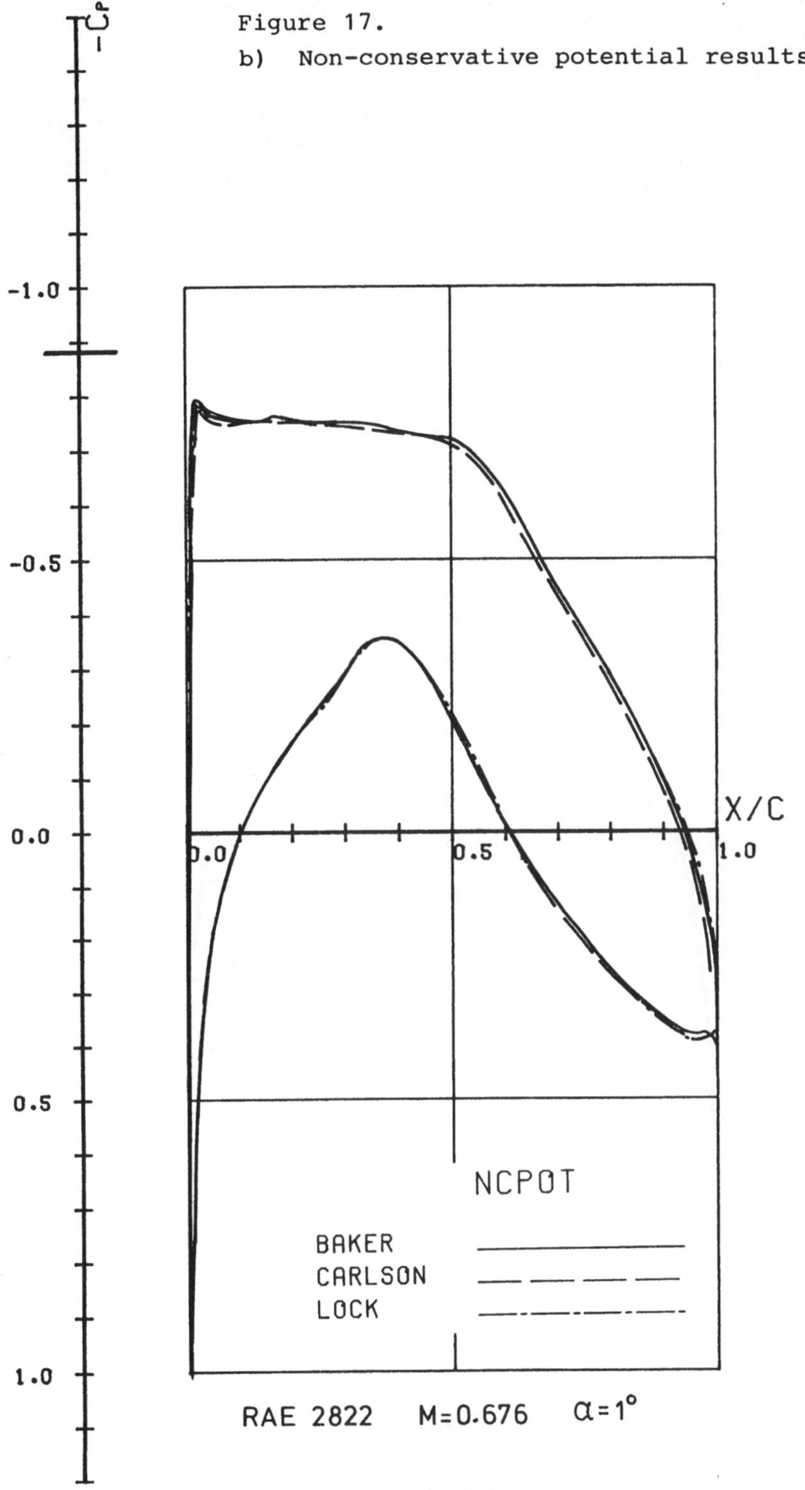

Figure 17.
b)   Non-conservative potential results.

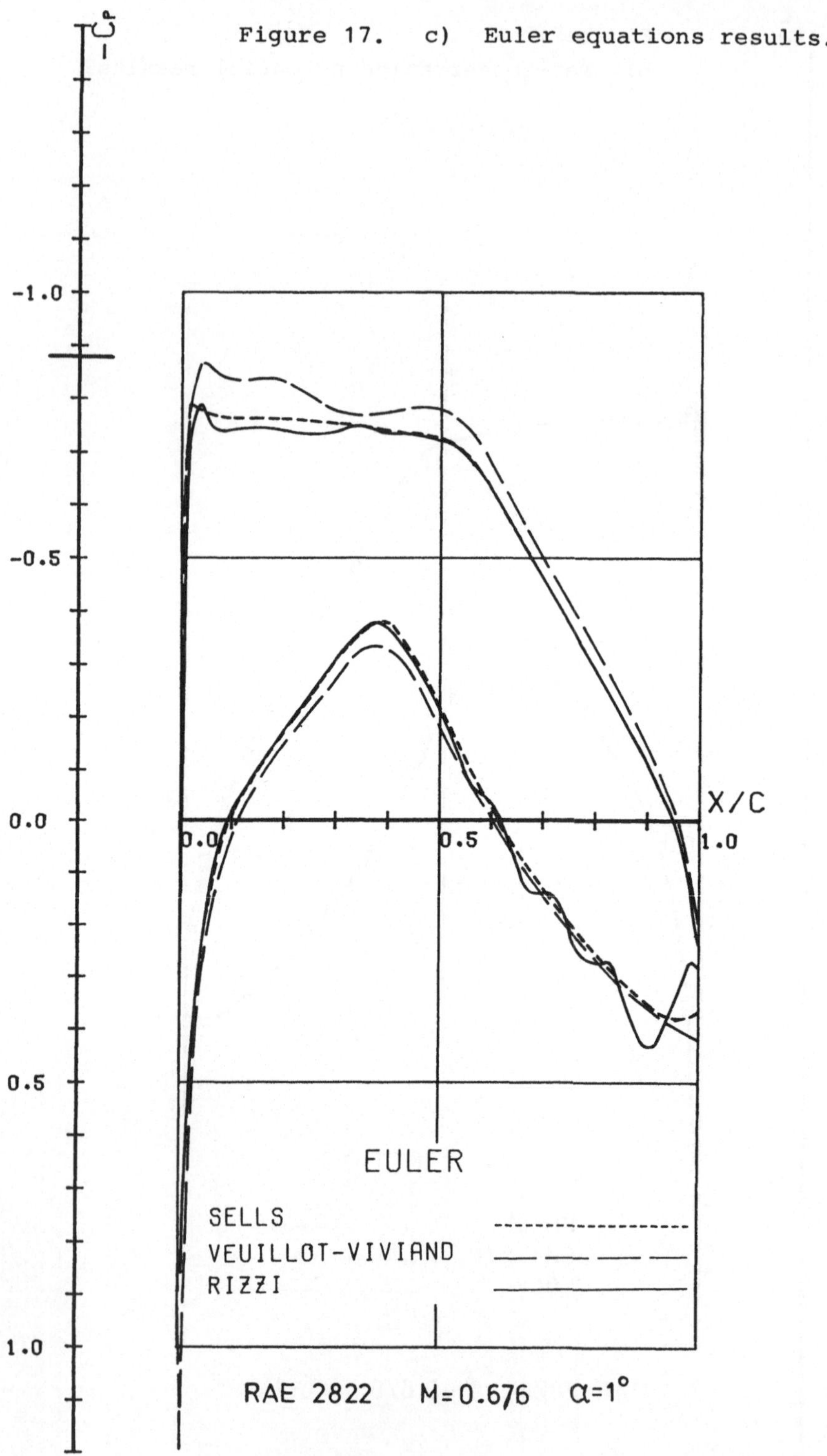

Figure 17.   c)   Euler equations results.

218

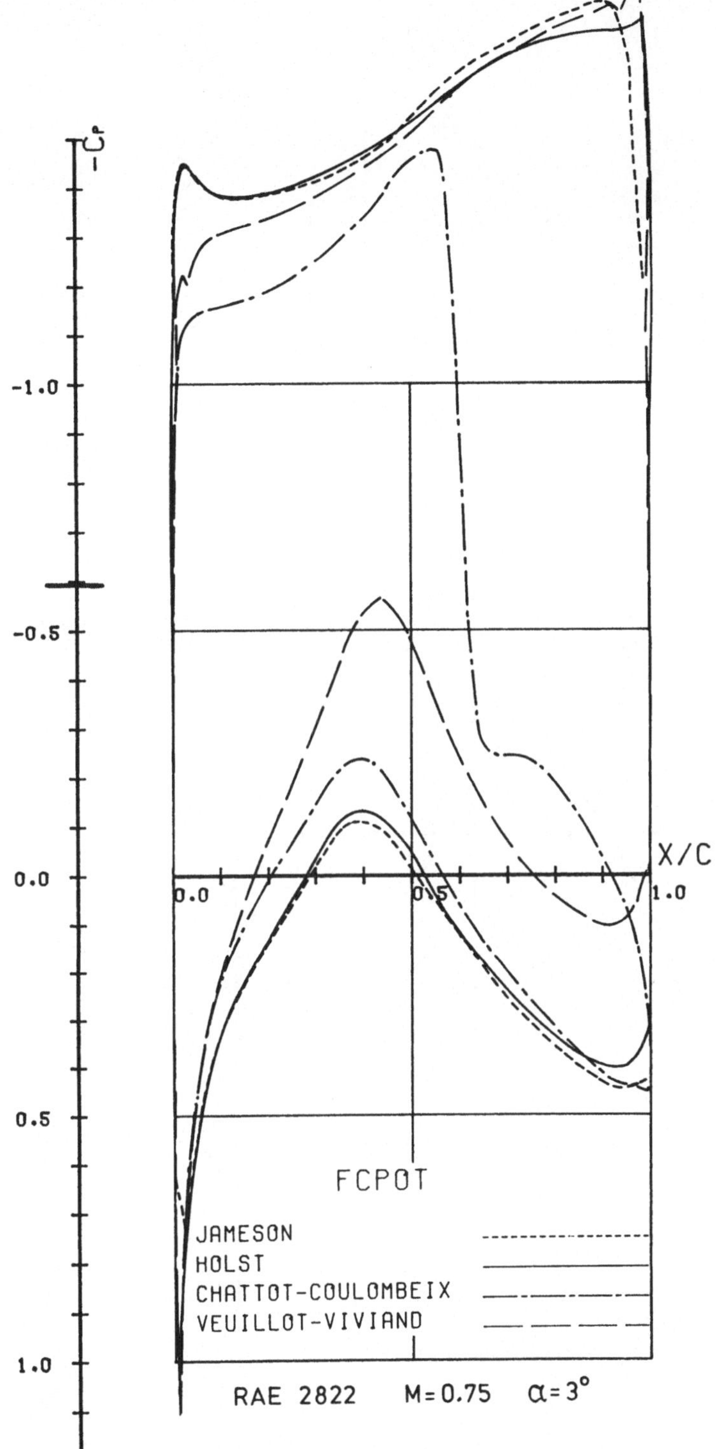

Figure 18. Comparison of the pressure coefficients computed on the airfoil. Problem C.II. Supercritical flow.

a) Fully-conservative potential results.

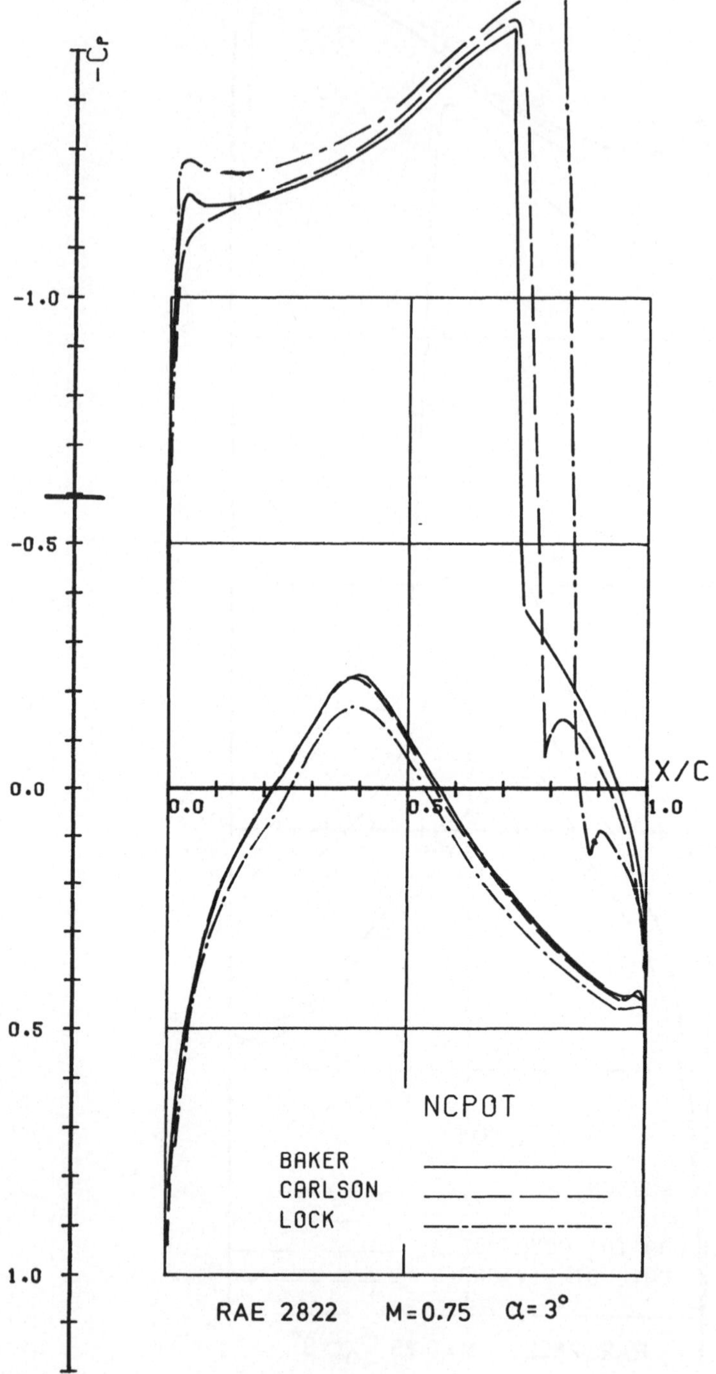

Figure 18.   b)   Non-conservative potential results.

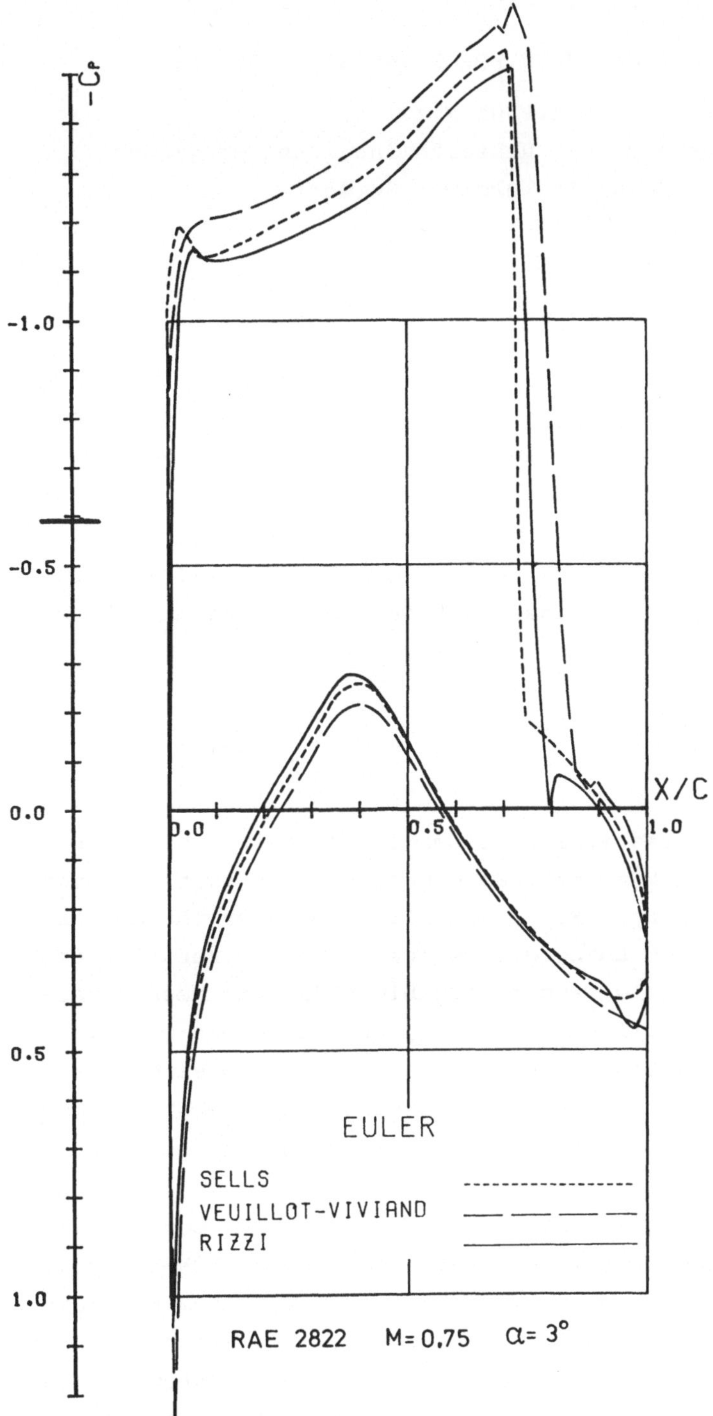

Figure 18.   c) Euler equations results.

221

COMPUTATIONAL MESH FOR TRANSONIC AIRFOILS

Arthur Rizzi

FFA The Aeronautical Research Institute of Sweden

S-161 11 Bromma, Sweden

SUMMARY

The standard mesh that we propose for Problems A, C, D, and G
may be generated by the program described here in conjunction
with the accompanying data cards. It is a curvilinear body-
fitted type system produced by direct geometrical construction,
where normally the body coincides with the first mesh line.
Provision was made, however, to locate the body half way between the first two mesh lines in order to accommodate those
methods which require it.

INTRODUCTION

The study of many practical problems often rests upon the numerical solution to a mathematical boundary-value problem having
a complicated geometry. For a numerical solution the domain of
the problem has first to be discretized, and the need for an
adaptable, controllable, and reasonably efficient mesh generation procedure is well recognized by practitioners of computational methods in fluid dynamics. Underlying most of the procedures is a transformation to a coordinate system which con-
forms to the boundaries and spans the interior of the domain.
A coordinate transformation maps a simply-connected domain one-
to-one onto a rectangular domain, and therefore a combination
of these may be used for more general domains. With branch cuts
and other methods for smoothly connecting one system to another,
a wide variety of problems can be studied. The success of such
an approach is aided if the transition from one system to the
next is done smoothly, which calls for some degree of control
over where the transition is to be located, what type of trans-

formation is to be used, and what the final distribution of
mesh points shall be, particularly near the bounding regions.

Three different types of procedures, primarily two-dimensional,
are currently in use for the generation of a mesh about a single
or multiple airfoil, with some extension to simple three-dimensional shapes. The first of these is based upon complex variables and conformal mapping[1,2], but is somewhat constrained in
its ability to prescribe desirable mesh properties and grid-
point distributions along the boundaries, and is not readily
extendable to three dimensions. In order to overcome these
limitations the second procedure[3] was developed, based on the
numerical solution of two coupled Poisson equations, one for
each coordinate direction of the problem domain, the source
terms of which are arbitrarily chosen to control the locations
of coordinate curves, and hence mesh spacing, in the interior.
The distribution of the boundary grid points must still be determined, however, by other means, and is used as boundary conditions in the generation procedure. The source terms may be
located anywhere in the domain, but exact control of the resulting mesh spacing is missing. For example in the construction of an airfoil mesh Sorenson and Steger[4] found it was particularly difficult to find two source functions which would
yield a predetermined and uniform spacing between the airfoil
and the first adjacent grid line. Instead, they set these
terms to zero, solved the two corresponding Laplace equations,
and of the two families of resulting coordinate curves, they
retained only that one (ξ = constant) which intersects the airfoil, and discarded the other. By a direct subdivision of each
of the retained lines, they achieved an acceptable mesh,
although one that still lacks differentiability across the line
following the trailing edge, a feature inherent to this type
of approach because of the boundary conditions imposed on branch
cuts.

The third way to generate a network of grid points is by direct
geometrical construction[5]. It possesses none of the mathematical sophistication of the previous two, but does afford a de-

gree of control that is constrained only be the various geometrical elements which one builds into it. The procedure proposed herein falls into this category. The idea is to offer,
in a FORTRAN program, direct control of the grid spacing and
point distribution near the airfoil, while providing an adequate coordinate system in the interior with a minimal degree
of complication.

DIRECT CONSTRUCTION OF THE MESH

The construction of the mesh is founded on the curvilinear coordinate system ξ, η, illustrated in Fig. 1, for which the line
$\eta = \eta_{min}$ coincides with the airfoil and the curvilinear coordinate cut which follows thereafter. Like Sorenson and Steger it
begins by adopting one family of coordinate curves that should
intersect the airfoil (ξ = constant), but instead of solving
numerically two partial differential equations, the choice here
is a family of hyperbolas (see Figs. 2 and 3), given in the
parameter θ by

$$x = B + A \cosh(\eta) \cos \theta$$
$$y = A \sinh(\eta) \sin \theta$$

In addition to its simplicity, the primary reason for this
choice is that far away from the airfoil, where its effect is
almost a point disturbance, the hyperbolas approach their
straight-line asymptotes, and the resulting grid system is practically a radial one. Furthermore the existence of a degener-
ate (straight-line) hyperbola when $\theta = \pi/2$ can simplify the
adjoining of this mesh to another.

The origins of the hyperbolas ξ_i, η_{min} are set by the desired
distribution of points on the airfoil, parameterized by its arc
length (SFOIL(I)). The user of PROGRAM MESH specifies the interval of arc length on the airfoil between the first two grid
points at the leading edge and the last two at the trailing
edge (Fig. 4 a). A blended distribution function (SSTRCH) for

the arc length (Fig. 4 b) is then achieved, and a SPLINE fit
versus x and y yields the coordinates of the grid points on the
airfoil. It solves for the parameter θ which is a good indicator of the suitability of the global distribution of ξ = constant lines. At the leading and trailing edges the first mesh
interval in the η direction is specified by the user (Fig. 4 c)
and the remaining are then calculated from a quadratic relation
between these two. Subsequent intervals are prescribed by an
exponential stretching function (STRECH). That part of the
mesh past the trailing edge is obtained by translating the hyperbola at the trailing edge downstream, according to some
prescribed function, to the last desired position XDOWN.

In order that the computational domain be simply connected, a
cut must be made in the physical domain (see Fig. 5 and Table 1).
The program offers some generality for this in that the cut may
be curvilinear. It is an exponential curve that leaves the
trailing edge at an angle THECUT with the x axis and attains
the value y = YCUT at x = XDOWN.

For those methods like the usual relaxation procedures the program can also construct a mesh in which the body is located
half-way between the first two mesh lines (J= 1 and 2, see Fig.6).
This is accomplished simply by removing the COMMENT label from
statements MESH2500-2520, MESH2850-2870, and MESH3030-3050.

The nodal points x, y which are generated are stored in the
arrays $X(I,J)$ and $Y(I,J)$.

An outline of the structure of the program is illustrated by
the flow chart in Fig. 7, and an entire listing of the FORTRAN
statements comprising this program is given in Table 2.

The standard Workshop meshes proposed for Problems A, C, D,
and G are generated by the use of PROGRAM MESH together with

the input data*) listed in Table 3. These meshes are plotted
in Figs. 8. A copy of the program and input data may be obtained by direct request to the author.

ACKNOWLEDGEMENT

It is a pleasure to acknowledge and publicly thank Drs.H.Viviand
and J.P. Veuillot and others at ONERA for their help in the
development of and improvement in this mesh-generating program.

REFERENCES

1.   Sells, C.C.L.: Plane Subcritical Flow past a Lifting Air-
     foil, Proc. Roy. Soc. (London), Vol. 308 A, 1968,
     pp. 377-401.

2.   Caughey, D.A.: A Systematic Procedure for Generating Use-
     ful Conformal Mappings. Intl. J. Numerical Methods in
     Engr., Vol. 12, pp. 1651-57, 1978.

3.   Thompson, J.F.; Thames, F.C., and Mastin, C.W.: Automatic
     Numerical Generation of Body-fitted Curvilinear Coordinate
     System for Field Containing and Number of Arbitrary Two-
     Dimensional Bodies. J. Comp. Phys., Vol. 15, pp. 299-319,
     1974.

4.   Sorenson, R.L.; and Steger, J.L.: Simplified Clustering
     of Non-orthogonal Grids Generated by Ellystic Partial Dif-
     ferential Equations. NASA TM 73252, 1977.

5.   Eiseman, P.R.: A Coordinate System for A Viscous Tran-
     sonic Cascade Analysis. J. Comp. Phys., Vol. 26,
     pp. 307-338, 1978.

-----

*) During the Workshop Antony Jameson pointed out that the coor-
   dinates of the Korn 1 airfoil listed here and taken from
   Kacprzynski et al. (NRC Aero. Rep. LR-554, 1971) are not entirely accurate and may give rise to a weak shock wave even
   at design conditions. Garabedian has subsequently corrected
   this inaccuracy.

Table 1. Definition of some of the input parameters.

```
NPTWK       NUMBER OF POINTS IN THE WAKE
IL          TOTAL NUMBER OF POINTS IN THE PSE(I) DIRECTION FROM THE
            DOWNSTREAM LOWER EDGE TO THE DOWNSTREAM UPPER EDGE OF
            THE COORDINATE CUT  (MUST BE ODD)
JL          TOTAL NUMBER OF POINTS IN THE ETA(J) DIRECTION FROM THE
            AIRFOIL SURFACE TO FARFIELD
BMINA       B-A  (A AND B ARE CONSTANTS FOR THE HYPERBOLAS)
THETE       VALUE OF THETA AT THE TRAILING EDGE (DEG.)
CL          CHORD LENGTH - NORMALLY EQUAL TO ONE
XUP,XDOWN   THE COMPUTATIONAL DOMAIN EXTENDS FROM XUP TO XDOWN
YFAR        THE UPSTREAM PART OF FARFIELD BOUNDARY IS AN ELLIPSE
            CENTERED AT ! X=B Y=0 - AND YFAR IS THE SEMIMINOR AXIS
            PARALLEL TO OY
YCUT        THE COORDINATE CUT HAS AN EXPONENTIAL SHAPE WHICH
            BECOMES PARALLEL TO THE X-AXIS AT LARGE VALUES OF X. AT
            X=XDOWN IT PASSES THROUGH THE POINT Y=YCUT
THECUT      THE ANGLE THAT THE CUT MAKES WITH THE X-AXIS AT THE
            TRAILING EDGE   (DEG.)
DSLE        THE LENGTH OF THE FIRST MESH CELL ON THE AIRFOIL SURFACE
            AT THE LEADING EDGE (LE)
DSTE        THE LENGTH OF THE MESH CELL ON THE AIRFOIL SURFACE AT
            THE TRAILING EDGE (TE)
DETLE       THE THICKNESS OF THE MESH CELL AT THE LEADING EDGE
DETTE       THE THICKNESS OF THE MESH CELL AT THE TRAILING EDGE
            OBSERVE! DSLE,DSTE,DETLE,DETTE ARE SCALED BY THE CHORD
                LENGTH (CL)
XSL1        THE RELATIVE VALUE OF ARC LENGTH AT  N=ISL1  ON THE
            LOWER SURFACE WHERE THE TRAILING EDGE STRETCHING ENDS
XSL2        THE RELATIVE VALUE OF ARC LENGTH AT  N=ISL2  ON THE
            LOWER SURFACE WHERE THE LEADING EDGE STRETCHING BEGINS
                OBSERVE ! XSL1 AND XSL2 ARE MEASURED FROM THE TRAILING
                    EDGE AND ARE SCALED BY THE ARC LENGTH OF THE
                    LOWER SURFACE (STOTL)
XSU1        THE RELATIVE VALUE OF ARC LENGTH AT  N=ISU1  ON THE
            UPPER SURFACE WHERE THE LEADING EDGE STRETCHING ENDS
XSU2        THE RELATIVE VALUE OF ARC LENGTH AT  N=ISU2  ON THE
            UPPER SURFACE WHERE THE TRAILING EDGE STRETCHING BEGINS
                OBSERVE ! XSU1 AND XSU2 ARE MEASURED FROM THE LEADING
                    EDGE AND ARE SCALED BY THE ARC LENGTH OF THE
                    UPPER SURFACE (STOTU)
NSL1        NUMBER OF MESH CELLS IN THE STRETCHED REGION 1<N<ISL1
            WHICH DETERMINES THAT   ISL1=NSL1+1
            (THE TRAILING EDGE CORRESPONDS TO N=1 OR N=NPW AND THE
            LEADING EDGE TO N=N0)
NSL2        NUMBER OF MESH CELLS IN THE STRETCHED REGION ISL2<N<N0
            WHICH DETERMINES THAT   ISL2=N0-NSL2
NSU1        NUMBER OF MESH CELLS IN THE STRETCHED REGION  N0<N<ISU1
            WHICH DETERMINES THAT   ISU1=N0+NSU1
NSU2        NUMBER OF MESH CELLS IN THE STRETCHED REGION ISU2<N<NPW
            WHICH DETERMINES THAT   ISU2=NPW-NSU2
TEXT        TITLE
NIN         NUMBER OF INPUT COORDINATES THAT DEFINE THE AIRFOIL
N00         INDEX OF THE NOSE INPUT COORDINATES (I.E. XIN=0 YIN=0)
XIN,YIN     VALUES OF INPUT COORDINATES THAT DEFINE THE AIRFOIL
```

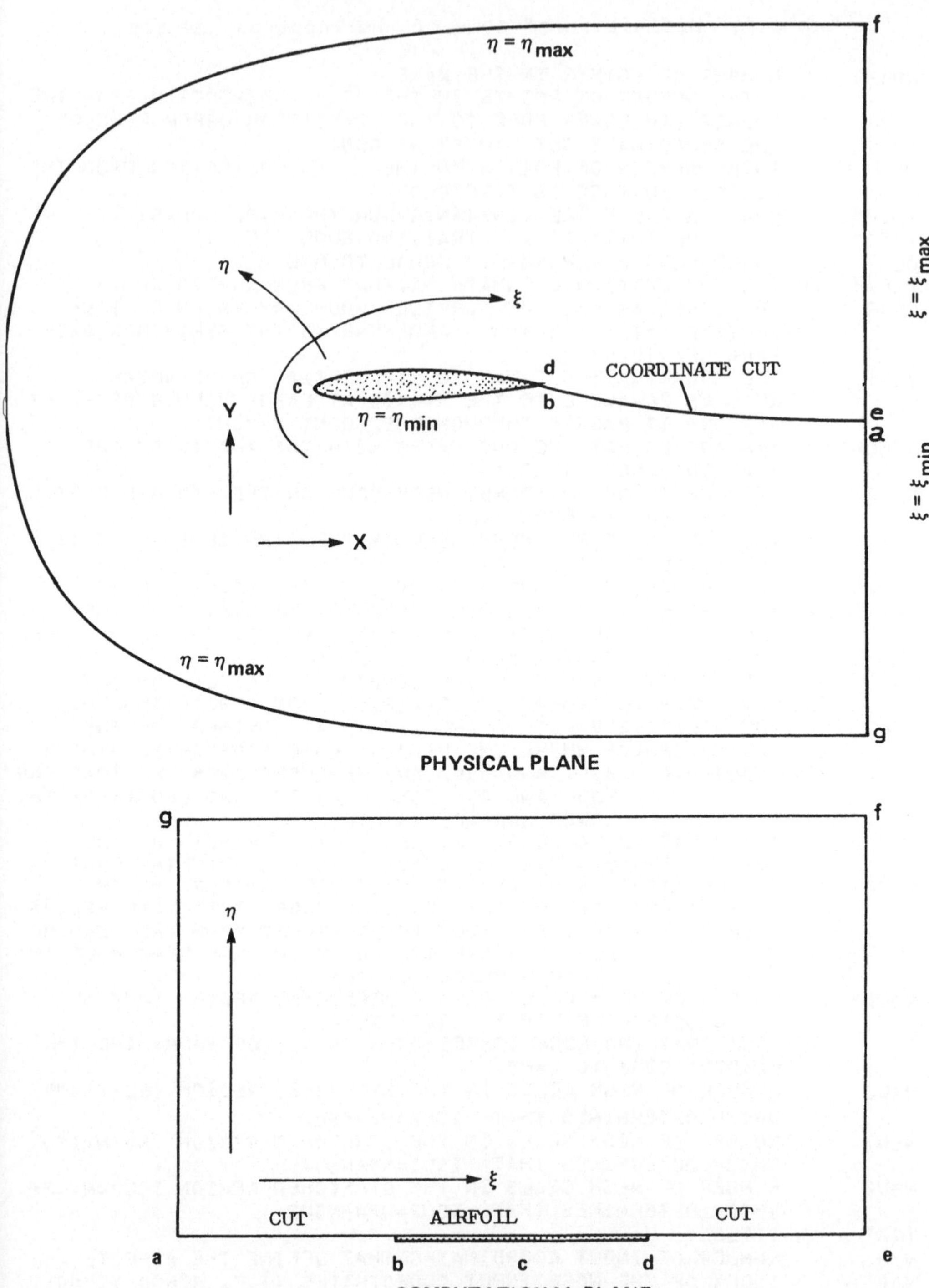

Fig. 1 Coordinate transformation of the domain of the problem to a rectangular computational domain.

228

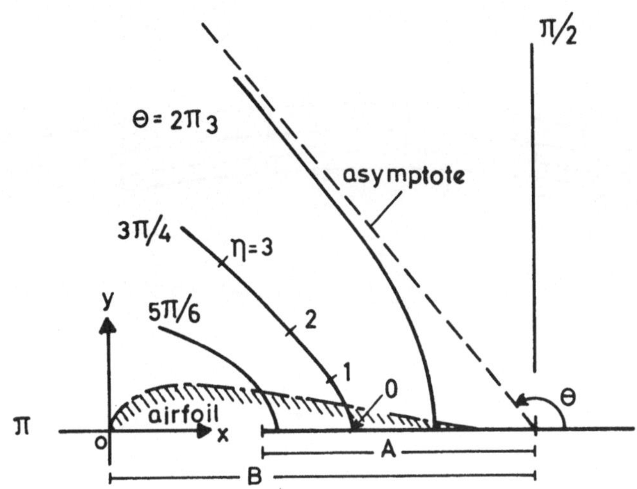

Fig. 2   The family of hyperbolas used for
the basic construction of the mesh.

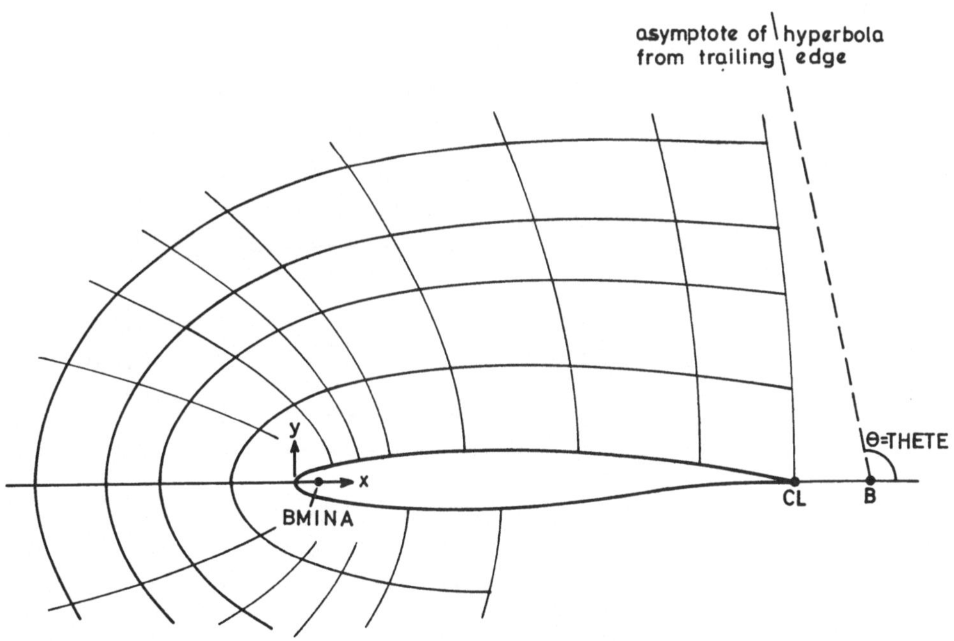

Fig. 3   Partial control of the global distribution of Theta
achieved by the two parameters   THETE, BMINA.

DSLE
DSTE

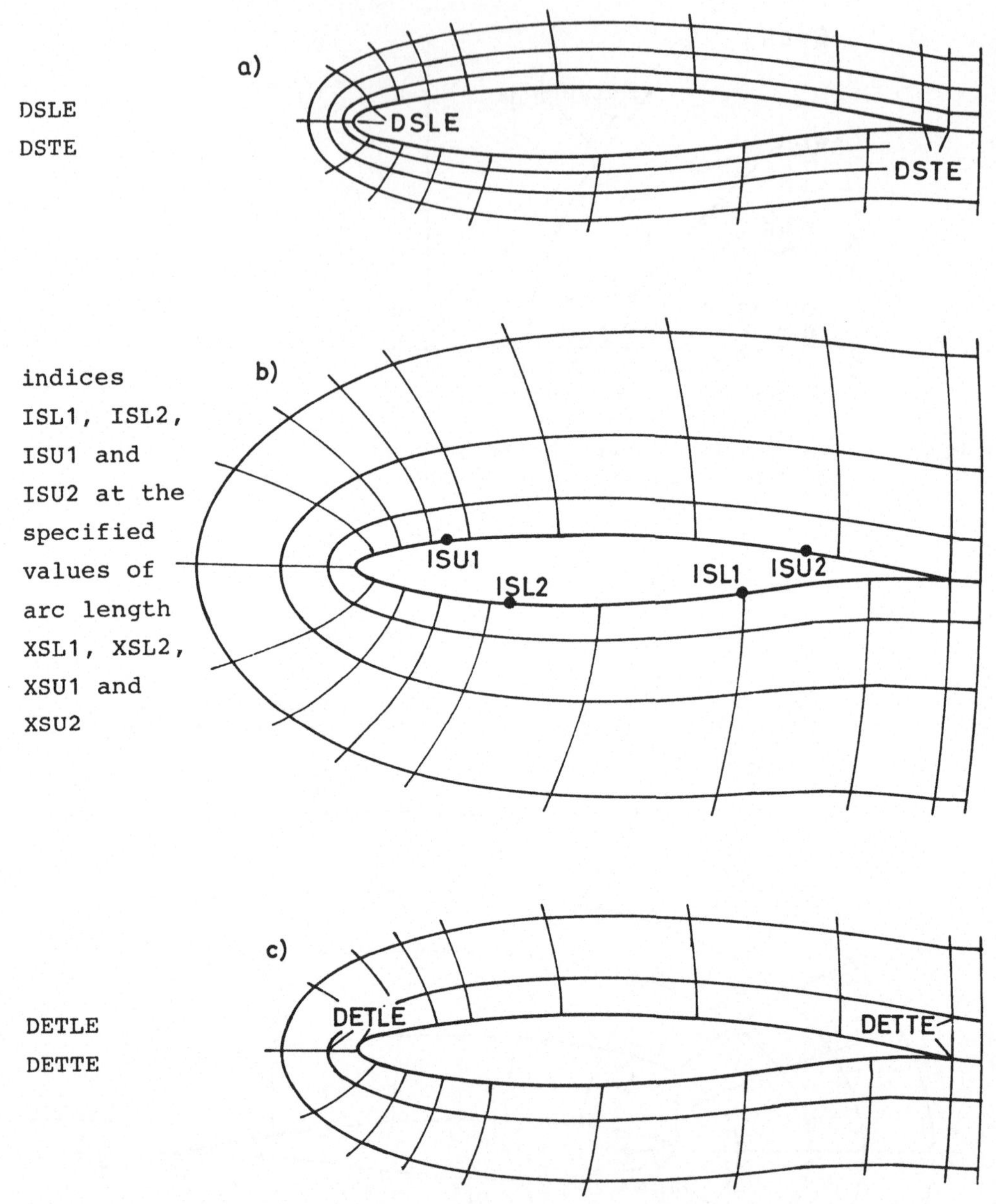

indices
ISL1, ISL2,
ISU1 and
ISU2 at the
specified
values of
arc length
XSL1, XSL2,
XSU1 and
XSU2

DETLE
DETTE

Control of these properties is managed by the
parameters defined in the illustrations above

Fig. 4 Specification of the location of grid points
and length of Mesh intervals near the airfoil.

230

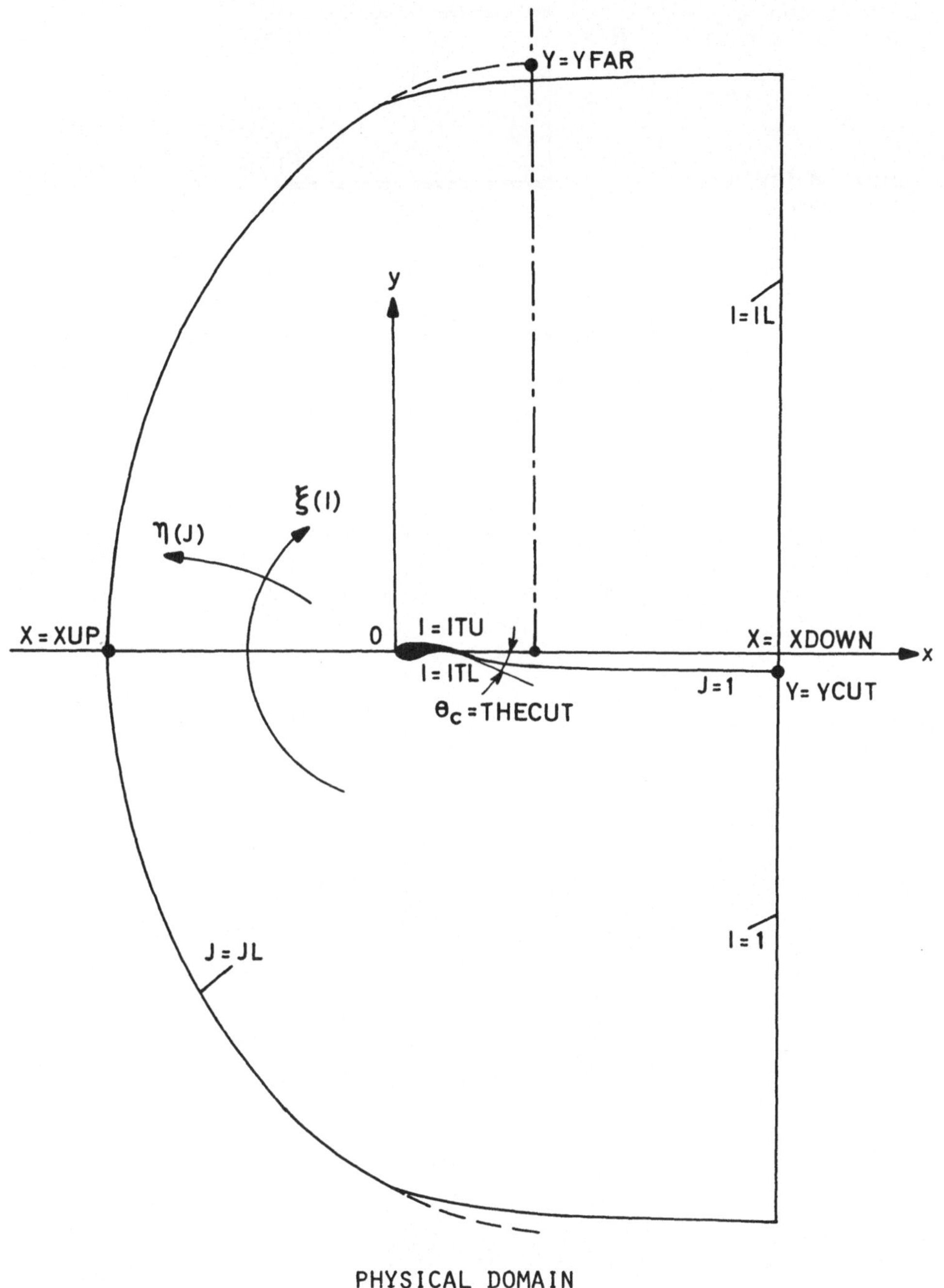

Fig. 5 a Definition of some of the input parameters.

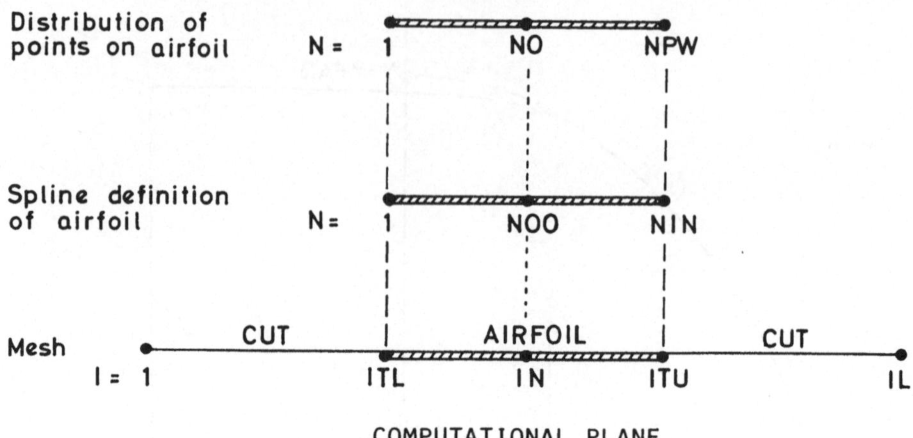

Fig. 5 b Various indexing systems are used
for the definition of the airfoil.

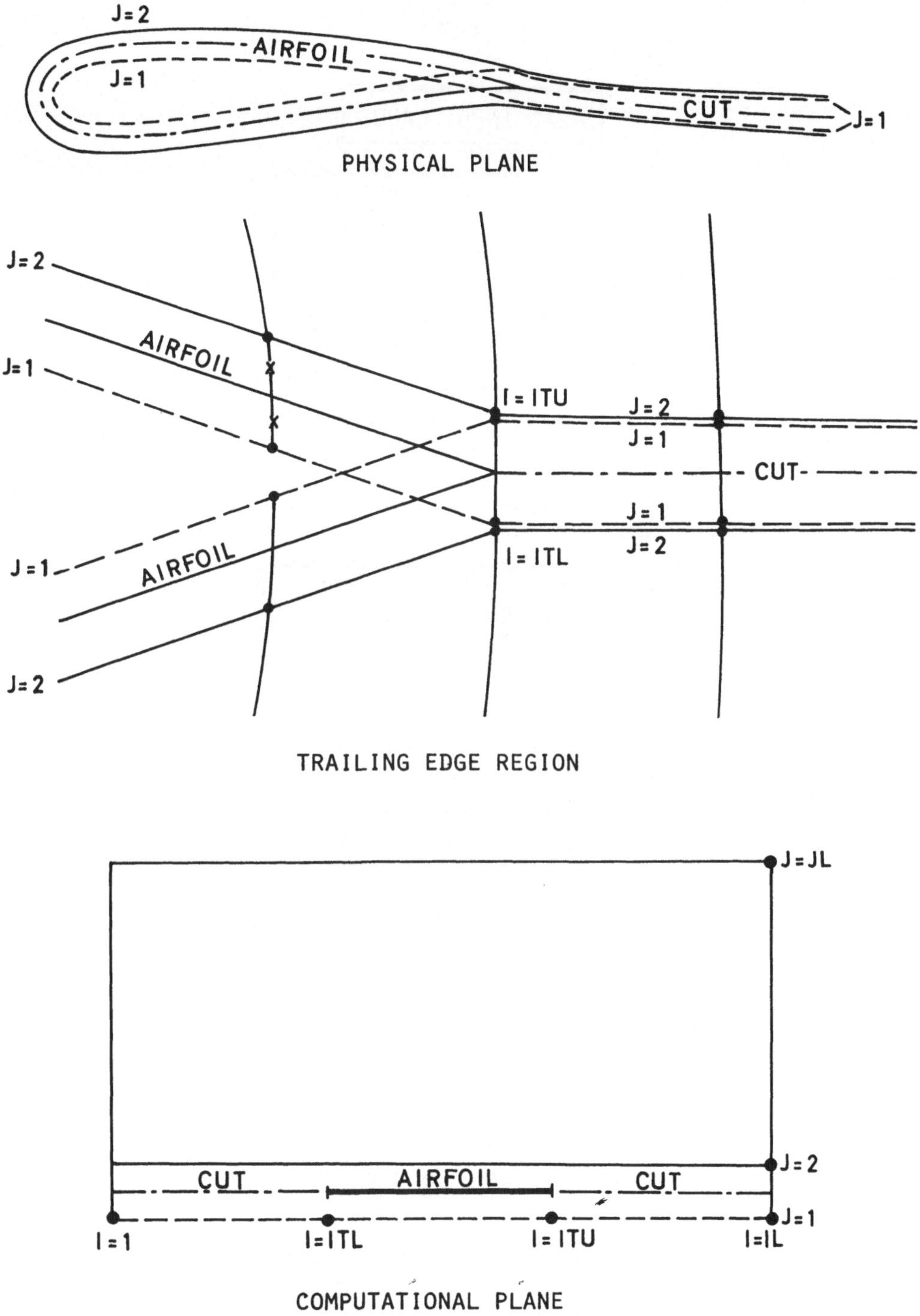

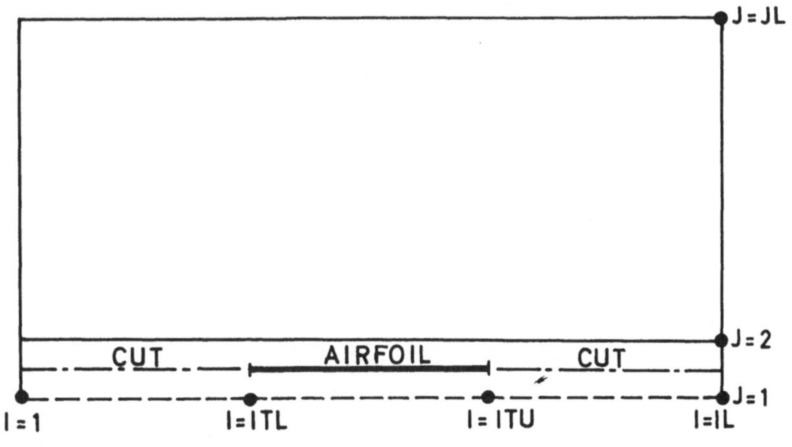

Fig. 6   Mesh lines J = 1 and J = 2 in the
case of a half-mesh at the wall.

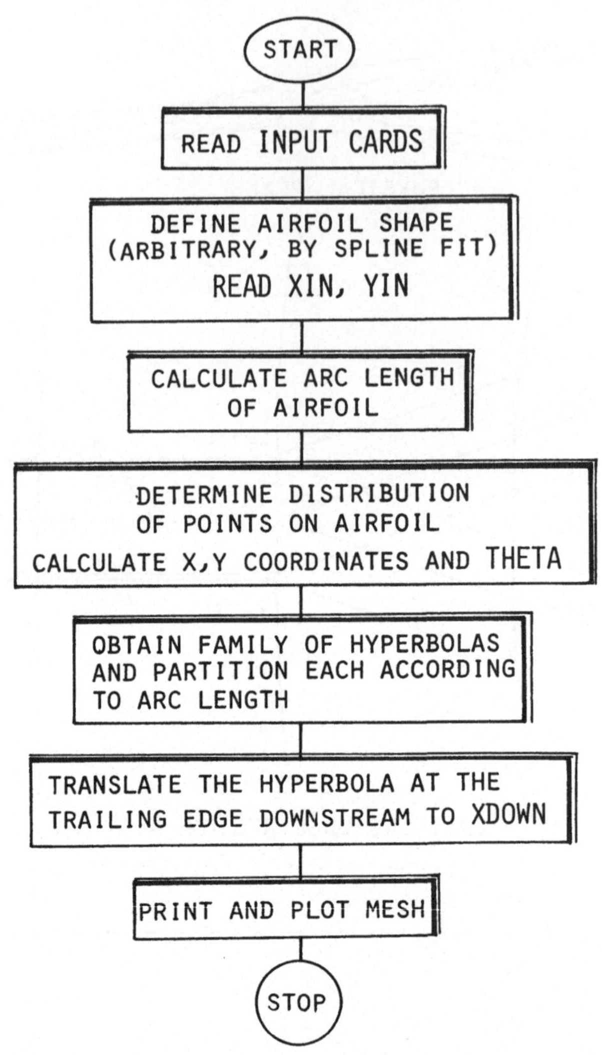

NOTE: Subroutines SPLINE and STRECH make use of
        several tolerance limits when finding roots.
        These may depend on the accuracy of the par-
        ticular machine being used.

Fig. 7 Block diagram for PROGRAM MESH.

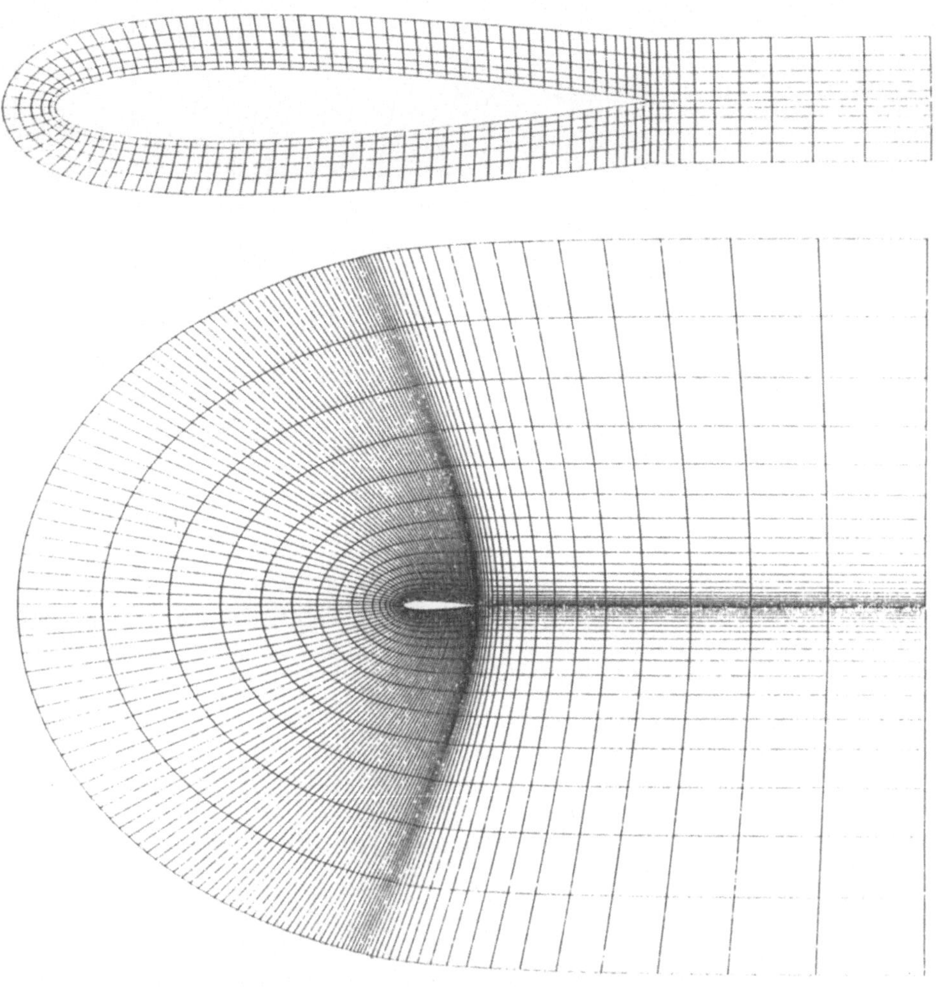

Fig. 8 The standard meshes for the workshop,
Problems A, C, D and G.

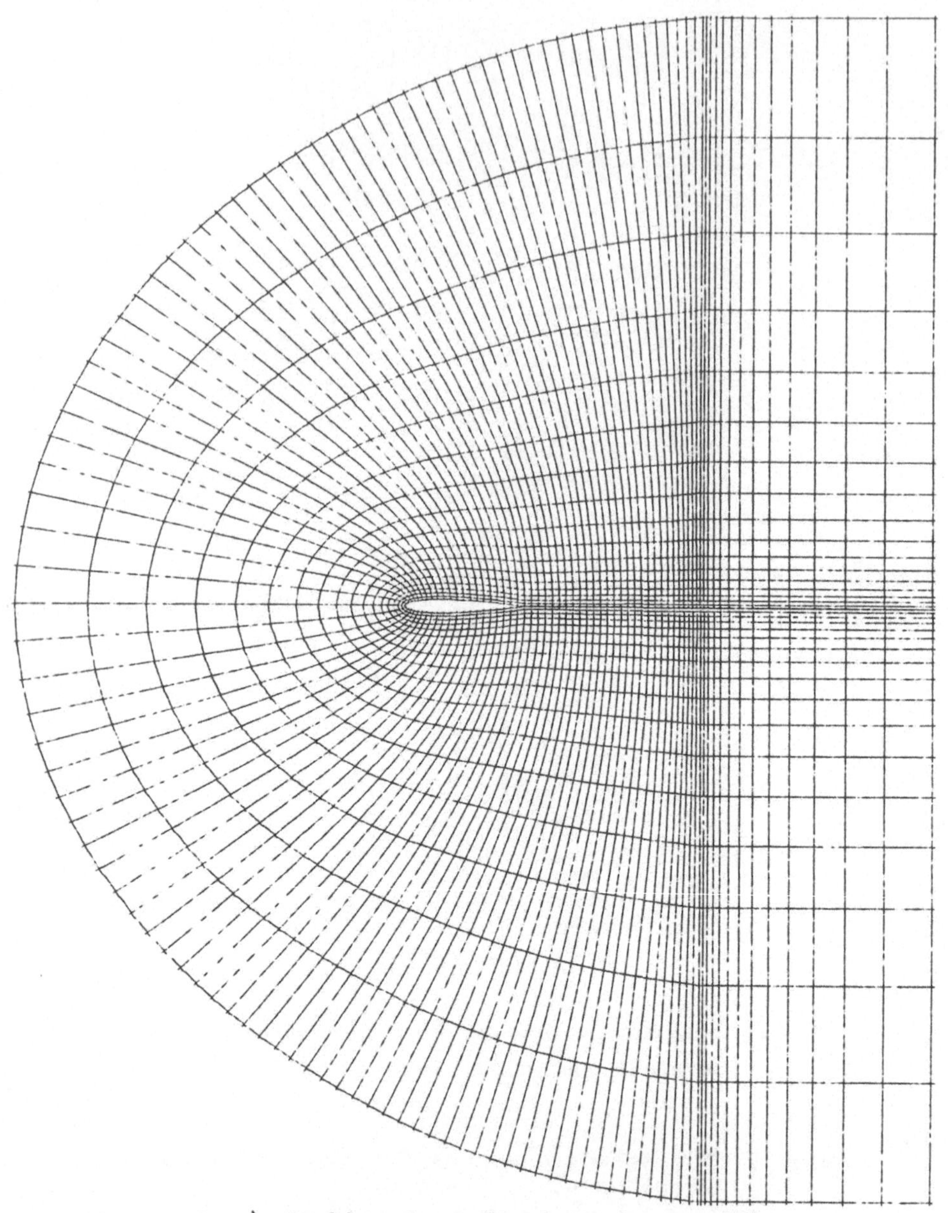

a_{ii}) Problem A.   (Part II-3, M = .95).

Airfoil NACA 0012 141x25 points
(only partially drawn). The outer
boundary extends 24 chords above
and below the airfoil and 12 chords
downstream of it.

Fig. 8 Cont.

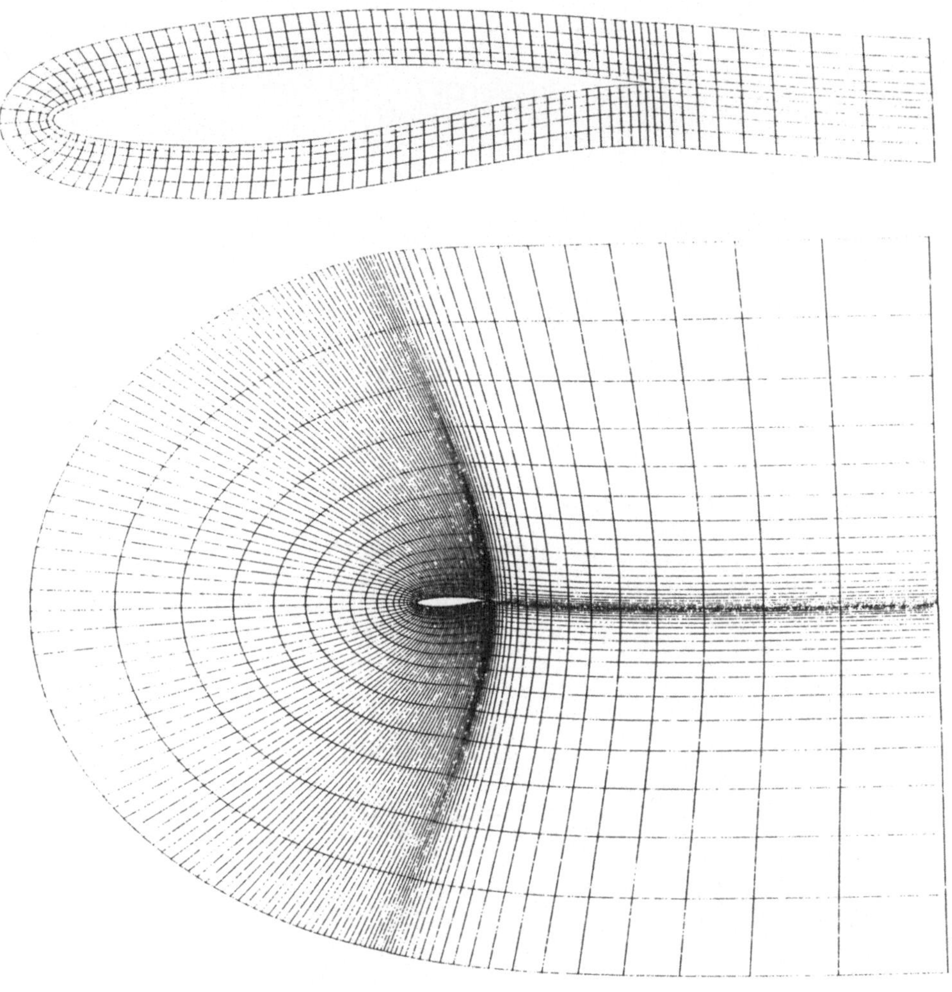

b) Problem C.
    Airfoil RAE 2822. 141x21 points.

Fig. 8 Cont.

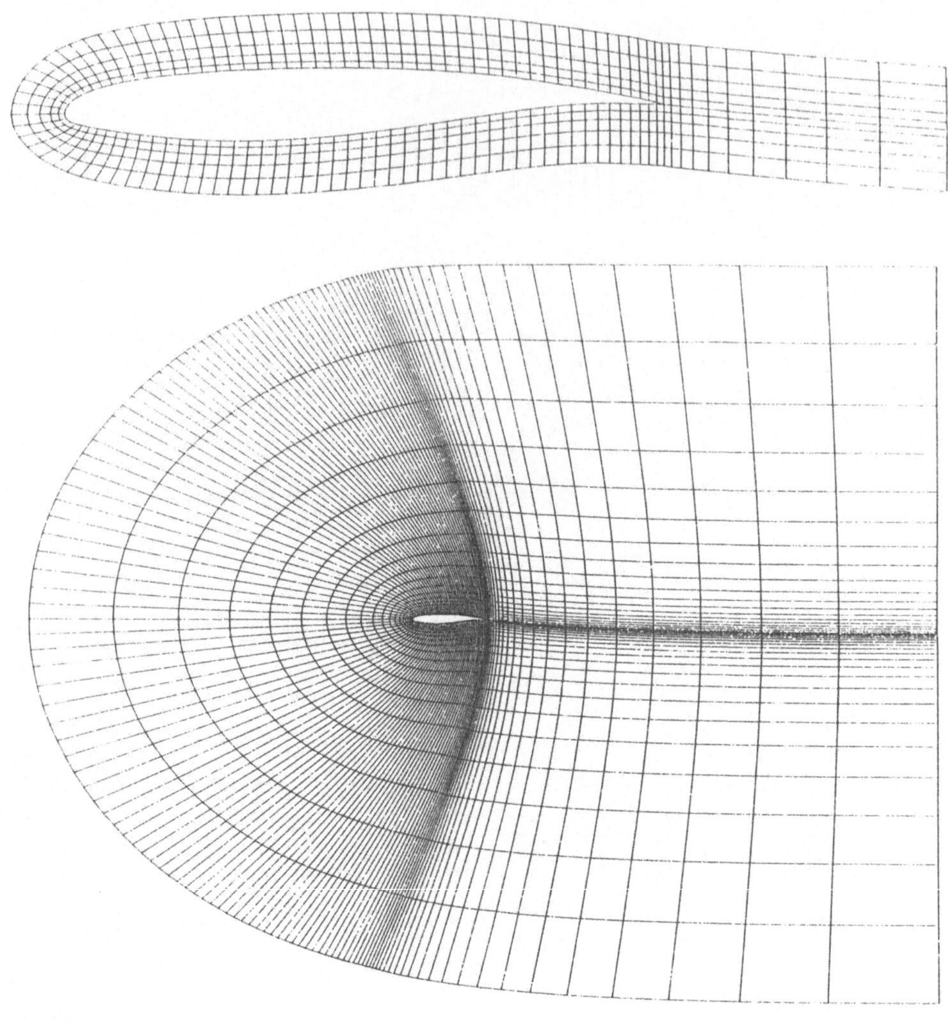

c) Problem D.
    Airfoil CAST 7 (closed) 141x21 points.

    Fig. 8 Cont.

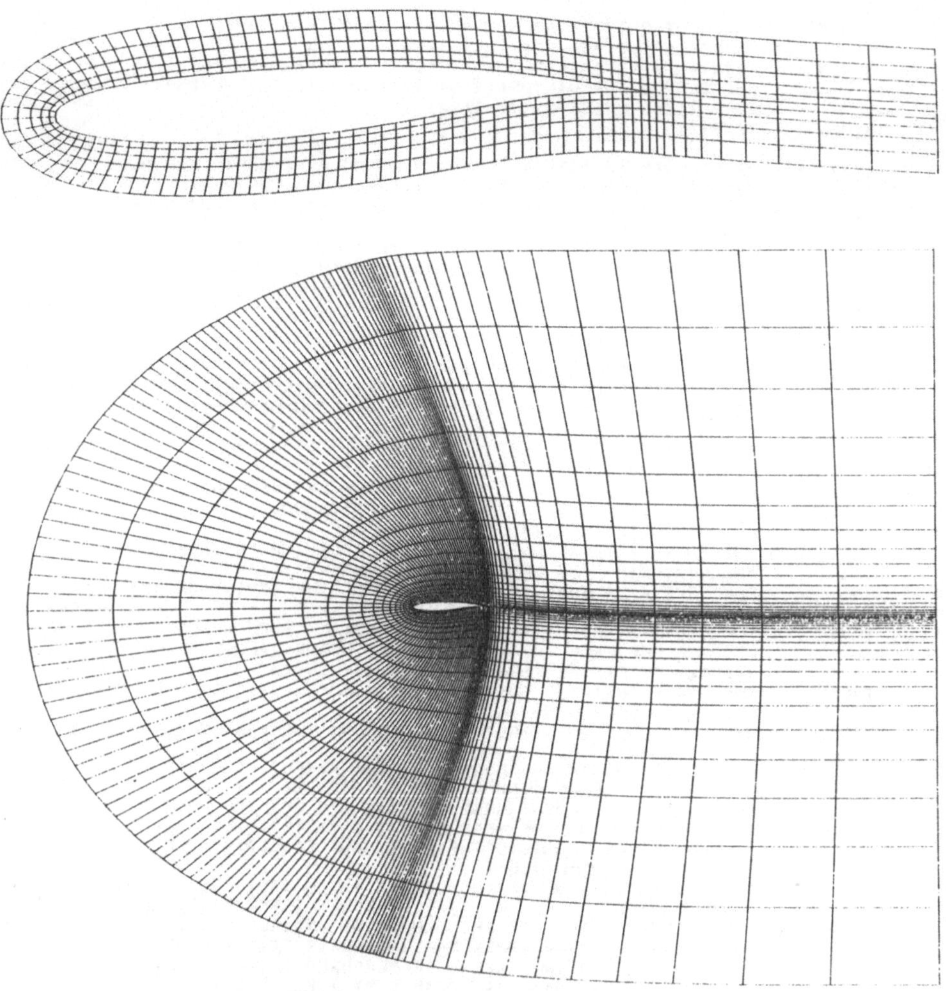

d) Problem G.
    Airfoil KORN 1 (closed) 141x21 points.

Fig. 8 Cont.

```
      PROGRAM MESH(INPUT,OUTPUT,TAPE5=INPUT,TAPE6=OUTPUT,TAPE7)          MESH   10
      DIMENSION  X(256,32),Y(256,32),THETA(256)                         MESH   20
      DIMENSION  XIN(256),YIN(256), XOUT(256),YOUT(256), SEN(256),      MESH   30
     .           SFOIL(256),DSFOIL(256), D1Y(256),D2Y(256),D3Y(256),    MESH   40
     .           SINT(10), XX(8192),YY(8192)                            MESH   50
      DIMENSION TEXT(8)                                                 MESH   60
C                                                                       MESH   70
C THIS PROGRAM DETERMINES AN AIRFOIL MESH BY USING HYPERBOLAS FOR ONE   MESH   80
C FAMILY OF THE GENERAL COORDINATES THAT FIT THE BODY. THE PARAMETERS   MESH   90
C THAT DESCRIBE THESE HYPERBOLAS ARE  THETA  AND  ETA. THE CARTESIAN    MESH  100
C COORDINATES OF THE POINTS ARE GIVEN BY!                              MESH  110
C                         X  =   B + A*COSH(ETA)*COS(THETA)            MESH  120
C                         Y  =       A*SINH(ETA)*SIN(THETA)            MESH  130
C                                                                       MESH  140
C INPUT PARAMETERS !                                                    MESH  150
C INPUT PARAMETERS !                                                    MESH  160
C                                                                       MESH  170
C    NPTWK      NUMBER OF POINTS IN THE WAKE                            MESH  180
C    IL         TOTAL NUMBER OF POINTS IN THE PSE(I) DIRECTION FROM THE MESH  190
C               DOWNSTREAM LOWER EDGE TO THE DOWNSTREAM UPPER EDGE OF   MESH  200
C               THE COORDINATE CUT  (MUST BE ODD)                       MESH  210
C    JL         TOTAL NUMBER OF POINTS IN THE ETA(J) DIRECTION FROM THE MESH  220
C               AIRFOIL SURFACE TO FARFIELD                            MESH  230
C    BMINA      B-A  (A AND B ARE CONSTANTS FOR THE HYPERBOLAS)        MESH  240
C    THETE      VALUE OF THETA AT THE TRAILING EDGE (DEG.)             MESH  250
C    CL         CHORD LENGTH - NORMALLY EQUAL TO ONE                   MESH  260
C    XUP,XDOWN  THE COMPUTATIONAL DOMAIN EXTENDS FROM XUP TO XDOWN     MESH  270
C    YFAR       THE UPSTREAM PART OF FARFIELD BOUNDARY IS AN ELLIPSE   MESH  280
C               CENTERED AT ! X=B Y=0 - AND YFAR IS THE SEMIMINOR AXIS MESH  290
C               PARALLEL TO OY                                         MESH  300
C    YCUT       THE COORDINATE CUT HAS AN EXPONENTIAL SHAPE WHICH      MESH  310
C               BECOMES PARALLEL TO THE X-AXIS AT LARGE VALUES OF X. AT MESH  320
C               X=XDOWN IT PASSES THROUGH THE POINT Y=YCUT            MESH  330
C    THECUT     THE ANGLE THAT THE CUT MAKES WITH THE X-AXIS AT THE    MESH  340
C               TRAILING EDGE  (DEG.)                                  MESH  350
C    DSLE       THE LENGTH OF THE FIRST MESH CELL ON THE AIRFOIL SURFACEMESH  360
C               AT THE LEADING EDGE (LE)                               MESH  370
C    DSTE       THE LENGTH OF THE MESH CELL ON THE AIRFOIL SURFACE AT  MESH  380
C               THE TRAILING EDGE (TE)                                 MESH  390
C    DETLE      THE THICKNESS OF THE MESH CELL AT THE LEADING EDGE     MESH  400
C    DETTE      THE THICKNESS OF THE MESH CELL AT THE TRAILING EDGE    MESH  410
C                  OBSERVE! DSLE,DSTE,DETLE,DETTE ARE SCALED BY THE CHORDMESH 420
C                        LENGTH (CL)                                   MESH  430
C    XSL1       THE RELATIVE VALUE OF ARC LENGTH AT  N=ISL1  ON THE    MESH  440
C               LOWER SURFACE WHERE THE TRAILING EDGE STRETCHING ENDS  MESH  450
C    XSL2       THE RELATIVE VALUE OF ARC LENGTH AT  N=ISL2  ON THE    MESH  460
C               LOWER SURFACE WHERE THE LEADING EDGE STRETCHING BEGINS MESH  470
C                  OBSERVE ! XSL1 AND XSL2 ARE MEASURED FROM THE TRAILINGMESH 480
C                        EDGE AND ARE SCALED BY THE ARC LENGTH OF THEMESH  490
C                        LOWER SURFACE (STOTL)                         MESH  500
C    XSU1       THE RELATIVE VALUE OF ARC LENGTH AT  N=ISU1  ON THE    MESH  510
C               UPPER SURFACE WHERE THE LEADING EDGE STRETCHING ENDS   MESH  520
C    XSU2       THE RELATIVE VALUE OF ARC LENGTH AT  N=ISU2  ON THE    MESH  530
C               UPPER SURFACE WHERE THE TRAILING EDGE STRETCHING BEGINS MESH 540
C                  OBSERVE ! XSU1 AND XSU2 ARE MEASURED FROM THE LEADING MESH 550
C                        EDGE AND ARE SCALED BY THE ARC LENGTH OF THEMESH  560
C                        UPPER SURFACE (STOTU)                         MESH  570
C    NSL1       NUMBER OF MESH CELLS IN THE STRETCHED REGION 1<N<ISL1  MESH  580
C               WHICH DETERMINES THAT  ISL1=NSL1+1                     MESH  590
C               (THE TRAILING EDGE CORRESPONDS TO N=1 OR N=NPW AND THE MESH  600
C               LEADING EDGE TO N=N0)                                  MESH  610
C    NSL2       NUMBER OF MESH CELLS IN THE STRETCHED REGION ISL2<N<N0 MESH  620
```

Cont.

```
C                    WHICH DETERMINES THAT   ISL2=NO-NSL2              MESH 630
C       NSU1         NUMBER OF MESH CELLS IN THE STRETCHED REGION   NO<N<ISU1 MESH 640
C                    WHICH DETERMINES THAT   ISU1=NO+NSU1             MESH 650
C       NSU2         NUMBER OF MESH CELLS IN THE STRETCHED REGION ISU2<N<NPW MESH 660
C                    WHICH DETERMINES THAT   ISU2=NPW-NSU2            MESH 670
C       TEXT         TITLE                                            MESH 680
C       NIN          NUMBER OF INPUT COORDINATES THAT DEFINE THE AIRFOIL MESH 690
C       NOO          INDEX OF THE NOSE INPUT COORDINATES (I.E. XIN=0 YIN=0) MESH 700
C       XIN,YIN      VALUES OF INPUT COORDINATES THAT DEFINE THE AIRFOIL MESH 710
C                                                                     MESH 720
C     INTEGRATION BY SIMPSONS RULE                                    MESH 730
      SIMSON( X1, X2) = A*(X2-X1)*(     SQRT(SINH(X1)**2+STHESQ)       MESH 740
     .                           +4.*SQRT(SINH(.5*(X1+X2))**2+STHESQ)  MESH 750
     .                           +   SQRT(SINH(X2)**2+STHESQ) )/6.     MESH 760
C                                                                     MESH 770
      PI = 4.*ATAN( 1. )                                              MESH 780
      READ(5,103) TEXT                                                MESH 790
      READ(5,1) NPTWK, IL, JL                                         MESH 800
      READ(5,2) RMINA, THETE                                          MESH 810
      READ(5,2) CL, XUP, XDOWN, YFAR                                  MESH 820
      READ(5,2) YCUT, THECUT                                          MESH 830
      READ(5,2) DSLE,DSTE, DETLE,DETTE                                MESH 840
      READ(5,2) XSL1,XSL2, XSU1,XSU2                                  MESH 850
      READ(5,1) NSL1,NSL2, NSU1,NSU2                                  MESH 860
      READ(5,1) NIN, NOO                                              MESH 870
      READ(5,2)  (XIN(I),YIN(I), I=1,NIN)                             MESH 880
C                                                                     MESH 890
      ITL = NPTWK +1                                                  MESH 900
      ITU = IL -NPTWK                                                 MESH 910
      IN  = IL/2 +1                                                   MESH 920
      NO  = IN -ITL +1                                                MESH 930
      NPW = IL -2*NPTWK                                               MESH 940
      NPT = IL*JL                                                     MESH 950
      NM5 = 5*INT(FLOAT(NPW)/5.)                                      MESH 960
      NRM = NPW -NM5                                                  MESH 970
      N5P = NM5 +1                                                    MESH 980
      IM5 = 5*INT(FLOAT(IL)/5.)                                       MESH 990
      IRM = IL -IM5                                                   MESH1000
      I5P = IM5 +1                                                    MESH1010
      ISL1 = NSL1+1                                                   MESH1020
      ISL2 = NO-NSL2                                                  MESH1030
      ISU1 = NO+NSU1                                                  MESH1040
      ISU2 = NPW-NSU2                                                 MESH1050
      A = (CL-RMINA) /(1.+COS(PI*THETE/180.))                        MESH1060
      R = RMINA +A                                                    MESH1070
      XFAR = XUP-R                                                    MESH1080
      DYCUT = TAN( PI*THECUT/180. )                                   MESH1090
C                                                                     MESH1100
      WRITE(6,99)                                                     MESH1110
      WRITE(6,111) TEXT                                               MESH1120
      WRITE(6,106)                                                    MESH1130
      WRITE(6,114) CL, XUP, XDOWN, YFAR                               MESH1140
      WRITE(6,100) ITL, IN, ITU, IL                                   MESH1150
      WRITE(6,104) NPTWK, NPW, NO, JL                                 MESH1160
      WRITE(6,101) A, R, THETE, RMINA                                 MESH1170
      WRITE(6,120) YCUT, THECUT                                       MESH1180
      WRITE(6,110) DSLE, DSTE, DETLE, DETTE                           MESH1190
      WRITE(6,121) XSL1,XSL2, XSU1,XSU2                               MESH1200
      WRITE(6,122) NSL1,NSL2, NSU1,NSU2                               MESH1210
      WRITE(6,123) NPT                                                MESH1220
C                                                                     MESH1230
C---------------------------------------------------------------------MESH1240
```

Cont.

Table 2. Cont.

```
C       *   FIND BODY POINTS   *                                          MESH1250
C-------------------------------------------------------------------------MESH1260
C                                                                         MESH1270
C.. CALCULATE ARC LENGTH AND CURVATURE OF AIRFOIL                         MESH1280
C             (LOWER TO UPPER SURFACE, TRAILING-LEADING-TRAILING EDGE)    MESH1290
      SEN   (1) = 0.                                                      MESH1300
      SFOIL (1) = 0.                                                      MESH1310
      DO 4 I=2,NIN                                                        MESH1320
         SEN(I) = SEN(I-1) +SQRT( (XIN(I)-XIN(I-1))**2                    MESH1330
                                 +(YIN(I)-YIN(I-1))**2 )                  MESH1340
    4 CONTINUE                                                            MESH1350
      DO 10 NTIMES=1,2                                                    MESH1360
      DO 6 I=2,NIN                                                        MESH1370
         DSINT = (SEN(I)-SEN(I-1)) /(10-1)                               MESH1380
         DO 7 L=1,10                                                      MESH1390
            SINT(L) = SEN(I-1) + (L-1)*DSINT                              MESH1400
    7    CONTINUE                                                         MESH1410
         CALL SPLINE (SEN,XIN, SINT,THETA, XOUT, D1Y,D2Y,D3Y, NIN,10 )    MESH1420
         CALL SPLINE (SEN,YIN, SINT,THETA, YOUT, D1Y,D2Y,D3Y, NIN,10 )    MESH1430
         SUM = 0.                                                         MESH1440
         DO 9 L=2,10                                                      MESH1450
            SUM = SUM +DSINT*SQRT( (.5*(XOUT(L-1)+XOUT(L)))**2            MESH1460
                                  +(.5*(YOUT(L-1)+YOUT(L)))**2 )          MESH1470
    9    CONTINUE                                                         MESH1480
         SFOIL(I) = SFOIL(I-1) + SUM                                      MESH1490
    6 CONTINUE                                                            MESH1500
      DO 10 I=1,NIN                                                       MESH1510
         SEN(I) = SFOIL(I)                                                MESH1520
   10 CONTINUE                                                            MESH1530
C             DIMENSIONALIZE BY ARC LENGTH AND CHORD LENGTH               MESH1540
      STOTL = SEN(NOO)-SEN(1)                                             MESH1550
      STOTU = SEN(NIN)-SEN(NOO)                                           MESH1560
      STOT  = STOTU +STOTL                                                MESH1570
      XSL1  = SEN(1) +XSL1*STOTL                                          MESH1580
      XSL2  = SEN(1) +XSL2*STOTL                                          MESH1590
      XSU1  = SEN(NOO) +XSU1*STOTU                                        MESH1600
      XSU2  = SEN(NOO) +XSU2*STOTU                                        MESH1610
      DSLE  = DSLE *CL                                                    MESH1620
      DSTE  = DSTE *CL                                                    MESH1630
      DETLE = DETLE*CL                                                    MESH1640
      DETTE = DETTE*CL                                                    MESH1650
C             CURVATURE        (XOUT=CURVATURE K , YOUT=K*DS)             MESH1660
      CALL SPLINE (SEN,XIN, SEN,XOUT, YOUT, D1Y,D2Y,D3Y, NIN,NIN)         MESH1670
      CALL SPLINE (SEN,YIN, SEN,XOUT, YOUT, SFOIL,DSFOIL,D3Y, NIN,NIN)    MESH1680
      DO 12 N=1,NIN                                                       MESH1690
         XOUT(N) = D1Y(N)*DSFOIL(N) -SFOIL(N)*D2Y(N)                      MESH1700
   12 CONTINUE                                                            MESH1710
      YOUT(1) = 0.                                                        MESH1720
      DO 16 N=2,NIN                                                       MESH1730
         YOUT(N) = (SEN(N)-SEN(N-1))* XOUT(N)                             MESH1740
   16 CONTINUE                                                            MESH1750
C                                                                         MESH1760
C.. DEFINE DESIRED DISTRIBUTION OF ARC LENGTH    SFOIL                    MESH1770
C             LOWER SURFACE                                               MESH1780
      CALL SSTRCH(1,ISL1,ISL2,NO, 0.,XSL1,XSL2,SEN(NOO),DSTE,DSLE,SFOIL)  MESH1790
C             UPPER SURFACE                                               MESH1800
      CALL SSTRCH( NO,ISU1,ISU2,NPW, SEN(NOO),XSU1,XSU2,SEN(NIN),         MESH1810
                   DSLE,DSTE, SFOIL )                                     MESH1820
      DSFOIL(1) = 0.                                                      MESH1830
      DO 11 I=2,NPW                                                       MESH1840
         DSFOIL(I) = SFOIL(I) -SFOIL(I-1)                                 MESH1850
   11 CONTINUE                                                            MESH1860
```

Cont.

Table 2. Cont.

```
      WRITE(6,115)  STOTL, STOT                                          MESH1870
      WRITE(6,5)(NN,(DSFOIL(NN-1+II),SFOIL(NN-1+II),II=1,5), NN=1,NM5,5)MESH1880
      IF(NRM.GT.0) WRITE(6,5)N5P,(DSFOIL(NM5+II),SFOIL(NM5+II),II=1,NRM)MESH1890
C          CURVATURE       (XX=CURVATURE K ,  YY=K*DS)                   MESH1900
      CALL SPLINE (SEN,XOUT, SFOIL,  XX , DSFOIL, D1Y,D2Y,D3Y, NIN,NPW) MESH1910
      CALL SPLINE (SEN,YOUT, SFOIL,  YY , DSFOIL, D1Y,D2Y,D3Y, NIN,NPW) MESH1920
      WRITE(6,125)                                                       MESH1930
      WRITE(6,5)(NN,(YY      (NN-1+II),XX      (NN-1+II),II=1,5), NN=1,NM5,5)MESH1940
      IF(NRM.GT.0) WRITE(6,5)N5P,(YY      (NM5+II),XX      (NM5+II),II=1,NRM)MESH1950
C                                                                        MESH1960
C.. CALCULATE AIRFOIL COORDINATES AND ANGLES THETA                      MESH1970
      CALL SPLINE (SEN,XIN, SFOIL,XOUT, THETA, D1Y,D2Y,D3Y, NIN,NPW )   MESH1980
      CALL SPLINE (SEN,YIN, SFOIL,YOUT, THETA, D1Y,D2Y,D3Y, NIN,NPW )   MESH1990
      WRITE(6,116)                                                       MESH2000
      WRITE(6,5) (NN, (XOUT(NN-1+II),YOUT (NN-1+II),II=1,5), NN=1,NM5,5)MESH2010
      IF(NRM.GT.0) WRITE(6,5)N5P,(XOUT   (NM5+II),YOUT (NM5+II),II=1,NRM)MESH2020
      DO 15 N=1,NPW                                                      MESH2030
      I = N +ITL -1                                                      MESH2040
      ARG=((A-XOUT(N)+B)**2+YOUT(N)**2)*((A+XOUT(N)-B)**2+YOUT(N)**2)   MESH2050
      ZEE = ( A**2 +(XOUT(N)-B)**2 +YOUT(N)**2 -SQRT(ARG)) /(2.*A**2)   MESH2060
      THETA(I) = ACOS( -SQRT(ZEE) )                                     MESH2070
      IF (N.LT.NO)  THETA(I) = 2.*PI -THETA(I)                          MESH2080
      X(I,1) = XOUT(N)                                                   MESH2090
      Y(I,1) = YOUT(N)                                                   MESH2100
   15 CONTINUE                                                           MESH2110
      SUMTHE = .5*PI - THETA(ITU-1)                                     MESH2120
      DTH = THETA(ITU) -THETA(ITU-1)                                    MESH2130
      CALL STRECH (DTH, SUMTHE, IL-ITU+1, RDTH)                         MESH2140
      DO 20 I=ITU,IL                                                     MESH2150
      THETA(I) = THETA(I-1) +DTH                                        MESH2160
      IF (I.GT.ITU)  THETA(ITL+ITU-I) = 2.*PI -THETA(I)                MESH2170
      DTH = RDTH*DTH                                                    MESH2180
   20 CONTINUE                                                           MESH2190
      D1Y(1) = 180.*THETA(1)/PI                                         MESH2200
      D2Y(1) = 0.                                                        MESH2210
      DO 25 I=2,IL                                                       MESH2220
      D1Y(I) = 180.*THETA(I)/PI                                         MESH2230
      D2Y(I) = 180.*(THETA(I)-THETA(I-1))/PI                           MESH2240
   25 CONTINUE                                                           MESH2250
      WRITE(6,117)                                                       MESH2260
      WRITE(6,5) (NN, (D1Y (NN-1+II),D2Y  (NN-1+II),II=1,5), NN=1,IM5,5)MESH2270
      IF(IRM.GT.0) WRITE(6,5)I5P,(D1Y    (IM5+II),D2Y  (IM5+II),II=1,IRM)MESH2280
C                                                                        MESH2290
C-----------------------------------------------------------------------MESH2300
C      *    DEFINE THE ELLIPTIC-HYPERBOLIC COORDINATES OVER AIRFOIL   *MESH2310
C-----------------------------------------------------------------------MESH2320
C                                                                        MESH2330
      DO 30 I=ITL,ITU                                                    MESH2340
      DS = DETLE +(DETTE-DETLE)*(FLOAT(I-IN)/FLOAT(NO-1))**2            MESH2350
      RFAR = 1./SQRT((COS(THETA(I))/XFAR)**2+(SIN(THETA(I))/YFAR)**2)  MESH2360
      ETAJL = ALOG( 2.*RFAR/A )                                        MESH2370
      ETA1 = EFCNXY( X(I,1)-B, Y(I,1), A,THETA(I) )                    MESH2380
      DETA = (ETAJL -ETA1)/ 50.                                        MESH2390
      ETA = ETA1                                                        MESH2400
      STHESQ = SIN( THETA(I) )**2                                      MESH2410
      SUMDS = 0.                                                        MESH2420
      DO 35 L=1,50                                                      MESH2430
      SUMDS = SUMDS +SIMSON( ETA, ETA+DETA )                           MESH2440
      ETA = ETA +DETA                                                  MESH2450
   35 CONTINUE                                                          MESH2460
      CALL STRECH( DS, SUMDS, JL-1, RDS )                              MESH2470
      IF (I.EQ.ITU)  SUMDSU = SUMDS                                    MESH2480
```

Cont.

                    Table 2.   Cont.

```
        IF (I.EQ.ITL)  SUMDSL = SUMDS                                    MESH2490
C       ETA1 = EFCNS( -DS/2., 0., ETA1, A,THETA(I) )                     MESH2500
C       X(I,1) = B +A*COSH(ETA1)*COS(THETA(I))                          MESH2510
C       Y(I,1) =    A*SINH(ETA1)*SIN(THETA(I))                          MESH2520
        ETA = ETA1                                                      MESH2530
          ETA = EFCNS( S+DS, S, ETA, A,THETA(I) )                       MESH2560
          X(I,J) = B +A*COSH(ETA)*COS(THETA(I))                         MESH2570
          Y(I,J) =    A*SINH(ETA)*SIN(THETA(I))                         MESH2580
          S = S +DS                                                     MESH2590
          DS = DS*RDS                                                   MESH2600
   30 CONTINUE                                                          MESH2610
C                                                                       MESH2620
C-------------------------------------------------------------------------MESH2630
C          *    TRANSLATE HYPERBOLAS DOWNSTREAM   *                     MESH2640
C-------------------------------------------------------------------------MESH2650
C                                                                       MESH2660
      XTEU = X(ITU,1)                                                   MESH2670
      YTEU = Y(ITU,1)                                                   MESH2680
      XTEL = X(ITL,1)                                                   MESH2690
      YTEL = Y(ITL,1)                                                   MESH2700
      DX = DSTE                                                         MESH2710
      SUMDX = XDOWN-CL                                                  MESH2720
      CALL STRECH( DX, SUMDX, NPTWK, RDX )                             MESH2730
      XTRANS = 0.                                                      MESH2740
      DO 50 N=1,NPTWK                                                  MESH2750
        XTRANS = XTRANS +DX                                            MESH2760
        I = ITU +N                                                     MESH2770
        B = CL -A*COS(THETA(I)) +XTRANS                                MESH2780
        XCUT = CL +XTRANS                                              MESH2790
        X(I,1) = XCUT                                                  MESH2800
        Y(I,1) = YCUT +(YTEU-YCUT)*EXP( -DYCUT*(XCUT-XTEU)/YCUT )      MESH2810
        ETA1 = EFCNXY( XCUT-B, Y(I,1), A,THETA(I) )                   MESH2820
        DS = DETTE                                                     MESH2830
        CALL STRECH( DS, SUMDSU, JL-1, RDS )                          MESH2840
C       ETA1 = EFCNS( -DS/2., 0., ETA1, A,THETA(I) )                  MESH2850
C       X(I,1) = B +A*COSH(ETA1)*COS(THETA(I))                        MESH2860
C       Y(I,1) =    A*SINH(ETA1)*SIN(THETA(I))                        MESH2870
        ETA = ETA1                                                    MESH2880
        S = 0.                                                        MESH2890
        DO 51 J=2,JL                                                  MESH2900
          ETA = EFCNS( S+DS, S, ETA, A,THETA(I) )                     MESH2910
          X(I,J) = B +A*COSH(ETA)*COS(THETA(I))                       MESH2920
          Y(I,J) =    A*SINH(ETA)*SIN(THETA(I))                       MESH2930
          S = S +DS                                                   MESH2940
          DS = DS*RDS                                                  MESH2950
   51     CONTINUE                                                    MESH2960
        I = ITL -N                                                    MESH2970
        X(I,1) = XCUT                                                 MESH2980
        Y(I,1) = YCUT +(YTEL-YCUT)*EXP( -DYCUT*(XCUT-XTEL)/YCUT )     MESH2990
        ETA1 = EFCNXY( XCUT-B, Y(I,1), A,THETA(I) )                  MESH3000
        DS = DETTE                                                    MESH3010
        CALL STRECH( DS, SUMDSL, JL-1, RDS )                         MESH3020
C       ETA1 = EFCNS( -DS/2., 0., ETA1, A,THETA(I) )                 MESH3030
C       X(I,1) = B +A*COSH(ETA1)*COS(THETA(I))                       MESH3040
C       Y(I,1) =    A*SINH(ETA1)*SIN(THETA(I))                       MESH3050
        ETA = ETA1                                                   MESH3060
        S = 0.                                                       MESH3070
        DO 52 J=2,JL                                                 MESH3080
          ETA = EFCNS( S+DS, S, ETA, A,THETA(I) )                    MESH3090
          X(I,J) = B +A*COSH(ETA)*COS(THETA(I))                      MESH3100
          Y(I,J) =    A*SINH(ETA)*SIN(THETA(I))                      MESH3110
          S = S +DS                                                  MESH3120
```

                                                              Cont.

Table 2. Cont.

```
          DS = DS*RDS                                                  MESH3130
   52     CONTINUE                                                     MESH3140
          DX = DX*RDX                                                  MESH3150
   50 CONTINUE                                                         MESH3160
C                                                                      MESH3170
C------------------------------------------------------------------------MESH3180
C        *    PRINT THE NODAL POINTS    *                              MESH3190
C------------------------------------------------------------------------MESH3200
C                                                                      MESH3210
      REWIND 7                                                         MESH3220
      WRITE(7) TEXT                                                    MESH3230
      WRITE(7) IL, JL, ITL, ITU                                        MESH3240
      WRITE(7) CL,  ( (X(I,J),Y(I,J), I=1,IL), J=1,JL)                 MESH3250
      ENDFILE 7                                                        MESH3260
      REWIND 7                                                         MESH3270
C                                                                      MESH3280
      WRITE(6,107)                                                     MESH3290
      WRITE(6,108)                                                     MESH3300
      N=(JL+3)/4                                                       MESH3310
      I=0                                                              MESH3320
      DO 13 L=1,NPT,JL                                                 MESH3330
        I = I+1                                                        MESH3340
        MM = (I-1)*JL                                                  MESH3350
        LL=L+JL-1                                                      MESH3360
        LK=L+N-1                                                       MESH3370
        DO 14 K=L,LK                                                   MESH3380
          WRITE(6,109)   (J,X(I,J-MM),Y(I,J-MM), J=K,LL,N)            MESH3390
   14     CONTINUE                                                     MESH3400
        WRITE(6,109)                                                   MESH3410
   13 CONTINUE                                                         MESH3420
C                                                                      MESH3430
    1 FORMAT(16I5)                                                     MESH3440
    2 FORMAT(8F10.5)                                                   MESH3450
    5 FORMAT(1H ,I4,3X,2E11.4,3X,2E11.4,3X,2E11.4,3X,2E11.4,3X,2E11.4 )MESH3460
   99 FORMAT(1H1,45X,40H*   M E S H   C O O R D I N A T E S   *,//)    MESH3470
  100 FORMAT( 20X,4X,10HITL          ,I4,11X,10HIN          ,I4,11X,   MESH3480
     .               10HITU          ,I4,11X,10HIL          ,I4)      MESH3490
  101 FORMAT(/5X,15HHYPERBOLA TERMS,4X,10HA          ,G11.4,4X,        MESH3500
     .            10HB          ,G11.4,4X,10HTHETE      ,G11.4,4X,     MESH3510
     .            10HRMINA      ,G11.4)                                MESH3520
  103 FORMAT(8A10)                                                     MESH3530
  104 FORMAT( 20X,4X,10HNPTWK       ,I4,11X,10HNPW         ,I4,11X,    MESH3540
     .               10HNO          ,I4,11X,10HJL          ,I4)       MESH3550
  106 FORMAT(/1H ,15HMESH PARAMETERS)                                  MESH3560
  107 FORMAT(/////60X,12HNODAL POINTS/60X,12(1H=))                     MESH3570
  108 FORMAT(//4(7H    NBR,8X,1HX,11X,1HY,5X)/4(9H     ---  ,24(1H-))) MESH3580
  109 FORMAT(4(I7,2X,1P2E12.4))                                        MESH3590
  110 FORMAT(/5X,15HMESH STRETCHING,4X,10HDSLE       ,G11.4,4X,        MESH3600
     .            10HDSTE       ,G11.4,4X,10HDETLE      ,G11.4,4X,     MESH3610
     .            10HDETTE      ,G11.4)                                MESH3620
  111 FORMAT(/,(/20X,8A10),//)                                         MESH3630
  114 FORMAT( /5X,15HEXTENT OF MESH ,4X,10HCL         ,G11.4,4X,       MESH3640
     .            10HXUP        ,G11.4,4X,10HXDOWN      ,G11.4,4X,     MESH3650
     .            10HYFAR       ,G11.4)                                MESH3660
  115 FORMAT(///10X,45HARC LENGTH OF AIRFOIL    (DS,S)             ,   MESH3670
     .     25H    S..LOWER (TE TO LE) =,G11.4,12H  S..TOTAL =,G11.4/)  MESH3680
  116 FORMAT(///10X,45HMESH POINTS ON AIRFOIL    (X,Y)             /)  MESH3690
  117 FORMAT(///10X,47HANGLE OF ASYMPTOTES OF HYPERBOLAS   (THETA,DTH)/)MESH3700
  120 FORMAT(/5X,15HCOORDINATE CUT ,4X,10HYCUT       ,G11.4,4X         MESH3710
     .                             ,10HTHECUT     ,G11.4)             MESH3720
  121 FORMAT( 20X,4X,10HXSL1       ,G11.4,4X,10HXSL2       ,G11.4,4X,  MESH3730
     .               10HXSU1       ,G11.4,4X,10HXSU2       ,G11.4)    MESH3740
```

Cont.

Table 2. Cont.

```
  122 FORMAT( 20X,4X,10HNSL1        ,I4,11X,10HNSL2       ,I4,11X,      MESH3750
     .                 10HNSU1       ,I4,11X,10HNSU2       ,I4)         MESH3760
  123 FORMAT(//5X,15HNODAL POINTS    ,I5)                              MESH3770
  125 FORMAT(///10X,45HCURVATURE  K   OF AIRFOIL     ( K*DS, K )   /)  MESH3780
      STOP                                                            MESH3790
      END                                                            MESH3800


      SUBROUTINE STRECH (DX1,SUMDX,N,R)                              STRC  10
C                                                                    STRC  20
C     THIS SUBROUTINE PRODUCES A COORDINATE STRETCHING ACCORDING TO A  STRC  30
C     GEOMETRIC PROGRESSION. THE COORDINATE DIRECTION  X   BEGINS AT   STRC  40
C     X(IFIRST) AND ENDS AT X(ILAST). THE INTERVAL  DXI   THEN IS    STRC  50
C     DXI  =  X(I+1) - X(I)                                          STRC  60
C     THE ARGUMENTS ARE THE FOLLOWING:                              STRC  70
C                                                                    STRC  80
C        DX1 - A SPECIFIED VALUE OF THE FIRST INTERVAL              STRC  90
C              X(IFIRST+1) - X(IFIRST)                              STRC 100
C        SUMDX - THE SUM OF ALL THE INTERVALS ,THE DISTANCE XLAST-XFIRSTSTRC 110
C        N  -  THE NUMBER OF INTERVALS  ILAST-IFIRST BETWEEN XFIRST AND STRC 120
C              XLAST                                                STRC 130
C        R  -  RATIO (CONSTANT) OF SUCCESSIVE INTERVALS. (DX I+1)/(DX I)STRC 140
C                                                                    STRC 150
C                                                                    STRC 160
      F (R)  =  (R-1.)*SUMDX -DX1*(R**N-1.)                          STRC 170
      FP(R)  =   SUMDX -N*DX1*R**(N-1)                               STRC 180
      F1 = 1.E-5                                                     STRC 190
      F2 = 1.E-4                                                     STRC 200
      R = 1.5                                                        STRC 210
      DO 10 L=1,50                                                   STRC 220
        R ITER  =  R -F(R)/FP(R)                                     STRC 230
        IF (1.-F2.LT.RITER.AND.RITER.LT.1.)  RITER = 1.E+1          STRC 240
        IF (1..LT.RITER.AND.RITER.LT.1.+F2)  RITER = 1.E-1          STRC 250
        IF (ABS(RITER-R).LT.R*E1)  GO TO 1                          STRC 260
        R = RITER                                                   STRC 270
   10 CONTINUE                                                      STRC 280
      WRITE(6,5) N,R,RITER,SUMDX,DX1                                STRC 290
    5 FORMAT(25H NO CONVERGENCE IN STRECH/                          STRC 300
     .      21H N,R,RITER,SUMDX,DX1,I5,4F12.4,/                     STRC 310
     .      38H PROCEED WITH EQUAL SPACING.    R = 1.   )           STRC 320
      R = 1.00001                                                   STRC 330
      RETURN                                                        STRC 340
    1 R  =  R ITER                                                  STRC 350
      RETURN                                                        STRC 360
      END                                                           STRC 370


      SUBROUTINE SSTRCH (IS,I1,I2,IE, XS,X1,X2,XF, DXS,DXE, X)       SSTR  10
C                                                                    SSTR  20
C     THIS SUBROUTINE PRODUCES AN  "S-TYPE"  FUNCTION FOR STRETCHING A  SSTR  30
C     COORDINATE AT BOTH ENDS OF AN INTERVAL. A SEPARATE GEOMETRIC   SSTR  40
C     PROGRESSION IS USED AT EACH END TO GIVE EITHER AN EXPANDING OR SSTR  50
C     CONDENSING GRID DENSITY. THEY ARE MATCHED IN THE INTERIOR BY A CUBICSSTR  60
C     POLYNOMIAL.                                                   SSTR  70
C                                                                    SSTR  80
      DIMENSION X(2)                                                SSTR  90
      XM(I) = X1 +X1P*FLOAT(I-I1) +(3.*(X2-X1)-(X2P+2.*X1P)*(I2-I1)) SSTR 100
     .                    *( FLOAT(I-I1)/FLOAT(I2-I1) )**2          SSTR 110
     .                    -(2.*(X2-X1)-(X2P+   X1P)*(I2-I1))        SSTR 120
     .                    *( FLOAT(I-I1)/FLOAT(I2-I1) )**3          SSTR 130
C                                                                    SSTR 140
      ISP1  = IS +1                                                 SSTR 150
      IEMI2 = IE -I2                                                SSTR 160
C                                                                    SSTR 170
```

Cont.

Table 2. Cont.

```
C.. STARTING SECTOR                                                 SSTR 180
      DX = DXS                                                      SSTR 190
      CALL STRECH (DX, X1-XS, I1-IS, RS)                           SSTR 200
      X(IS) = XS                                                    SSTR 210
      DO 1 I=ISP1,I1                                                SSTR 220
         X(I) = X(I-1) +DX                                          SSTR 230
         DX = RS*DX                                                 SSTR 240
    1 CONTINUE                                                      SSTR 250
C                                                                   SSTR 260
C.. ENDING SECTOR                                                   SSTR 270
C         (REVERSE ORDER)                                           SSTR 280
      DX = -DXE                                                     SSTR 290
      CALL STRECH (DX, X2-XF, IE-I2, RE)                           SSTR 300
      X(IE) = XE                                                    SSTR 310
      DO 2 N=1,IEMI2                                                SSTR 320
         X(IE-N) = X(IE+1-N) +DX                                    SSTR 330
         DX = RE*DX                                                 SSTR 340
    2 CONTINUE                                                      SSTR 350
C                                                                   SSTR 360
C.. MIDDLE SECTOR     (MATCHING CUBIC POLYNOMIAL)                   SSTR 370
      DX1 = X(I1) -X(I1-1)                                          SSTR 380
      DX2 = X(I2+1) -X(I2)                                          SSTR 390
      X1P = DX1*RS*ALOG(RS) /(RS-1.)                               SSTR 400
      X2P = DX2*RE*ALOG(RE) /(RE-1.)                               SSTR 410
      DO 3 I=I1,I2                                                  SSTR 420
         X(I) = XM(I)                                               SSTR 430
    3 CONTINUE                                                      SSTR 440
C                                                                   SSTR 450
      RETURN                                                        SSTR 460
      END                                                           SSTR 470

      SUBROUTINE SPLINE (XIN,YIN, XOUT,YOUT, DYDX,D1Y,D2Y,D3Y, NIN,NOUT)SPLI  10
C                                                                   SPLI  20
C     COMPUTE A CUBIC SPLINE THROUGH THE SET OF POINTS  XIN(I),YIN(I) SPLI  30
C     XIN  MUST BE MONOTONIC                                        SPLI  40
C                                                                   SPLI  50
C     XIN,YIN        INDEPENDENT AND DEPENDENT INPUT VARIABLES      SPLI  60
C     XOUT,YOUT      INDEPENDENT AND DEPENDENT OUTPUT VARIABLES     SPLI  70
C     D1Y,D2Y,D3Y    1ST, 2ND, AND 3RD DERIVATIVES AT SPLINE POINTS XIN SPLI  80
C     DYDX           DERIVATIVE AT XOUT                             SPLI  90
C     NIN,NOUT       NUMBER OF INPUT AND OUTPUT VALUES  (NIN .GE. 4) SPLI 100
C     D1YI           VALUE OF 1ST DERIVATIVE AT INITIAL SPLINE POINT SPLI 110
C     D1YF           VALUE OF 1ST DERIVATIVE AT  FINAL  SPLINE POINT SPLI 120
C                                                                   SPLI 130
      DIMENSION  XIN(4),YIN(4), XOUT(2),YOUT(2), DYDX(2), D1Y(2),D2Y(2),SPLI 140
     .           D3Y(2)                                             SPLI 150
      YP1 (X1,X2,X3,X4, Y1,Y2,Y3,Y4) = ((X1-X2)*(X1-X3) +(X1-X3)*(X1-X4)SPLI 160
     .                                 +(X1-X2)*(X1-X4))*Y1/((X1-X2) SPLI 170
     .                                         *(X1-X3)*(X1-X4)) SPLI 180
     .                 -(X1-X3)*(X1-X4)*Y2/((X1-X2)*(X2-X3)*(X2-X4))SPLI 190
     .                 +(X1-X2)*(X1-X4)*Y3/((X1-X3)*(X2-X3)*(X3-X4))SPLI 200
     .                 -(X1-X2)*(X1-X3)*Y4/((X1-X4)*(X2-X4)*(X3-X4))SPLI 210
      EPSI1 = -1.E-10                                               SPLI 220
      EPSI2 = -EPSI1                                                SPLI 230
      NIM1 = NIN-1                                                  SPLI 240
      D1YI =YP1(XIN(1),XIN(2),XIN(3),XIN(4),YIN(1),YIN(2),YIN(3),YIN(4))SPLI 250
      D1YF =YP1( XIN(NIN),XIN(NIN-1),XIN(NIN-2),XIN(NIN-3),          SPLI 260
     .           YIN(NIN),YIN(NIN-1),YIN(NIN-2),YIN(NIN-3) )        SPLI 270
      DX = XIN(2)-XIN(1)                                            SPLI 280
      IF (DX.EQ.0.) GO TO 35                                        SPLI 290
      DF = (YIN(2)-YIN(1)) /DX                                      SPLI 300
C                                                                   SPLI 310
```

Cont.

Table 2. Cont.

```
C..          FORWARD LOOP OF TRIDIAGONAL MATRIX COMPUTATION          SPLI 320
      D1Y(1) = -.5                                                   SPLI 330
      D2Y(1) = 3.*(DF-D1YI) /DX                                      SPLI 340
      DO 5 I=2,NIM1                                                  SPLI 350
         DX1 = XIN(I+1)-XIN(I)                                       SPLI 360
           IF (DX1.EQ.0.) GO TO 36                                   SPLI 370
         DF1 = (YIN(I+1)-YIN(I)) /DX1                                SPLI 380
         B = 2.*(DX+DX1)                                             SPLI 390
         F = 6.*(DF1-DF)                                             SPLI 400
         DENOM = B +DX*D1Y(I-1)                                      SPLI 410
         D2Y(I) = (F-DX*D2Y(I-1)) /DENOM                             SPLI 420
         D1Y(I) = -DX1/DENOM                                         SPLI 430
         DX = DX1                                                    SPLI 440
         DF = DF1                                                    SPLI 450
    5 CONTINUE                                                       SPLI 460
C                                                                    SPLI 470
C..          BACK SUBSTITUTION OF TRIDIAGONAL MATRIX COMPUTATION     SPLI 480
      DENOM = 1. +.5*D1Y(NIM1)                                       SPLI 490
      D2Y(NIN) = ( -3.*(DF1-D1YF)/DX1 -.5*D2Y(NIM1) )/DENOM          SPLI 500
      D1Y(NIN) = 0.                                                  SPLI 510
      K = NIN                                                        SPLI 520
      DO 10 I=1,NIM1                                                 SPLI 530
         K = K-1                                                     SPLI 540
         D2Y(K) = D2Y(K) +D1Y(K)*D2Y(K+1)                           SPLI 550
         DX1 = XIN(K+1) -XIN(K)                                      SPLI 560
         DF1 = (YIN(K+1)-YIN(K)) /DX1                                SPLI 570
         D1Y(K+1) = DF1 +DX1/6.*(D2Y(K)+2.*D2Y(K+1))                SPLI 580
         D3Y(K+1) = (D2Y(K+1)-D2Y(K)) /DX1                           SPLI 590
   10 CONTINUE                                                       SPLI 600
      D1Y(1) = DF1 -DX1/6.*(2.*D2Y(1)+D2Y(2))                        SPLI 610
      D3Y(1) = D3Y(2)                                                SPLI 620
C                                                                    SPLI 630
C..          INTERPOLATE FOR GIVEN VALUES OF XOUT                    SPLI 640
      DO 15 J=1,NOUT                                                 SPLI 650
         DO 12 I=1,NIN                                               SPLI 660
            DX = XIN(I)-XOUT(J)                                      SPLI 670
            IF (DX.GE.EPSI1.AND.DX.LE.EPSI2)   GO TO 13              SPLI 680
              IF (DX.GE.EPSI2)  GO TO 14                             SPLI 690
   12    CONTINUE                                                    SPLI 700
         GO TO 37                                                    SPLI 710
   13       YOUT(J) = YIN(I)                                         SPLI 720
            DYDX(J) = D1Y(I)                                         SPLI 730
            GO TO 15                                                 SPLI 740
   14       DX = XOUT(J)-XIN(I)                                      SPLI 750
            YOUT(J) = YIN(I) +DX*(D1Y(I)+DX*.5*(D2Y(I)+DX/3.*D3Y(I)))SPLI 760
            DYDX(J) = D1Y(I) +DX*(D2Y(I)+DX*.5*D3Y(I))               SPLI 770
   15 CONTINUE                                                       SPLI 780
      RETURN                                                         SPLI 790
C                                                                    SPLI 800
   35 WRITE(6,100)                                                   SPLI 810
      WRITE(6,101) XIN(1),XIN(2)                                     SPLI 820
      CALL EXIT                                                      SPLI 830
   36 WRITE(6,100)                                                   SPLI 840
      WRITE(6,102) I,XIN(I),XIN(I+1)                                 SPLI 850
      CALL EXIT                                                      SPLI 860
   37 WRITE(6,100)                                                   SPLI 870
      WRITE(6,103) J,XOUT(J),XIN(NIN)                                SPLI 880
      CALL EXIT                                                      SPLI 890
C                                                                    SPLI 900
  100 FORMAT (/5X,18HSUBROUTINE SPLINE /)                            SPLI 910
  101 FORMAT (/5X,24HERROR IN INPUT   XIN(1)=,E12.4,5X,7HXIN(2)=,E12.4/)SPLI 920
  102 FORMAT (/5X,19HERROR IN INPUT   I=,I5,5X,7HXIN(I)=,F12.4,5X,   SPLI 930
```

Cont.

Table 2. Cont.


```
      .          9HXIN(I+1)=,E12.4/)                                     SPLI 940
 103 FORMAT (/5X,28HXOUT(J) IS OUT OF RANGE    J=,I5,5X,8HXOUT(J)=,F12.4SPLI 950
      .          ,5X,9HXIN(NIN)=,E12.4/)                                 SPLI 960
      END                                                                SPLI 970


      FUNCTION EFCNXY( X, Y,  A,THETA )                                   E(X)  10
C                                                                        E(X)  20
C  COMPUTES  ETA  FOR A GIVEN POINT  X,Y  ON THE I-TH HYPERBOLA (THETA)  E(X)  30
C                                                                        E(X)  40
         IF (ABS(TAN( THETA )).LE.1.)  GO TO 1                           E(X)  50
      SINHE = Y/( A*SIN(THETA) )                                         E(X)  60
      EFCNXY = ALOG( SINHE +SQRT(SINHE**2+1.) )                          E(X)  70
      RETURN                                                             E(X)  80
  1      COSHE = X/( A*COS(THETA) )                                      E(X)  90
         EFCNXY = ALOG( COSHE +SIGN( SQRT(COSHE**2-1.), Y*SIN(THETA)) )  E(X) 100
      RETURN                                                             E(X) 110
      END                                                                E(X) 120


      FUNCTION EFCNS( S2, S1, ETA1, A,THETA )                            F(S)  10
C                                                                        E(S)  20
C  COMPUTES THE UPPER LIMIT  ETA  OF THE GIVEN ARC LENGTH INTEGRAL  S2   F(S)  30
C                                                                        E(S)  40
      STHESQ = SIN( THETA )**2                                           E(S)  50
      DETA = SIGN( .002, S2-S1 )                                         E(S)  60
      S = S1                                                             E(S)  70
      ETA = ETA1                                                         E(S)  80
      DO 1 L=1,5000                                                      E(S)  90
         DS = A*SQRT( SINH(ETA)**2 +STHESQ )*DETA                        E(S) 100
         S = S +DS                                                       F(S) 110
         ETA = ETA +DETA                                                 E(S) 120
         IF( (S-S1)*(S-S2).GE.0. )  GO TO 10                             F(S) 130
  1 CONTINUE                                                             E(S) 140
     WRITE(6,5)                                                          E(S) 150
  5 FORMAT(22H NO SOLUTION IN EFCNS )                                    F(S) 160
     RETURN                                                              E(S) 170
 10 EFCNS = ETA +DETA*(S2-S)/DS                                          E(S) 180
     RETURN                                                              F(S) 190
     END                                                                 F(S) 200
```

Table 3. Input data.

```
     A I R F O I L    A N D    M F S H    D A T A
                          F O R
     I N P U T    T O    P R O G R A M    M F S H


NACA 0012 AIRFOIL (CLOSED) *)  STD. MFSH PROB. A (EXCFPT PART II-3, M=.95 CASF)
   20   141    21     03                        NPTWK, IL, JL, IPLOT
.02            115.                             RMINA, THFTF
1.            -5.           7.          5.      CL, XUP, XDOWN, YFAR
-.2            0.                               YCUT, THECUT
.008          .012         .010        .012    DSLE, DSTF,  DETLE, DFTTE
.130          .865         .135        .870     XSL1, XSL2,  XSU1, XSU2
   08    10    10    08                         NSL1, NSL2,  NSU1, NSU2
  133    67                          NACA 0012 AIRFOIL (CLOSED)
1.000000   0.0         0.990000 -0.002200   0.970000 -0.005400   0.933914 -0.010170
0.864459 -0.018812   0.800129 -0.026217   0.740545 -0.032581   0.685356 -0.038049
0.634239 -0.042726   0.586893 -0.046696   0.543040 -0.050030   0.502422 -0.052787
0.464800 -0.055021   0.429054 -0.056782   0.397679 -0.058116   0.367785 -0.059065
0.340096 -0.059671   0.314450 -0.059970   0.290696 -0.059998   0.268695 -0.059788
0.248316 -0.059369   0.229441 -0.058769   0.211959 -0.058013   0.195766 -0.057124
0.180768 -0.056122   0.166876 -0.055026   0.154009 -0.053851   0.142092 -0.052613
0.131053 -0.051324   0.120829 -0.049996   0.111359 -0.048638   0.102588 -0.047259
0.094464 -0.045866   0.086939 -0.044467   0.079970 -0.043066   0.073514 -0.041668
0.067535 -0.040278   0.061997 -0.038897   0.056868 -0.037530   0.052117 -0.036178
0.047716 -0.034843   0.043640 -0.033527   0.039865 -0.032229   0.036368 -0.030952
0.033129 -0.029694   0.030129 -0.028457   0.027351 -0.027239   0.024777 -0.026041
0.022394 -0.024861   0.020186 -0.023699   0.018141 -0.022554   0.016247 -0.021423
0.014493 -0.020306   0.012868 -0.019200   0.011363 -0.018103   0.009969 -0.017012
0.008679 -0.015923   0.007482 -0.014831   0.006374 -0.013732   0.005348 -0.012617
0.004398 -0.011477   0.003518 -0.010297   0.002702 -0.009054   0.001947 -0.007713
0.001248 -0.006198   0.000600 -0.004318   0.0          0.0          0.000600  0.004318
0.001248  0.006198   0.001947  0.007713   0.002702  0.009054   0.003518  0.010297
0.004398  0.011477   0.005348  0.012617   0.006374  0.013732   0.007482  0.014831
0.008679  0.015923   0.009969  0.017012   0.011363  0.018103   0.012868  0.019200
0.014493  0.020306   0.016247  0.021423   0.018141  0.022554   0.020186  0.023699
0.022394  0.024861   0.024777  0.026041   0.027351  0.027239   0.030129  0.028457
0.033129  0.029694   0.036368  0.030952   0.039865  0.032229   0.043640  0.033527
0.047716  0.034843   0.052117  0.036178   0.056868  0.037530   0.061997  0.038897
0.067535  0.040278   0.073514  0.041668   0.079970  0.043066   0.084939  0.044467
0.094464  0.045866   0.102588  0.047259   0.111359  0.048638   0.120829  0.049996
0.131053  0.051324   0.142092  0.052613   0.154009  0.053851   0.166876  0.055026
0.180768  0.056122   0.195766  0.057124   0.211959  0.058013   0.229441  0.058769
0.248316  0.059369   0.268695  0.059788   0.290696  0.059998   0.314450  0.059970
0.340096  0.059671   0.367785  0.059065   0.397679  0.058116   0.429054  0.056782
0.464800  0.055021   0.502422  0.052787   0.543040  0.050030   0.586893  0.046696
0.634239  0.042726   0.685356  0.038049   0.740545  0.032581   0.800129  0.026217
0.864459  0.018812   0.933914  0.010170   0.970000  0.005400   0.990000  0.002200
1.000000   0.0

NACA 0012 AIRFOIL (PROLONGED FICTITIOUSLY)  FINE MESH PROB A  II-3.  M=.95 CASF
   20   141    25     03                        NPTWK, IL, JL, IPLOT
.02            90.                              RMINA, THFTF
2.5           -14.         12.         24.0     CL, XUP, XDOWN, YFAR
-.2            0.                               YCUT, THFCUT
.008          .012         .010        .012    DSLE, DSTF,  DETLF, DETTF
.130          .865         .135        .870     XSL1, XSL2,  XSU1, XSU2
   08    10    10    08                         NSL1, NSL2,  NSU1, NSU2
  141    71                          NACA 0012  (FOR MACH=.95)
2.500000   .000000   1.500000   .000000   1.050000   .000000   1.010000    .000000
1.000000   .000000    .990000  -.002200    .970000  -.005400    .933914   -.010170


                                                    Cont.
```

Table 3. Cont.

.864459	-.018812	.800129	-.026217	.740545	-.032581	.685356	-.038049
.634239	-.042726	.586893	-.046696	.543040	-.050030	.502422	-.052787
.464800	-.055021	.429954	-.056782	.397679	-.058116	.367785	-.059065
.340096	-.059671	.314450	-.059970	.290696	-.059998	.268695	-.059788
.248316	-.059369	.229441	-.058769	.211959	-.058013	.195766	-.057124
.180768	-.056122	.166876	-.055026	.154009	-.053851	.142092	-.052613
.131053	-.051324	.120829	-.049996	.111359	-.048638	.102588	-.047259
.094464	-.045866	.086939	-.044467	.079970	-.043066	.073514	-.041668
.067535	-.040278	.061997	-.038897	.056868	-.037530	.052117	-.036178
.047716	-.034843	.043640	-.033527	.039865	-.032229	.036368	-.030952
.033129	-.029694	.030129	-.028457	.027351	-.027239	.024777	-.026041
.022394	-.024861	.020186	-.023699	.018141	-.022554	.016247	-.021423
.014493	-.020306	.012868	-.019200	.011363	-.018103	.009969	-.017012
.008678	-.015923	.007482	-.014831	.006374	-.013732	.005348	-.012617
.004398	-.011477	.003518	-.010297	.002702	-.009054	.001947	-.007713
.001248	-.006198	.000600	-.004318	.000000	.000000	.000600	.004318
.001248	.006198	.001947	.007713	.002702	.009054	.003518	.010297
.004398	.011477	.005348	.012617	.006374	.013732	.007482	.014831
.008678	.015923	.009969	.017012	.011363	.018103	.012868	.019200
.014493	.020306	.016247	.021423	.018141	.022554	.020186	.023699
.022394	.024861	.024777	.026041	.027351	.027239	.030129	.028457
.033129	.029694	.036368	.030952	.039865	.032229	.043640	.033527
.047716	.034843	.052117	.036178	.056868	.037530	.061997	.038897
.067535	.040278	.073514	.041668	.079970	.043066	.086939	.044467
.094464	.045866	.102588	.047259	.111359	.048638	.120829	.049996
.131053	.051324	.142092	.052613	.154009	.053851	.166876	.055026
.180768	.056122	.195766	.057124	.211959	.058013	.229441	.058769
.248316	.059369	.268695	.059788	.290696	.059998	.314450	.059970
.340096	.059671	.367785	.059065	.397679	.058116	.429954	.056782
.464800	.055021	.502422	.052787	.543040	.050030	.586893	.046696
.634239	.042726	.685356	.038049	.740545	.032581	.800129	.026217
.864459	.018812	.933914	.010170	.970000	.005400	.990000	.002200
1.000000	.000000	1.010000	.000000	1.050000	.000000	1.500000	.000000
2.500000	.000000						

```
          RAE 2822  AIRFOIL                    STANDARD MESH 141X21 FOR WORKSHOP PROBLEM C
   20  141    21                               NPTWK, IL, JL
 .01        115.                               BMINA, THETE
1.          -5.        7.        5.            CL, XUP, XDOWN, YFAR
-.2         -8.                                YCUT, THECUT
 .008       .012       .010      .012          DSLE, DSTE,  DETLE, DETTE
 .130       .865       .135      .870          XSL1, XSL2,  XSU1, XSU2
   08   10   10    08                          NSL1, NSL2,  NSU1, NSU2
  129   65                                     RAE 2822  AIRFOIL
```

1.	0.	.99940	+.00004	.99759	+.00014	.99459	+.00030
.90030	+.00050	.98502	+.00072	.97847	+.00093	.97077	+.00111
.96194	+.00171	.95200	+.00120	.94096	+.00103	.92886	+.00069
.91574	+.00016	.90160	-.00060	.88651	-.00159	.87048	-.00283
.85355	-.00431	.83578	-.00605	.81720	-.00803	.79785	-.01024
.77778	-.01264	.75705	-.01536	.73570	-.01823	.71378	-.02129
.69134	-.02449	.66845	-.02781	.64514	-.03121	.62149	-.03463
.59754	-.03804	.57336	-.04140	.54901	-.04464	.52453	-.04772
.50000	-.05056	.47547	-.05310	.45099	-.05526	.42663	-.05698
.40245	-.05822	.37851	-.05897	.35486	-.05924	.33156	-.05905
.30866	-.05846	.28622	-.05755	.26430	-.05638	.24295	-.05499
.22221	-.05343	.20215	-.05169	.18280	-.04991	.16422	-.04777
.14645	-.04561	.12952	-.04333	.11349	-.04093	.09840	-.03843
.08427	-.03584	.07114	-.03317	.05904	-.03042	.04801	-.02759
.03806	-.02469	.02923	-.02172	.02153	-.01871	.01498	-.01565
.00961	-.01256	.00541	-.00944	.00241	-.00631	.00060	-.00316
0.	0.	.00060	.00316	.00241	.00631	.00541	.00942

Cont.

Table 3. Cont.

```
.00961     .01248     .01498     .01549     .02153     .01844     .02923     .02135
.03806     .02422     .04801     .02706     .05904     .02987     .07114     .03264
.08427     .03536     .09840     .03801     .11349     .04059     .12952     .04307
.14645     .04546     .16422     .04773     .18280     .04987     .20215     .05188
.22221     .05375     .24295     .05547     .26430     .05703     .28622     .05841
.30866     .05963     .33156     .06066     .35486     .06150     .37851     .06213
.40245     .06256     .42663     .06278     .45099     .06277     .47547     .06253
.50000     .06203     .52453     .06125     .54901     .06019     .57336     .05885
.59754     .05722     .62149     .05534     .64514     .05326     .66845     .05099
.69134     .04857     .71378     .04603     .73570     .04338     .75705     .04064
.77778     .03785     .79785     .03502     .81720     .03218     .83578     .02935
.85355     .02655     .87048     .02382     .88651     .02115     .90160     .01858
.91574     .01611     .92886     .01377     .94096     .01156     .95200     .00951
.96194     .00762     .97077     .00592     .97847     .00440     .98502     .00309
.99039     .00200     .99459     .00114     .99759     .00051     .99940     .00013
1.         0.

      CAST 7  AIRFOIL (CLOSED)        STANDARD MESH 141X21 FOR WORKSHOP PROBLEM D
  20  141  21                         NPTWK, IL, JL
.02      115.                         BMINA, THETE
1.         -5.        7.        5.    CL, XUP, XDOWN, YFAR
-.2      -7.                          YCUT, THECUT
.008       .012       .010      .012  DSLE, DSTE, DETLE, DETTE
.130       .865       .135      .870  XSL1, XSL2, XSU1, XSU2
  08  10  10  08                      NSL1, NSL2, NSU1, NSU2
  61  32                                  CAST 7  AIRFOIL  (CLOSED)
.999998 -.002151   .977500 -.001230   .950000 -.000400   .920000  .000787
.875000  .001038   .815000 -.002087   .755000 -.008096   .695000 -.015660
.635000 -.023715   .575000 -.031525   .515000 -.038542   .455000 -.044304
.395000 -.048363   .335000 -.050268   .275000 -.049645   .215000 -.046321
.155000 -.040398   .115000 -.035181   .087500 -.031022   .065000 -.027131
.047500 -.023626   .035000 -.020850   .025000 -.018526   .017500 -.016627
.012500 -.015053   .007500 -.012695   .005000 -.010822   .003500 -.009271
.002000 -.007133   .001000 -.005060   .000400 -.003189   0.0      0.0
.0004    .003367   .0010    .005306   .0020    .007536   .0035    .010124
.0050    .012293   .0075    .015341   .0125    .020110   .0350    .032730
.0475    .037702   .0650    .043075   .0875    .048272   .1150    .053034
.1550    .058089   .2150    .063124   .2750    .066167   .3350    .067865
.3050    .068511   .4550    .068165   .5150    .066764   .5750    .064155
.6350    .060079   .6950    .054228   .7550    .046385   .8150    .036622
.8750    .025408   .9200    .016511   .9500    .010500   .9775    .003850
1.0000  -.002151

    **) KORN 1  AIRFOIL              STANDARD MESH 141X21 FOR WORKSHOP PROBLEM G
  20  141  21                         NPTWK, IL, JL
.015     115.                         BMINA, THETE
1.         -5.        7.        5.    CL, XUP, XDOWN, YFAR
-.2      -6.8                         YCUT, THECUT
.008       .012       .010      .012  DSLE, DSTE, DETLE, DETTE
.130       .865       .135      .870  XSL1, XSL2, XSU1, XSU2
  08  10  10  08                      NSL1, NSL2, NSU1, NSU2
  131  59                                  KORN 1  AIRFOIL
1.       0.         .99658 +.00037   .98618  .00133   .97362  .00224
.95866   .00307    .94364  .00366    .92650  .00407   .90371  .00421
.87942   .00389    .85881  .00322    .83854  .00223   .81919  .00096
.80115  -.00052    .78490 -.00209    .76851 -.00387   .74670 -.00654
.72670  -.00923    .70892 -.01177    .69037 -.01453   .67125 -.01743
.65758  -.01952    .64220 -.02185    .62353 -.02466   .60536 -.02733
.57867  -.03110    .54991 -.03492    .52191 -.03834   .47547 -.04329
.45011  -.04560    .42640 -.04749    .40459 -.04900   .38774 -.05001
```

Cont.

Table 3. Cont.

.36360	-.05121	.34498	-.05195	.32418	-.05256	.31025	-.05285
.29715	-.05303	.28215	-.05312	.26410	-.05306	.24800	-.05286
.22861	-.05241	.20304	-.05144	.16728	-.04930	.11865	-.04448
.08537	-.03946	.06254	-.03484	.03915	-.02863	.02538	-.02384
.01539	-.01934	.01000	-.01618	.00706	-.01398	.00477	-.01181
.00307	-.00978	.00189	-.00793	.00102	-.00613	.00046	-.00438
.00025	-.00333	.00005	-.00138	0.	0.	.00005	.00184
.00026	.00384	.00057	.00539	.00107	.00695	.00153	.00797
.00267	.00985	.00425	.01174	.00686	.01419	.01104	.01736
.01959	.02249	.02826	.02668	.03773	.03055	.05261	.03557
.05536	.03640	.05761	.03706	.05914	.03749	.06063	.03790
.06210	.03830	.06435	.03890	.06730	.03964	.07007	.04031
.07269	.04092	.07459	.04134	.07598	.04162	.07715	.04189
.07840	.04215	.07969	.04241	.08104	.04267	.08245	.04294
.08394	.04322	.08565	.04353	.08764	.04388	.08997	.04428
.09273	.04473	.09617	.04528	.10049	.04594	.10581	.04672
.11235	.04762	.12052	.04870	.13068	.04995	.14294	.05134
.15753	.05286	.17482	.05450	.19506	.05620	.21813	.05792
.24393	.05956	.27225	.06107	.30286	.06238	.33543	.06344
.36946	.06421	.40400	.06464	.43858	.06473	.47312	.06448
.50711	.06387	.53941	.06295	.56970	.06173	.59828	.06023
.62479	.05850	.64832	.05665	.66888	.05479	.68691	.05297
.70683	.05072	.73855	.04668	.77693	.04105	.80900	.03577
.83689	.03077	.87557	.02336	.91365	.01575	.93978	.01057
.97178	.00456	.99786	.00029	1.	0.		

---

*) It is essential that no ambiguity arises from the defi-
nition of these test problems. For this reason those
airfoils (Problems A and D), which have a thick trailing
edge have been very slightly modified so as to make the
trailing edge pointed (i.e. singled valued).

**) During the Workshop Antony Jameson pointed out that the
coordinates of the Korn 1 airfoil listed here and taken
from Kacprzynski et al. (NRC Aero. Rep. LR-554, 1971)
are not entirely accurate and may give rise to a weak
shock wave even at design conditions. Garabedian has subsequently corrected this inaccuracy.

OTHER MESH SYSTEMS USED BY THE PARTICIPANTS

The majority of grid networks used by the participants can be
classified as being either of the elliptic or parabolic type,
so termed because of their kinship to either the classical
elliptic or parabolic coordinates systems. Our standard Workshop mesh, however, as well as Lerat and Siclès's mesh are
hybrids of these two types in that ahead of the trailing edge
they resemble an elliptic mesh, but behind it they are more
like the parabolic type.

If provided by the participant, further details like the number
of nodal points and a plot of his mesh are given below, together with the label (a, b, c, etc.) we use in our comparison
to distinguish between several different meshes when the
participant used more than one.

(A. Rizzi)

Holst provided no plot, but used an elliptic mesh of increasing
nodal density: a) 105 x 28,   b) 149 x 39,   c) 205 x 56

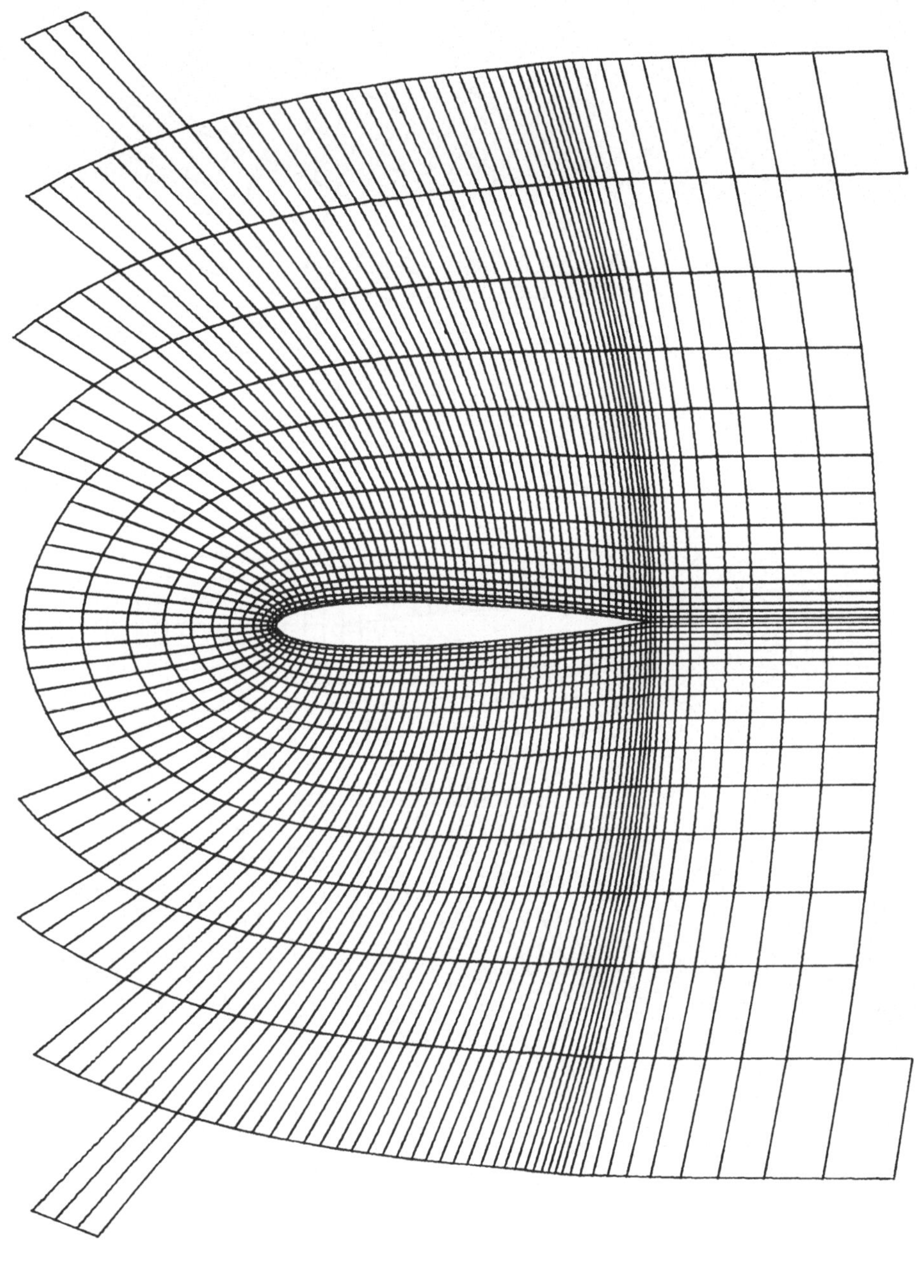

Figure 1. Jameson i). (almost) standard Workshop mesh a

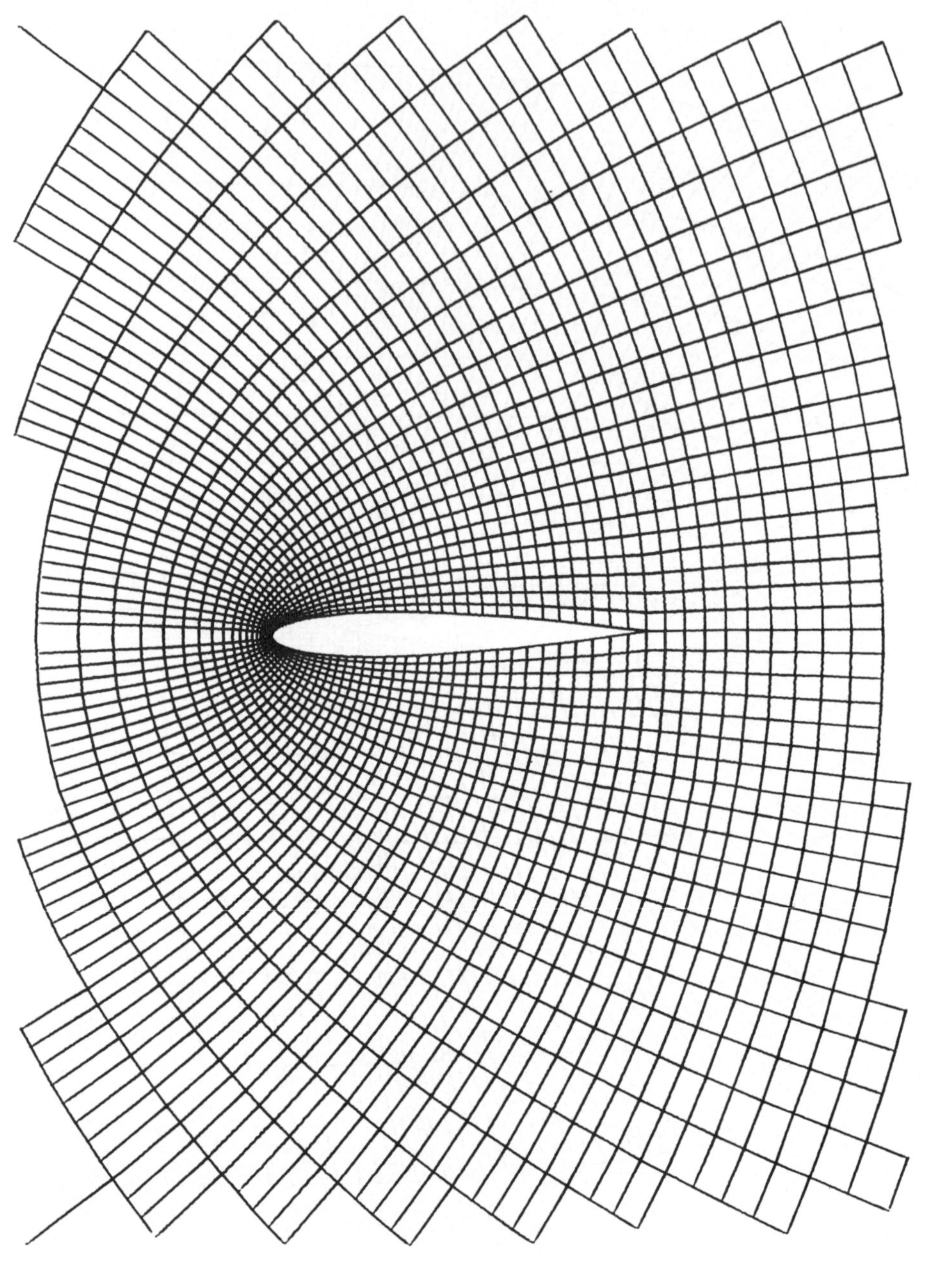

Figure 1ii. Jameson. parabolic mesh b .

Figure 1iii. Jameson parabolic mesh c .

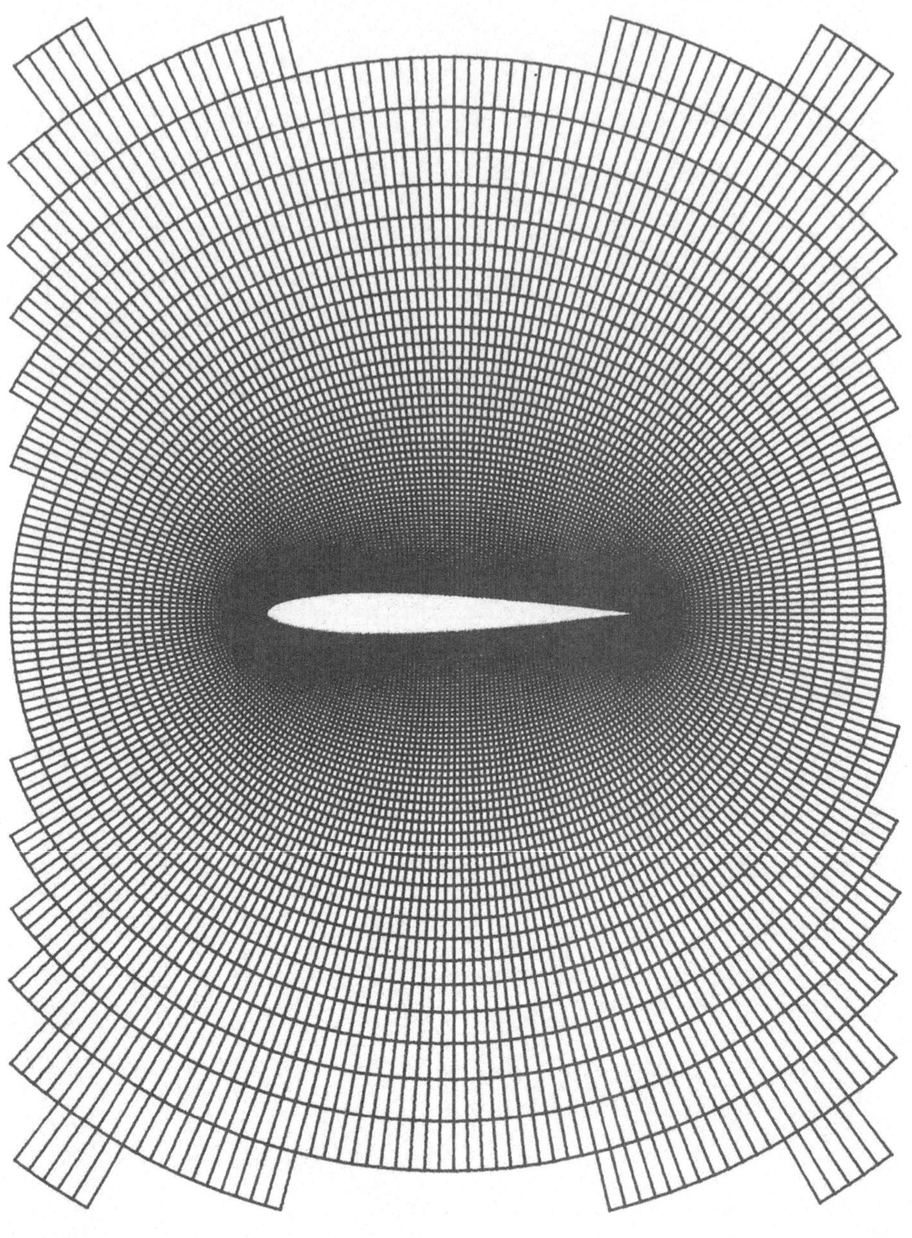

Figure 1iv Jameson elliptic mesh c´ .

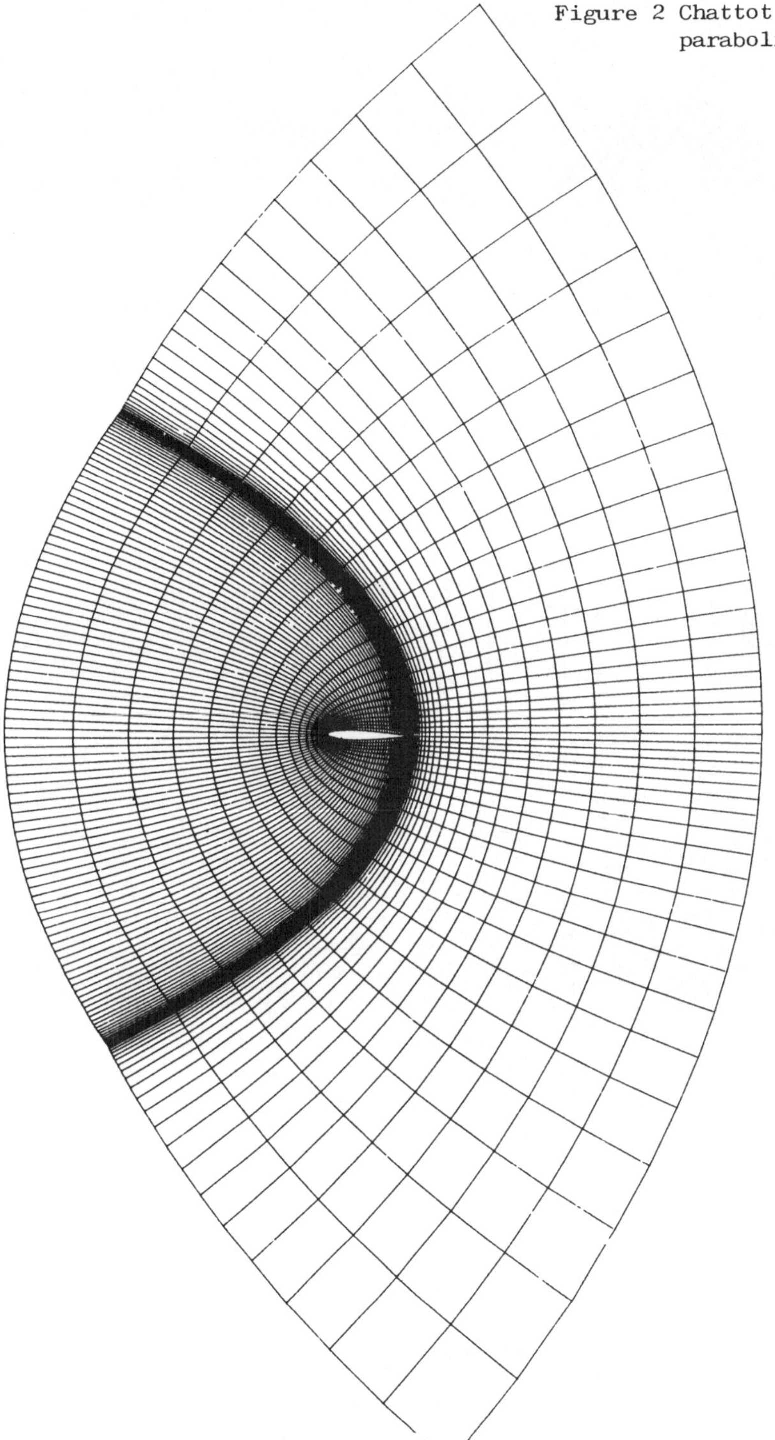

Figure 2 Chattot and Veuillot parabolic mesh a´ .

Fig. 2(con'd) expanded scale

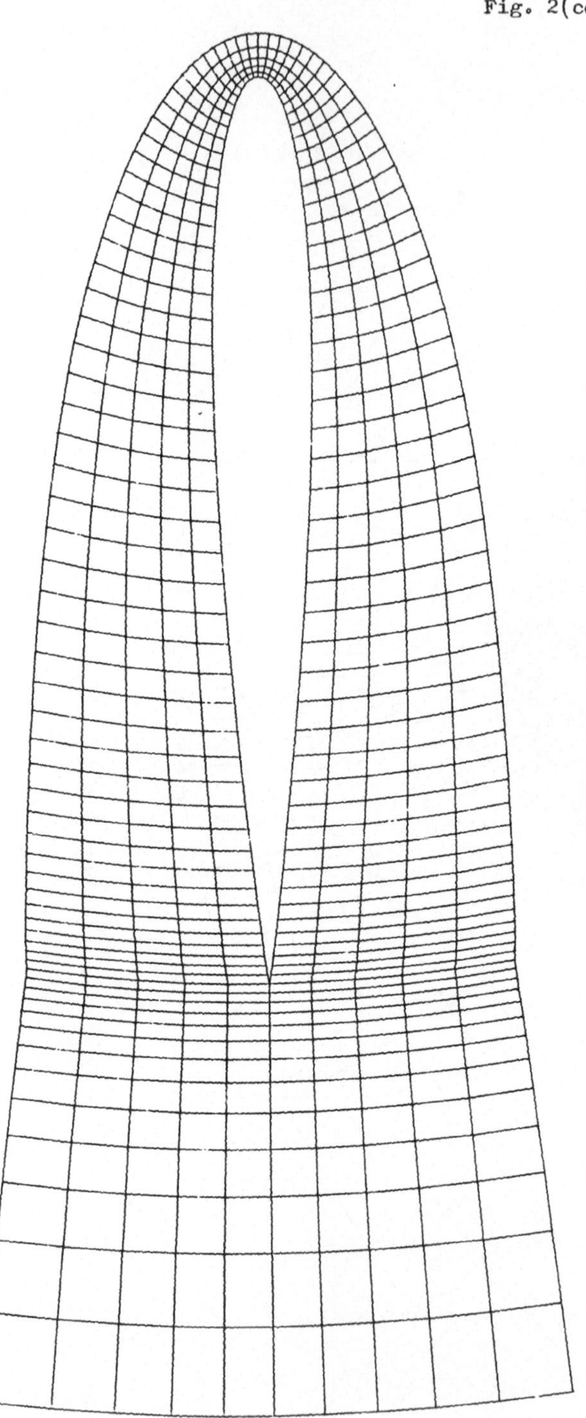

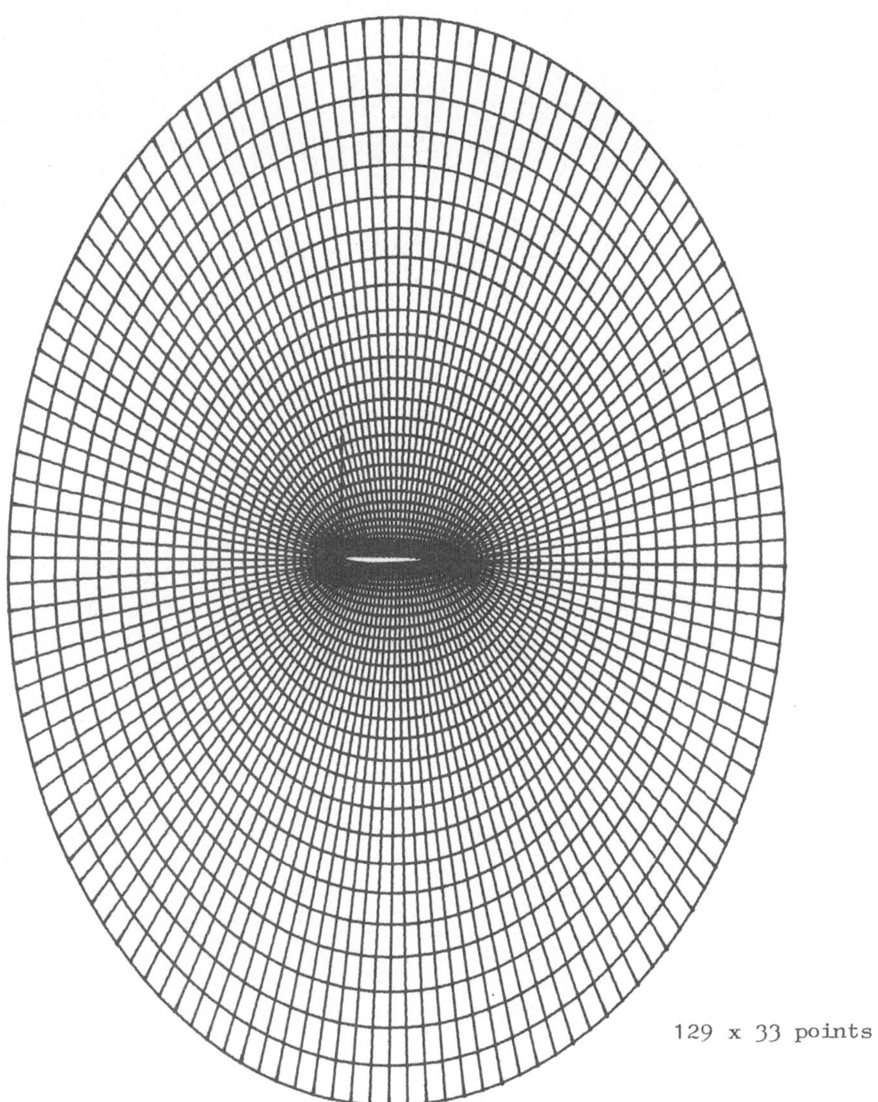

Figure 3. Eberle elliptic mesh

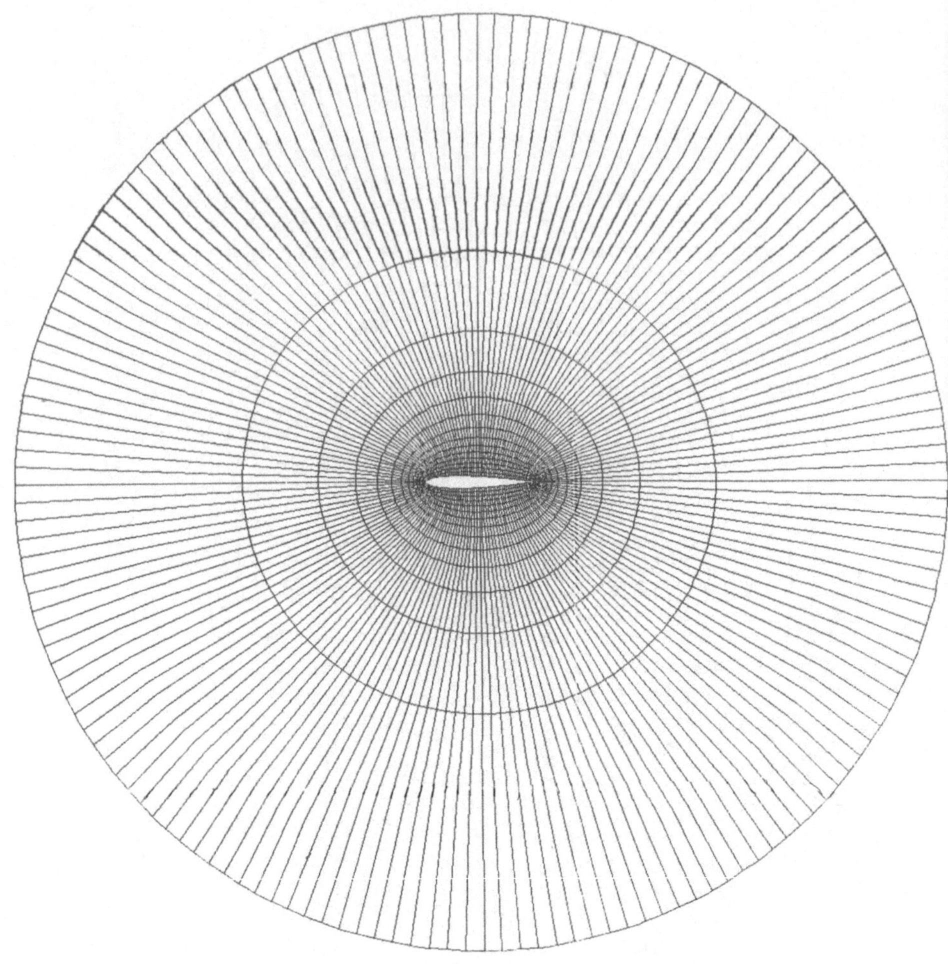

Figure 4 Baker elliptic mesh: a - 106 x 28, b - 210 x 56 .

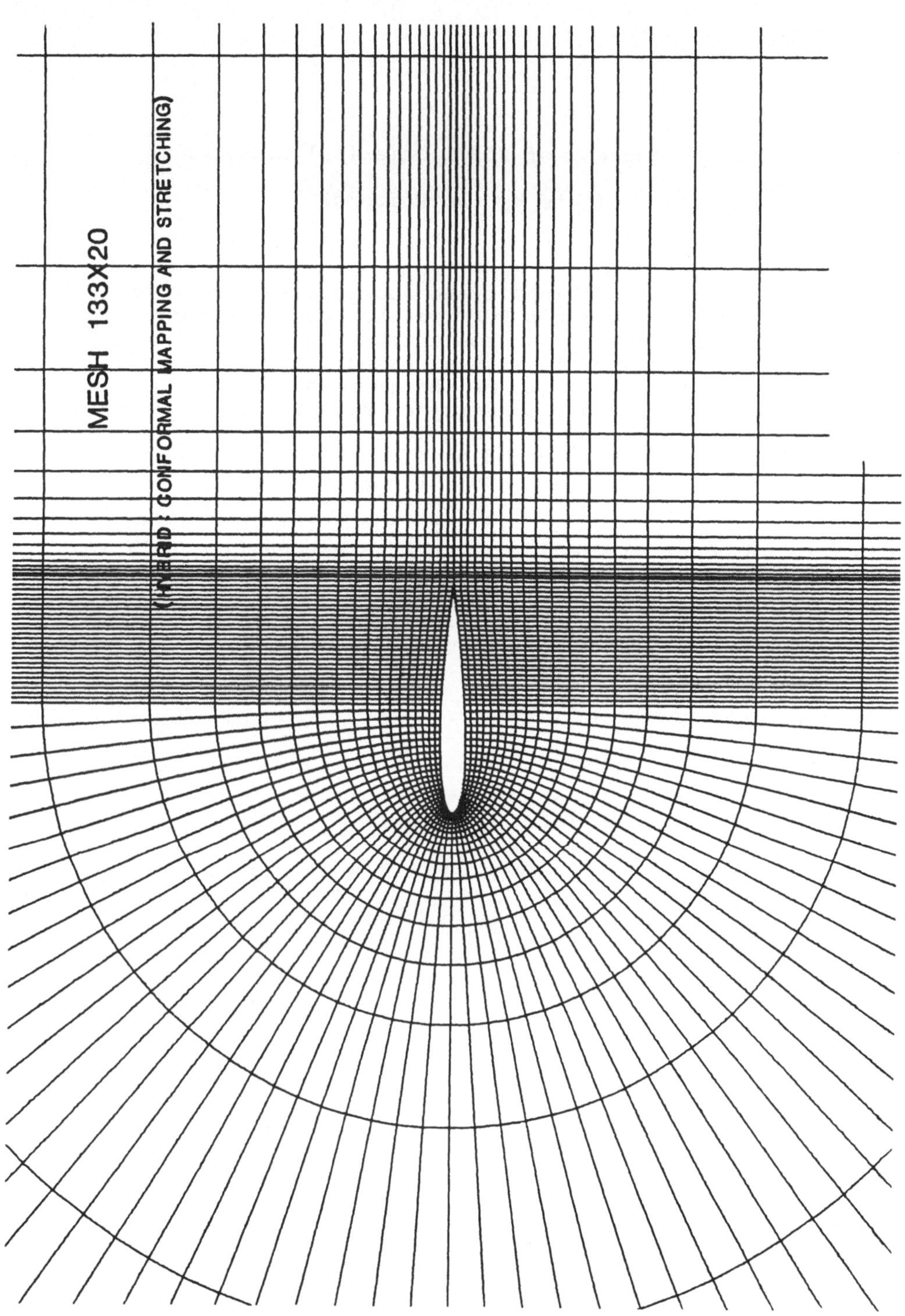

Figure 5. Lerat hybrid mesh b .

Nonuniqueness of FCPOT Numerical Solution

(Note by the Editors)

After the Workshop Steinhoff and Jameson devised an example
to test whether Jameson's FCPOT method could in fact produce
a nonunique solution. Jameson informs us that when he applies
his method to an airfoil with a cusped trailing edge (They
claim a similar result can be obtained on the NACA 0012 airfoil also) AND inforces symmetry on his trailing edge condition he obtains the symmetric solution displayed in Fig. 1.
But if he applies his method as if it were an asymmetric
problem, i.e., allows his usual trailing edge procedure to
operate, he obtains not the expected symmetric solution, but
the one shown in Fig. 2. In both cases the log of the residues is under -11, evidently both are bona fide solutions
to the difference equations.

JOUKOWSKI AIRFOIL
MACH 0.840   ALPHA 0.000
GRID 256X64   RESO.622E-12
CL 0.0000   CO 0.0636   CM 0.0000
C-type grid
CP
X/C
1.200 1.000 0.800 0.600 0.400 0.200 0.000 -0.200 -0.400 -0.600 -0.800 -1.000 -1.200 -1.400
0.000 0.200 0.400 0.600 0.800 1.000

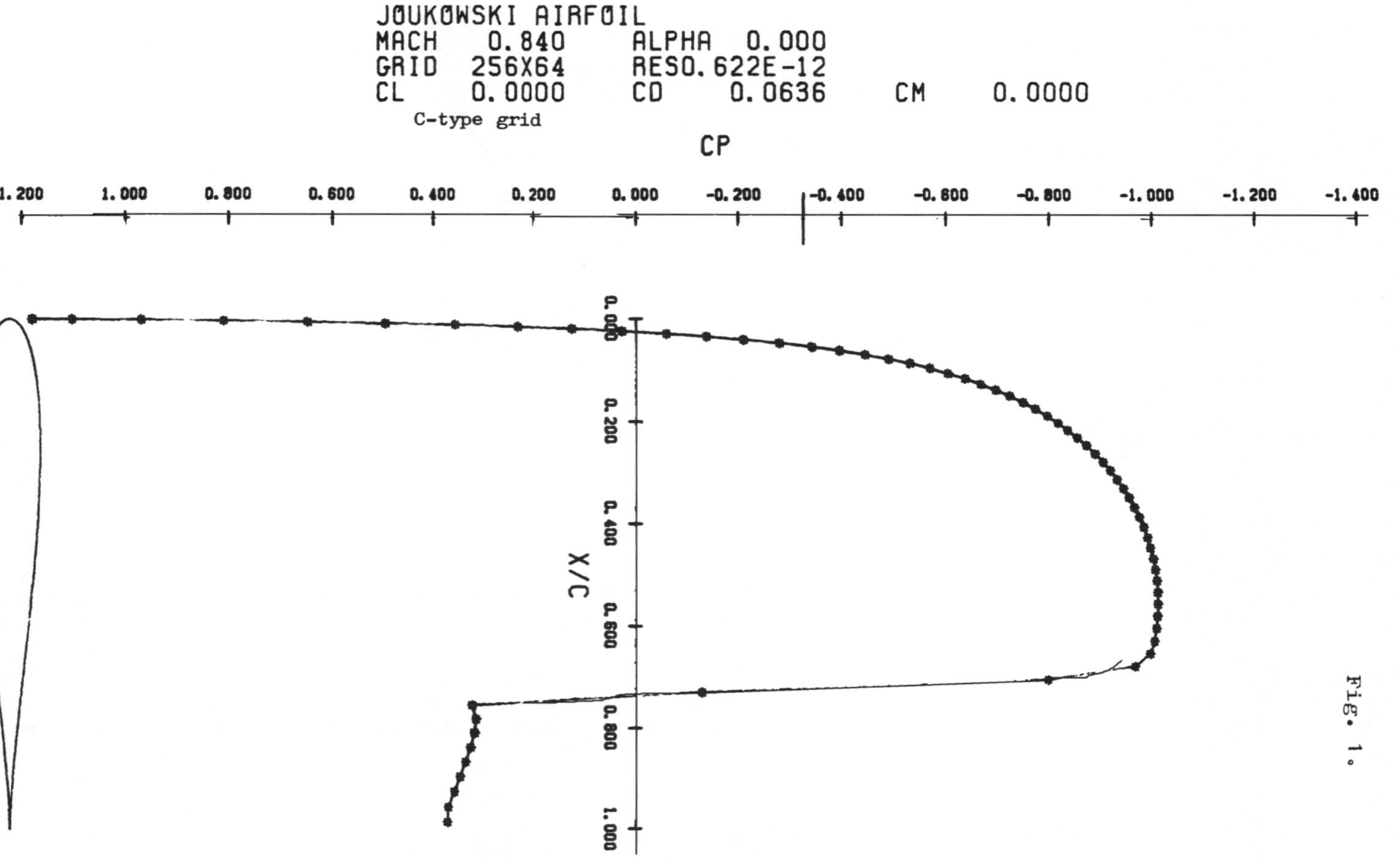

Fig. 1.

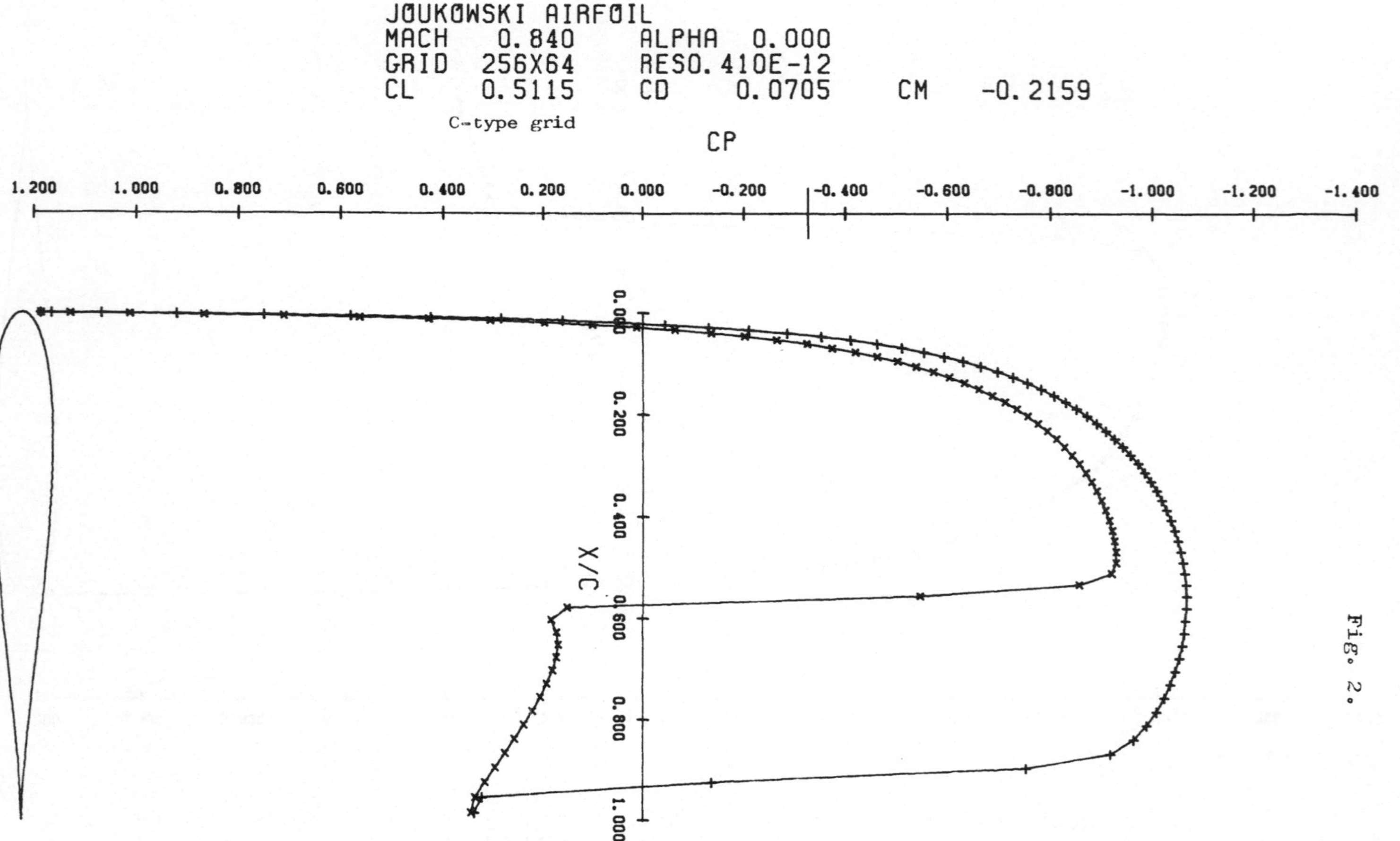
JOUKOWSKI AIRFOIL
MACH 0.840 ALPHA 0.000
GRID 256X64 RESO. 410E-12
CL 0.5115 CD 0.0705 CM -0.2159
C-type grid
CP
1.200 1.000 0.800 0.600 0.400 0.200 0.000 -0.200 -0.400 -0.600 -0.800 -1.000 -1.200 -1.400
X/C
0.000
0.200
0.400
0.600
0.800
1.000
Fig. 2.